Thermoelectricity in Metallic Conductors

Thermoelectricity in Metallic Conductors

Edited by

Frank J. Blatt and Peter A. Schroeder
Michigan State University

Springer Science+Business Media, LLC

Library of Congress Cataloging in Publication Data

International Conference on Thermoelectric Properties of Metallic Conductors, 1st,
  Michigan State University, 1977.
  Thermoelectricity in metallic conductors.

  Includes index.
  1. Electric conductors—Congresses. 2. Thermoelectricity—Congresses. 3. Metals—
Electric properties—Congresses. I. Blatt, Frank J. II. Schroeder, Peter A. III. Title.
QC610.9.I58 1977   537.6'2   78-6010
  ISBN 978-1-4757-6832-9   ISBN 978-1-4757-6830-5 (eBook)
  DOI 10.1007/978-1-4757-6830-5

Proceedings of the First International Conference
on Thermoelectric Properties of Metallic Conductors
held at Michigan State University, East Lansing, Michigan,
August 10—12, 1977

© 1978 Springer Science+Business Media New York
Originally published by Plenum Press, New York in 1978
Softcover reprint of the hardcover 1st edition 1978

Preface

The first International Conference on Thermoelectric Properties of Metallic Conductors was held at Michigan State University on August 10-12, 1977. The conference was sponsored and supported by the National Science Foundation, the Office of Naval Research and the Ford Motor Company.

Although the topic may appear, at first glance, rather narrow and of limited interest, it impacts significantly on numerous fields of research, in each instance providing a unique and fruitful technique for securing important data that is frequently difficult to obtain by other means. Thus, though thermoelectricity is the thread that binds these pages together, the papers constitute a patchwork quilt that includes critical phenomena, superconductivity, many-body theory, quasi one-dimensional systems, liquid metals, to mention only a few.

This volume contains the 12 invited and 31 contributed papers, arranged in the order in which they were presented, as well as much of the frequently spirited and always illuminating discussion that followed these papers. Regrettably, not all of the discussion is included. Difficulties with the recording system during the first session (Wednesday morning) did not become apparent before the end of that session, and, consequently, none of the discussion--some of it fairly heated--appears in the proceedings; other remarks were lost to posterity through occasional malfunctioning of the recording facilities and/or failure of speakers to come near a microphone.

We have viewed our role more as acquisition rather than copy editors. We have taken the liberty of rephrasing in a few rare instances, but have done this only when the meaning would otherwise have been obscure. The reader will find that some of the papers submitted by contributors from non-English speaking countries retain unfamiliar idioms that may help remind participants of the international flavor of the meeting.

We take this opportunity to thank the authors for their contributions, assistance and cooperation and all participants whose
interest and enthusiasm contributed so much to the success of the
conference. Special thanks go to Drs. Barnard, Crisp, Foiles,
Guntherodt, Henry, and Pearson who served as chairmen of the six
sessions. Without the able, careful and patient work of Ms. Maxine
Pennington, Ms. Delores Sullivan, Ms. Jean Strachan and Ms. Helen
Bartlett, these proceedings would never have seen the light of day.

Finally, it is perhaps noteworthy that this conference convened
roughly fifty years after Borelius and his collaborators at Leiden
embarked on the project of determining the scale of absolute thermoelectric power. Their results have served as the ultimate foundation
of nearly all subsequent experimental and, consequently, much
theoretical research for half a century. This first International
Conference on Thermoelectric Properties of Metallic Conductors marks
the debut of a new scale of absolute thermopower, due to Roberts,
which, we are confident, will serve the scientific community for at
least another fifty years.

                                          F. J. Blatt and
East Lansing, Michigan   P. A. Schroeder, Editors
January, 1978

Contents

REMINISCENCES OF WORK WITH D.K.C. MacDONALD AT THE NATIONAL RESEARCH COUNCIL

I. M. Templeton

National Research Council

Ottawa, Canada

In January 1953 the following note,[1] under the heading 'Thermoelectricity at Low Temperatures' was sent to Acta Metallurgica:

> 'It has been suggested[2] that the foundations of electron theory in metals are well-established and that at best only "refinements of theory" may be looked for. However......it now appears that this belief may well be unfounded......We are at present carrying out an experimental study of thermoelectricity......Our experiments are already adequate to show that little detailed agreement can be found with the present theory......We believe that systematic experiments on thermoelectricity at low temperatures will yield much information of value on the processes of electron scattering in metals.'*

That note was the forerunner of over 40 publications which were to appear over the next 12 years or so, reaching a peak of 8 in 1961, two years before Keith MacDonald's death. They included a series of 9 major papers,[3-11] all but one published in Proceedings of the Royal Society, under the general heading 'Thermoelectricity at Low Temperatures', a group of 4 letters[12-15] on 'Thermoelectricity below 1°K', 22 other original publications[16-35,49,50] and 10 reviews and conference papers.[36-45] I consider myself very lucky to have

---

*D.K.C. MacDonald and W.B. Pearson, Physics Division, National Research Council, Ottawa

been in on the 'ground floor', so that I was carried along by this
remarkable tide of publications and became co-author or author in
about half of them. I don't intend, nor could I hope, to attempt
an exhaustive review of the work, but rather I will try to highlight some of the more stimulating (and frustrating!) moments of
working with Keith MacDonald and Bill Pearson during those exciting
years, with particular emphasis on our 'pioneer' work below 1 K.

My own interests have always been in tackling the challenge
of making experimental measurements in difficult conditions, so it
is perhaps appropriate for us to look first at the development of
the measurement techniques. When I first started work with Keith
MacDonald at the Clarendon Lab., Oxford, in the fall of 1950, his
basic instrument for measuring small voltages was the feedback
galvanometer amplifier which he had developed while studying electrical resistivity at low temperatures with Mendelssohn. This was,
and still is, a very elegant device which can outperform the best
modern 'nanovoltmeter'.

Fig. 1 shows the basic circuit: a negligibly small current i
flowing through the primary galvanometer G_1 is sufficient to cause
the split selenium (or silicon) photocell S to drive a large current
I through the decade feedback resistor R. G_2 measures the feedback
current which has to be inversely proportional to R, so that IxR
is equal to E. The system is thus self-balancing, with a very high
input impedance. In the original system the 450Ω of G_1 constituted
the main resistance of the primary circuit, so the thermal noise
within the 10 Hz or so response bandwidth of the system was of
order 10^{-8} volts. Bob Chambers improved this, when he was a Post-
doctorate Fellow, by substituting a 5Ω primary galvanometer and
improving the feedback gain by an optical grid arrangement. He
also used the system as a null detector, thus removing the remanent
objection of slight non-linearity due to the reduced feedback at
maximum sensitivity.

The first thermoelectric measurements were a general survey
of the so-called 'simple' metals sodium, potassium and copper,
including some alloys of rubidium in potassium, and of nickel and
tin in copper. The techniques were a little startling compared
with those in use today. For measurements of total EMF, over a
range from room temperature to liquid helium temperatures, the
liquefying chamber of a Collins liquefier was used as the cryostat,
and measurements were made continuously during the cooling run.
The glass observation plate was replaced with a perspex plate carrying wires and black-wax or Q-compound seals as required. For both
copper and the alkalis thin lead (Pb) wires provided the reference
via an absolute scale of EMF derived from earlier work of Borelius,
Keesom, Johannson and Linde.[46]* Copper, of course, posed no

---

*However, see Roberts' paper at this conference!

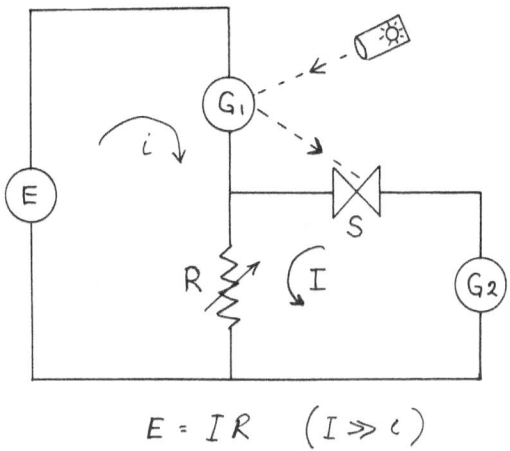

Fig. 1.   Feedback galvanometer amplifier.

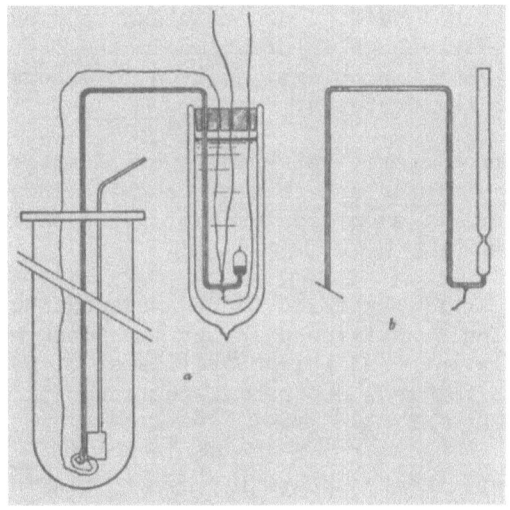

Fig. 2.   Measurement of thermo-E.M.F. in Alkali metals from 4.2 K
to 300 K.   (From Ref. 3).

particular installation problem, but the alkalis required a glass tube some 7 feet long in the shape of a Γ (Fig. 2)! Platinum wires provided the contacts, the top end was held in melting ice and the bottom was in thermal contact with the bulb of a gas thermometer, the temperature being read via a calibrated dial pressure gauge.

The next stage of development also used the Collins machine, this time simply to provide a liquid He reservoir. A copper tube with a heater provided the temperature gradient, the top of the tube being heated up to as much as 75 K. The temperatures were measured either by a gas thermometer and calibrated copper-constantan thermocouple (Fig. 3) or with a pair of gas thermometers.

Fig. 4 shows our chief technician at the time, Fred Richardson, and D.K.C. MacDonald operating the galvanometer amplifier.

The results were as shown in the next few figures (Figs. 5-7). For sodium, we appear to have the square-law EMF corresponding to a linear thermopower as predicted, but the magnitude is clearly appreciably larger than the theoretical prediction of Wilson. For potassium and dilute K-Rb alloys we appear to have a higher power law, while for copper not only is the effect positive, but there is apparently a gross sensitivity to the addition of tin. Borelius et al[47] had already found an even larger effect due to iron.

Now although as I said, I don't intend to discuss results in any detail, there is enough information in the last few figures to show up a number of the problems involved in extending these measurements to lower temperatures.

(1) The thermoelectric potentials were identified by reversing them at room temperature, and this of course allowed the inclusion of any spurious e.m.f.'s on the leads from the helium bath, which could be of order 1 μV.

(2) Even so, there appeared to be a discontinuity in slope at the superconducting transition-point of lead, suggesting that the Thomson heat scale for lead which Borelius et al had derived was wrong, and that S for lead dropped discontinuously rather than smoothly to zero at 7.2 K.[5]

(3) The giant effects apparently due to tin in copper might well, as we shall see, be described as a portrait of a red herring, one which was to occupy Bill Pearson, in particular, for some years to follow!

It was at this point that I made one of my more useful contributions to our work. We had been trying to make a mechanical reversing switch operate in liquid helium, with little success. I

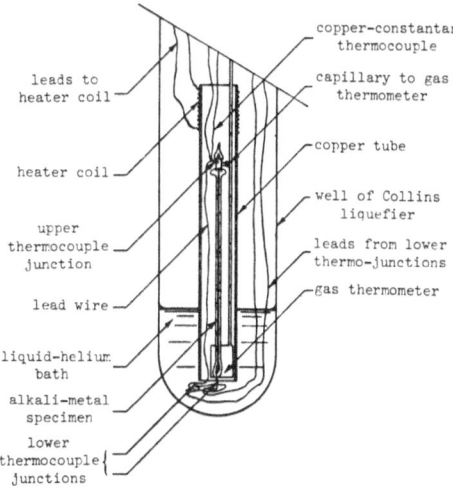

Fig. 3. Measurement of thermo-E.M.F. in alkali metals from 4.2 K to ~75 K. (From Ref. 3).

Fig. 4. MacDonald in characteristic pose at the galvanometer amplifier. Technician F.W. Richardson at left. (Photo: Canadian National Film Board).

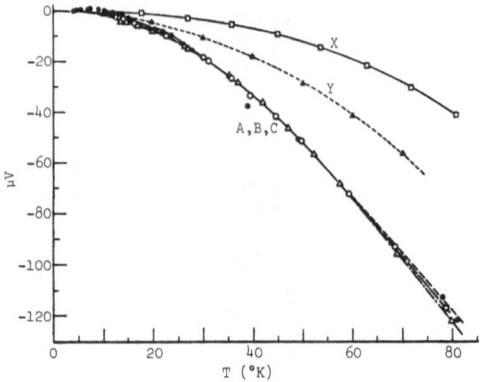

Fig. 5. Absolute thermo-E.M.F. in sodium. A,B,C experiment;
X,Y theory. (See Ref. 3).

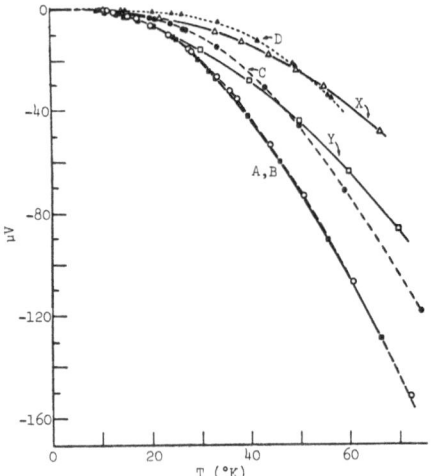

Fig. 6. Absolute thermo-E.M.F. in potassium (A,B) and potassium-
rubidium alloys (C,D). X,Y theory. (See Ref. 3).

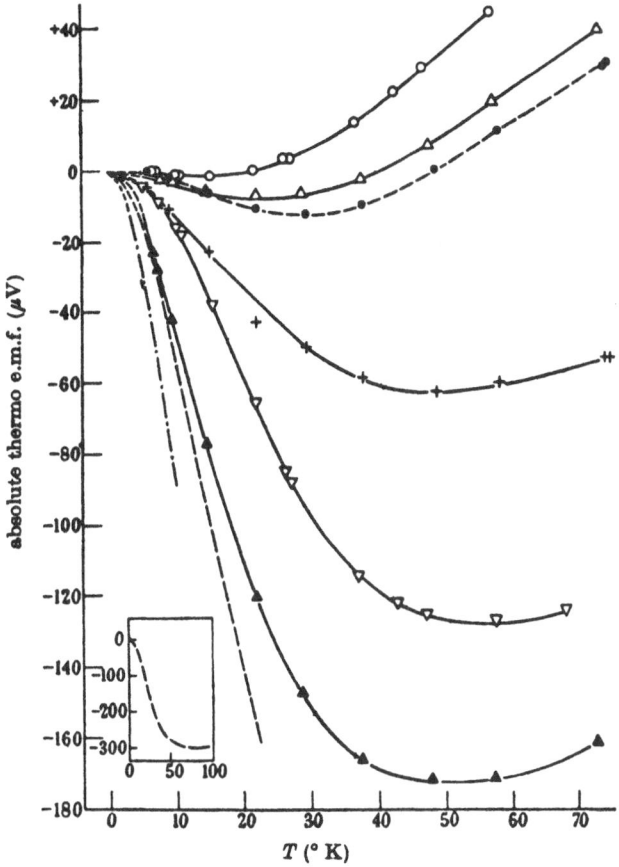

Fig. 7. Absolute thermo-E.M.F. in copper (top) and dilute copper-based alloys (mostly Cu(Sn)). Dashed lines are for Cu(Fe). (See Ref. 3).

had also been working for some time on a superconducting 'chopper' amplifier[20] for use in our experiments when it struck me that a reversing switch does not necessarily have to have any actual open circuits so long as the relative resistance of the 'open' and 'closed' parts have a large enough ratio, and the 'open' resistance is large compared with the source resistance. Hence (Fig. 8) if aa' were superconducting and bb' resistive (because of magnetic field or heating), then we have direct connection of E to D, with reversed connection if bb' is superconducting and aa' resistive. This is the principle of the superconducting reversing switch[19] which was used in one form or another in all our subsequent experiments. One of the later versions of these switches is shown in

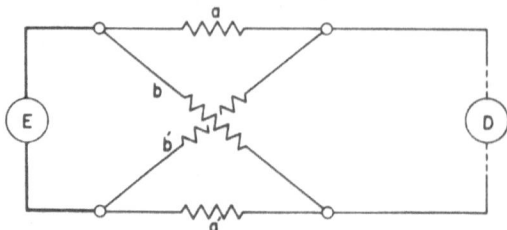

Fig. 8.   Superconducting reversing switch.   (Refs. 19 & 25).

Fig. 9 mounted on a cryostat: You can see the pair of magnetising coils used in switching the tantalum elements and the Pb wires coming from the specimen via platinum tube/soft glass seals.   We used this type for about 8 years, until Tony Guénault realised that a thermally operated type would avoid the inductive surges that tend to accompany changing magnetic fields however carefully the switching elements are wound.

Once the low-temperature e.m.f.'s could be identified cleanly and clearly, with a resolution approaching 10^{-9} V, things moved on

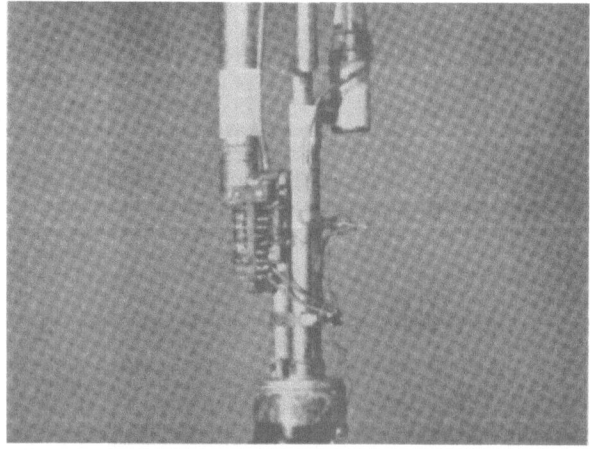

Fig. 9.   Later form of reversing switch in position on a cryostat. (Ref. 25.)

apace. The existing techniques of thermal conductivity measurements, with vacuum cans, controlled heaters, pumped helium pots
and differential gas thermometers were readily adapted for use in
thermoelectric measurements. We established a new low-temperature
end of the Pb scale by measuring against a stick of sintered Nb_3Sn
(incidentally, Bill Pearson was the only person I know who could
ever solder to that stuff, and then only when it was freshly prepared!), showing clearly that S was indeed discontinuous at the
superconducting transition point (Fig. 10). (We weren't the first
to do this - G. T. Pullan[48] had made some very elegant measurements
on tin in 1953 using a very sensitive superconducting galvanometer.)
Some alkali results are shown in Figs. 11-14.

The red herring of tin (and various other solutes) in copper
is perhaps worth a short comment. In addition to the rapidly
increasing negative thermopowers, there was also the appearance of
a resistance-minimum of increasing depth, until some 20 ppm of tin
had been added (Fig. 15). The reason for this is obvious with
hindsight from Fig. 16, where the effect of Sn is compared with
that of Fe. It can be seen that the initial change is equivalent
to adding iron, and the subsequent fall-off is as if from an iron-
dominated maximum. The problem arose because although the original
copper probably contained some 10-20 ppm of iron, it also contained

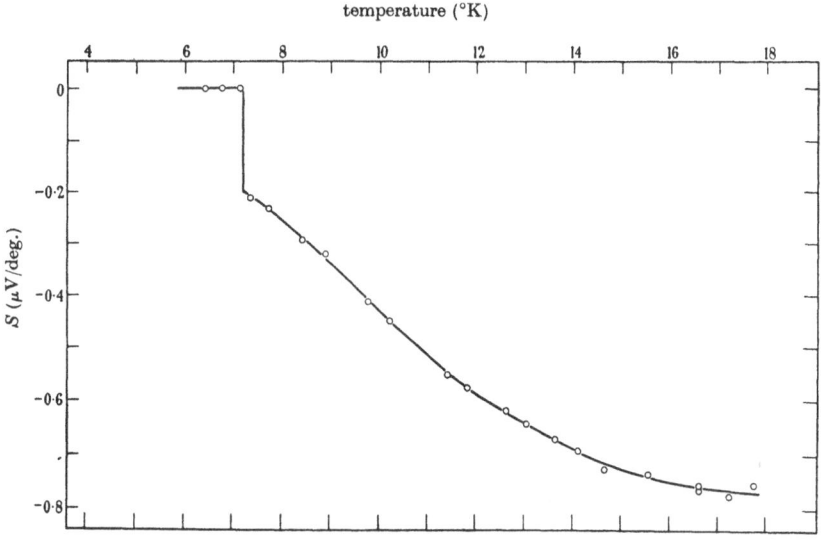

Fig. 10.   Thermopower of Pb vs. Nb_3Sn.   (Ref. 8).

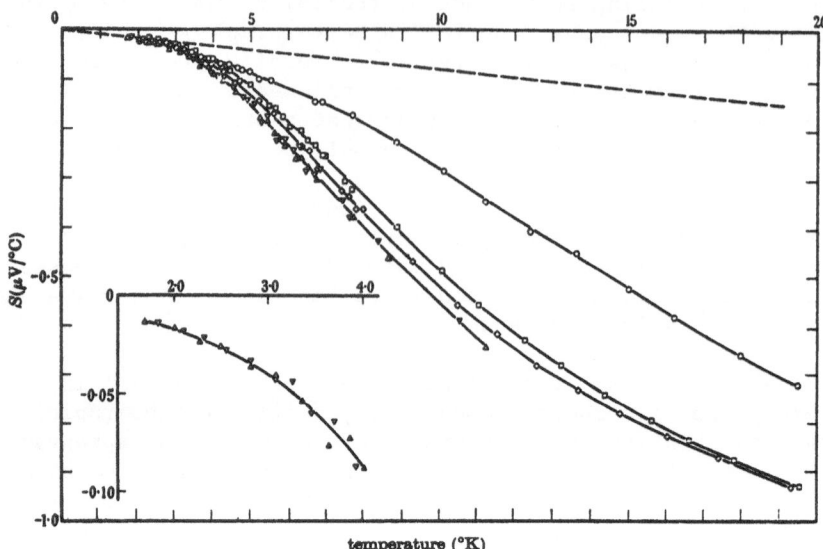

Fig. 11.   Absolute thermopower of sodium.   (Ref. 9).

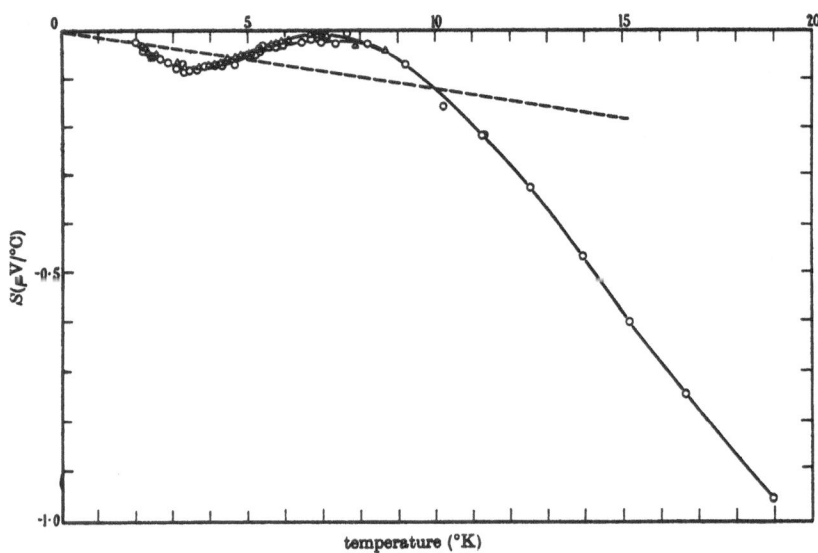

Fig. 12.   Absolute thermopower of potassium.   (Ref. 9).

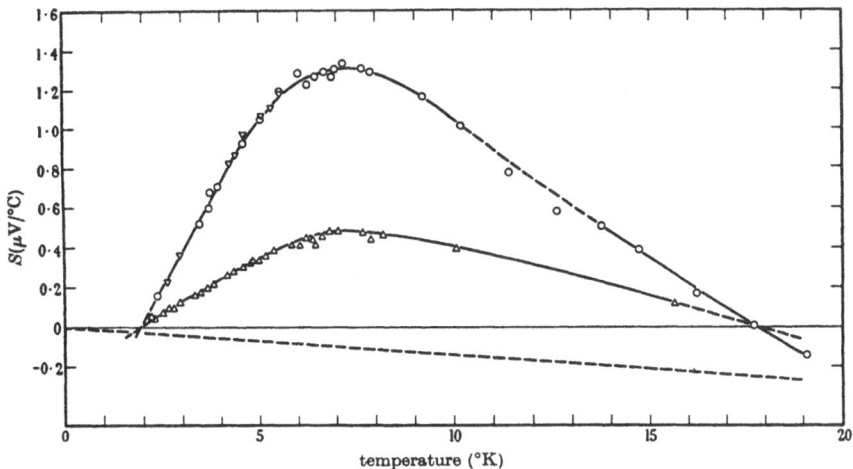

Fig. 13.   Absolute thermopower of rubidium.   (Ref. 9).

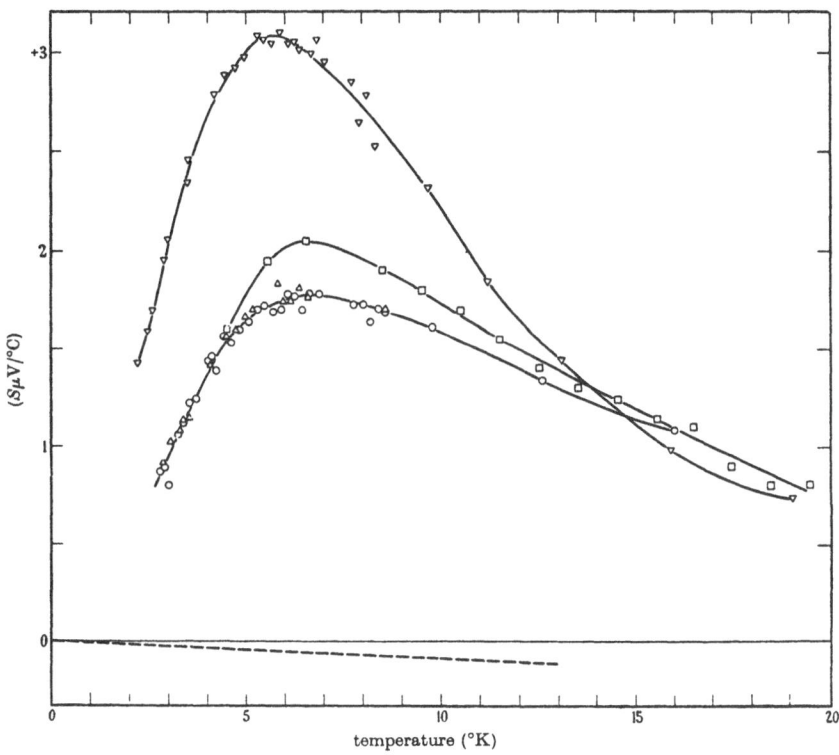

Fig. 14.   Absolute thermopower of caesium.   (Ref. 9).

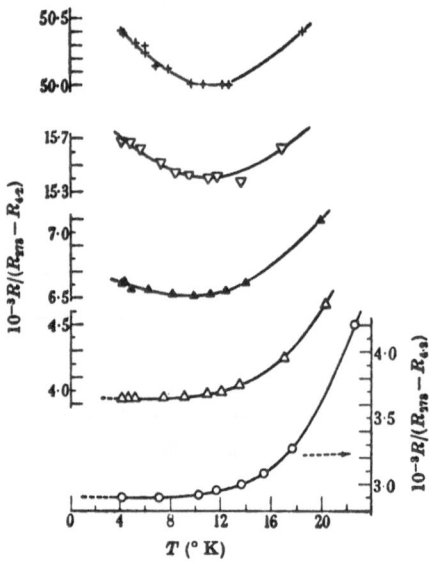

Fig. 15. Relative electrical resistances of copper and copper-tin alloys. (Ref. 3).

appreciable quantities of oxygen which took the iron out of solution as oxide. The added tin reduced the iron oxide, bringing the iron into solution and generating all the effects characteristic of iron. Thus (Fig. 17) tin added to oxidised copper (A) first brought iron into solution until all the oxide had been reduced (B), from which point the tin gradually established its own characteristic thermopower (C). If the Cu were reduced with hydrogen to (D), the addition of tin took thermopower straight to (C) while the addition of iron would take it to (E). As we found later, at very low temperatures iron has an even more catastrophic effect in gold, making Au(Fe) very useful as a low temperature thermocouple material. I often think we should have patented that idea!

I think I should go on now to introduce the subject of our film. Some time in 1957 we decided to attempt some measurements below 1 K - the best we had done until then was with a helium pot pumped to about 1.2 K at the 'cold' end and a heater at the 'hot' end of a specimen. We decided to use the demagnetisation of a paramagnetic salt pill to provide temperatures down to a few hundredths of a degree, and to solder the specimen between a metal mesh on which the salt was crystallised and a thick platinum wire sealed into a pumped helium pot. The susceptibility (and hence

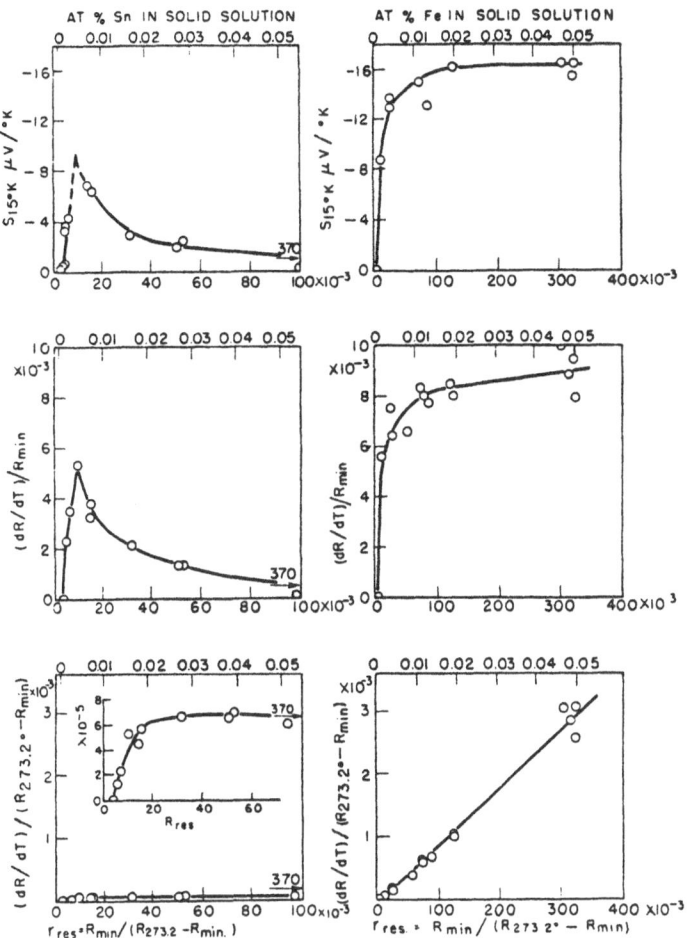

Fig. 16.   Thermopower and resistivities in copper-tin and copper-iron alloys.   (Ref. 21).

temperature) of the pill was measured by a mutual inductance/ fluxmeter type of system. We measured the decreasing thermoelectric potential, as the pill warmed up, via fine tantalum wires in the vacuum space, and Pb wires (leading to the reversing switch) within the helium pot. The pill in the first two pictures of Fig. 18 is rather lumpy and has an alkali metal specimen in place, while the second is rather more elegantly and correctly shaped for accurate susceptibility measurement and has a spiral of copper wire as the specimen. In choosing the salt to use we had to make a compromise between reaching the lowest possible temperature and having

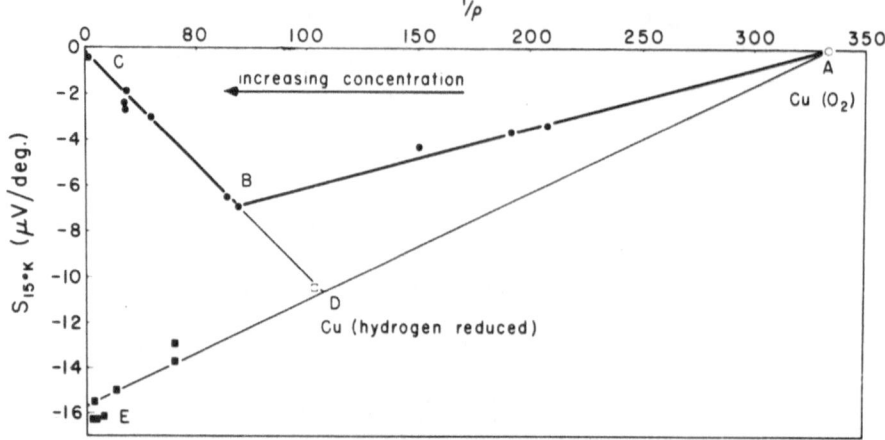

Fig. 17. Variation of thermopower at 15 K in copper containing
iron. (See text and Ref. 24).

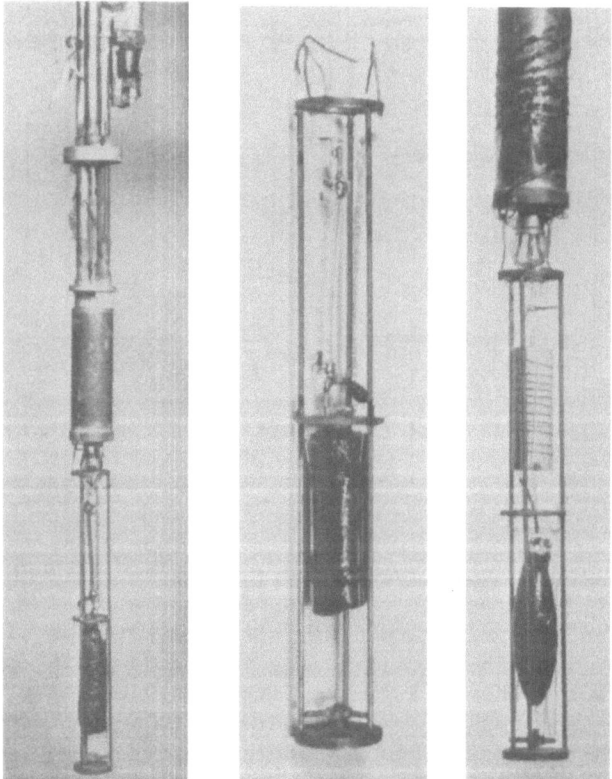

Fig. 18. Internal arrangements of cryostat for measurements below
1 K. (See text and Ref. 10).

reasonable thermal capacity – for most experiments we used either
ferric ammonium alum or ferric methylammonium alum. The susceptibility measuring coils were wound on the vacuum cans: the cans
therefore had to be of brass or of stainless steel with copper
strips for cooling, to avoid eddy current problems.

We published the first results of this work in a series of
notes to Philosophical Magazine[12-15], describing measurements on
gold, silver, the alkalis and, later, various transition metals.
The first major publication[10], giving full details of the technique,
was the eighth in the 'Thermoelectricity at Low Temperatures' series,
and dealt with the alkali metals below 2 K. Lithium and sodium
showed more ore less linear thermopowers, positive and negative,
respectively. Potassium (Fig. 19), combined with earlier results

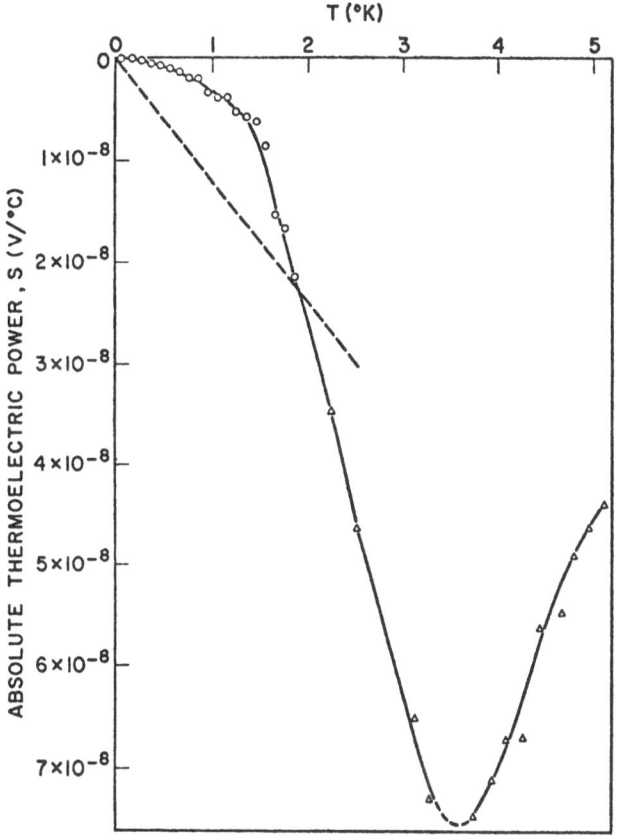

Fig. 19. Absolute thermopower of potassium. (Ref.10).

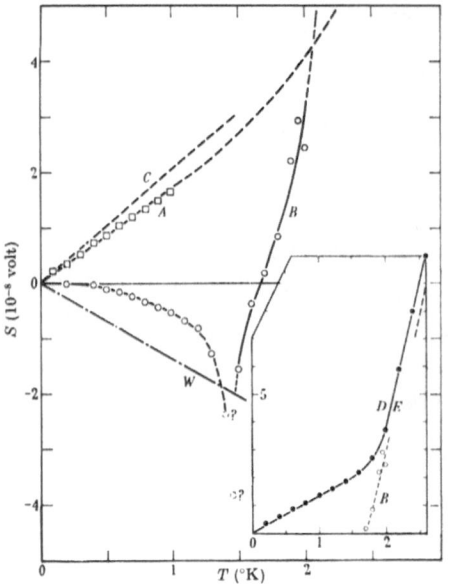

Fig. 20. Absolute thermopower
of rubidium. (Ref. 10)

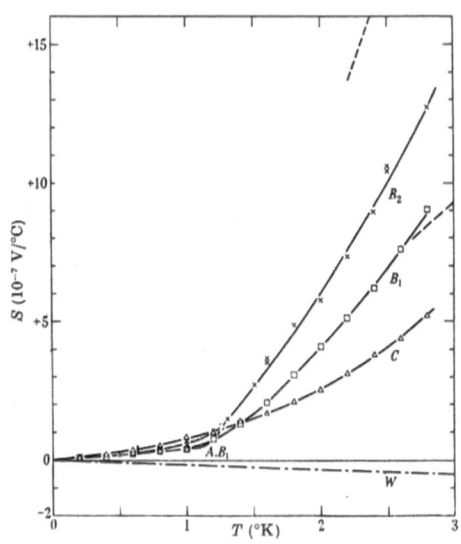

Fig. 21. Absolute thermopower
of caesium. (Ref. 10)

Fig. 22. W.B. Pearson, J.S. Dugdale & D.K.C. MacDonald during an
early experiment in the A.D. Little electromagnet. (Photo: Canadian
National Film Board.)

Fig. 23. The author controlling the demagnetization for an experiment below 1 K. (Photo: N.R.C.)

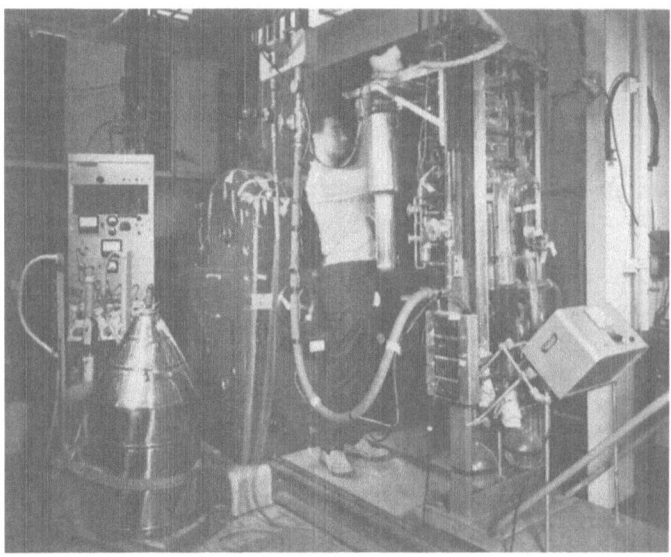

Fig. 24. P.R.F. Simon with a later version of a demagnetization cryostat. (Photo: Canadian National Film Board)

above 2 K, showed some rather fancy behaviour compared with the
simple Wilson formula, while rubidium (Fig. 20) and caesium
(Fig. 21) were even more startling. It was at this point that the
idea arose not only of normal phonon-drag, which would enhance the
diffusion term with a T^3 component, but of Umklapp phonon drag,
which given, for instance, a sufficiently distorted Fermi surface
could take over exponentially in the reverse sense at a point
depending on the Debye Θ of the metal concerned. All this area of
phonon drag was followed up by Keith in collaboration with
Tony Guénault[27,32].

While we were doing these experiments Keith became enthused
with the idea of making a film for high school and university
students which would attempt to illustrate the way in which
curiosity-oriented research operates, and emphasising the teamwork
involved in planning and carrying out an experiment. We decided
to base the film on our work below 1 K, and were most fortunate to
have the services of a skilled film-making crew in the Mechanical
Engineering Division of NRC, with John de Blois as the cameraman.

Most of the film was planned and 'acted', but the central
experiment was filmed during an actual run on a relatively high
conductivity specimen of gold. The warm-up time of the salt pill,
by conduction down the gold, was only about 5 minutes, so we forgot
completely about the cameraman and lights - whether this lent an
air of authenticity or simply of chaos I must leave you to judge!

At the conclusion of his talk, Dr. Templeton showed the film
'An Experiment in Physics', filmed in 1960 at the laboratories of
the National Research Council in Canada.

REFERENCES

1. D.K.C. MacDonald and W.B. Pearson, Acta Meta. 1, 242 (1953).
2. A.H. Wilson, Theory of Metals, (Cambridge University Press, 1936).
3. D.K.C. MacDonald and W.B. Pearson, Proc. Roy. Soc. A 219, 373, (1953).
4. D.K.C. MacDonald and W.B. Pearson, Proc. Roy. Soc. A 221, 534, (1954).
5. W.B. Pearson and I.M. Templeton, Proc. Roy. Soc. A 231, 534 (1955).
6. D.K.C. MacDonald and W.B. Pearson, Proc. Roy. Soc. A 241, 257, (1957).
7. J.P. Jan, W.B. Pearson and I.M. Templeton, Can. J. Phys. 36, 627, (1958).
8. J.W. Christian, J.P. Jan, W.B. Pearson and Templeton, Proc. Roy. Soc. A 245, 213, (1958).

9.   D.K.C. MacDonald, W.B. Pearson and I.M. Templeton, Proc. Roy.
     Soc. A 248, 107, (1958).
10. D.K.C. MacDonald, W.B.Pearson and I.M. Templeton, Proc. Roy.
     Soc. A 256, 334, (1960).
11. D.K.C. MacDonald, W.B.Pearson and I.M. Templeton, Proc. Roy.
     Soc. A 266, 161, (1962).
12. D.K.C. MacDonald, W.B. Pearson and I.M. Templeton, Phil. Mag.
     3, 657, (1958).
13. D.K.C. MacDonald, W.B. Pearson and I.M. Templeton, Phil. Mag.
     3, 917, (1958).
14. D.K.C. MacDonald, W.B.Pearson, and I.M. Templeton, Phil. Mag.
     4, 380, (1959).
15. D.K.C. MacDonald, W.B. Pearson, and I.M. Templeton, Phil. Mag.
     5, 867, (1960).
16. D.K.C. MacDonald and S.K. Roy, Phil. Mag. 44, 1364, (1953).
17. D.K.C. MacDonald and W.B. Pearson, Acta Met. 3, 392, (1955).
18. D.K.C. MacDonald and W.B. Pearson, Acta Met. 3, 403, (1955).
19. I.M. Templeton, J. Sci. Instrum. 32, 172, (1955).
20. W.B. Pearson, Can. J. Phys. 33, 473, (1955).
21. W.B. Pearson, Phil. Mag. 46, 911, (1955).
22. A.B. Bhatia and D.K.C. MacDonald, Can. J. Phys. 34, 419, (1956).
23. D.K.C. MacDonald, E. Mooser, W.B.Pearson, I.M. Templeton, and
     S.B. Woods, Phil. Mag. 4, 433, (1959).
24. D.K.C. MacDonald, A.V. Gold, W.B.Pearson, and I.M. Templeton,
     Phil. Mag. 5, 765, (1960).
25. I.M. Templeton, Solid State Elec. 1, 258, (1960).
26. W.B. Pearson and I.M. Templeton, Can. J. Phys. 39, 1034, (1961).
27. A.M. Guénault and D.K.C. MacDonald, Proc. Roy. Soc. A 264, 41,
     (1961).
28. D.K.C. MacDonald and W.B. Pearson, Proc. Phys. Soc. 78, 306,
     (1961).
29. A.M. Guénault and D.K.C. MacDonald, Phil. Mag. 6, 1201, (1961).
30. D.K.C. MacDonald, W.B. Pearson, and I.M. Templeton, Phil. Mag
     6, 1431, (1961).
31. A. Kjekshus and W.B. Pearson, Can. J. Phys. 40, 98, (1962).
32. A.M. Guénault and D.K.C. MacDonald, Proc. Roy. Soc. A. 274,
     154, (1963).
33. A.M. Guénault, Phil. Mag. 9, 331, (1964).
34. P.R.F. Simon, Proc. IXth Intl. Conf. Low Temp. Phys., August
     1964 (Pub. Plenum Press, 1964), p. 1045.
35. A.M. Guénault, Phil. Mag. 15, 17, (1967).
36. D.K.C. MacDonald, Physica 19, 841, (1953).
37. D.K.C. MacDonald and G.K. White, Int'l. Inst. Congress 1954,
     Philadelphia.
38. D.K.C. MacDonald, W.B.Pearson, and G.K. White, Bulletin Inter.
     du Froid, p. 107, 1955.
39. J.P. Jan, W.B. Pearson, and I.M. Templeton, Proc. Conf. de
     Basses Temperatures, p. 418, Sept. 1955, Paris.
40. D.K.C. MacDonald, Zeitschrift fur Physi. Chemie 16, 310, (1958).

41. D.K.C. MacDonald, W.B.Pearson, and I.M. Templeton, "Thermo-electricity at low temperatures", Proc. Madison Conf. on Low Temp. Phys., (1959).

42. D.K.C. MacDonald, Science 129, 943, (1959).

43. D.K.C. MacDonald and I.M. Templeton, Some comments of thermo-electric refrigeration at low temperatures, Proc. Intl. Conf., Prague, 1960, p. 650.

44. W.B. Pearson, Survey of thermoelectric studies of the Group 1 Metals at low temperatures carried out at the N.R.C. Labs, Ottawa, S.S.P. USSR, 3, 1411, (1961).

45. W.B. Pearson, Ultra High Purity Metals, p. 201, (1962).

46. G. Borelius, W.H. Keesom, C.H. Johansson, and J.O. Linde, 1932b Proc. Acad. Sci. Amst. 35, 10.

47. G.Borelius, W.H. Keesom, C.H. Johansson, and J.O. Linde, 1932a Proc. Acad. Sci. Amst. 35, 25.

48. G.T. Pullan, Proc. Roy. Soc. A, 217, 280 (1953).

49. W.B. Pearson, Can. J. Phys. 38, 1048, (1960).

50. W.B. Pearson, Phys. Rev. 119, 549, (1960).

THE ABSOLUTE THERMOPOWER OF LEAD

R. B. Roberts

National Measurement Laboratory, CSIRO

Sydney, Australia 2008

ABSTRACT

The absolute thermopower of lead has been calculated from the results[1] of direct measurement of its Thomson heat, μ, in the range from 10 to 350 K. The values for μ agree with those from the superconducting thermocouple experiment of Christian, Jan, Pearson and Templeton[2] from 10 to 17 K, and with those from the indirect measurements of Borelius, Keesom, Johansson and Linde[3] from about 80 to 300 K. From 17 to 80 K, however, the disagreement with the values used by Christian et al.[2] in constructing their absolute scale of thermoelectricity leads to a change in the scale of about 0.3 µV/K from 30 to 300 K. The abrupt nature of the change near 20 K makes a great difference to the shape of the calculated thermopowers of metals whose thermopowers are small ($\sim$ 1 µV/K). For example, the thermopowers of lead and the noble metals can be fitted to equations of the form $S = aT + b/T$ from just above the phonon drag peak to about the Debye θ. This paper will be concerned with the Thomson heat experiment, the validity of the new scale for thermoelectricity, and plans for extending the scale to 1000°C.

I. INTRODUCTION

Lead has long been used as the reference material required in a thermocouple measurement from which it is intended to

determine the absolute thermopower of some other material. Thus a
tabulation of the absolute thermopower of lead becomes a scale for
thermoelectricity.

The scale in present use was compiled by Christian, Jan,
Pearson and Templeton[2] from their own measurements of the behaviour
of a lead vs superconductor thermocouple from 7.2 to 17 K and
calculations of the Thomson heat, μ, of lead made by the Borelius,
Keesom, Johannsen and Linde group from 20 to 270 K. Because this
scale has had no experimental testing above 17 K it was decided to
measure directly the Thomson heat of lead and calculate its thermopower.[1] The region below 17 K would serve as a check on the present Thomson heat experiment and indeed on the Kelvin relation

$$S = \int \frac{\mu}{T} \, dT. \tag{1}$$

The Christian et al.[2] scale was verified to about 18 K, above
which temperature it was found to become progressively too negative
to 50 K where the discrepancy is -0.24 μV/K. At higher temperatures the discrepancy remains about constant. An internal cross
check was performed to verify the new scale.

II. METHOD

A full description of the apparatus is given by Roberts.[4]
Basically, a piece of lead wire was bent into a U-shape, electrically insulated and clamped between two isothermal blocks at the
top T_2 and bottom T_1 of the U, thereby exposing the working portion of the sample as two parallel wires of length L in a temperature gradient $(T_2-T_1)/L$. Provision was made to measure the change
in temperature δt of the mid-point of one wire occurring after the
reversal of a direct current i sent through the sample. It may
be shown that the Thomson heat may be calculated as

$$\mu = \frac{4kA}{iL} \frac{\delta t}{(T_2-T_1)} \tag{2}$$

where k is the thermal conductivity and A the cross-sectional area
of the sample.

Typically the difference, δt, amounted to 1 mK. It was detected by means of a differential thermocouple which drove a
Keithley nanovoltmeter. The differential junctions were placed
at the mid-points of the wires so the signal comprised an error

due to thermals in the leads, an error due to slight misplacement
of the junctions, an error in the zero point of the amplifier,
and the real difference between the temperatures of the mid-points.
Reversal of the sample current affects only the last component,
giving a signal change equivalent to twice the change in the temperature of the mid-point of one wire. The main features of this
arrangement are the symmetry of the sample and the absence of
switches in the detection circuit.

III.   PRELIMINARY TEST

The performance of the equipment was tested by changing the
experimental variables in Eq. (2). For variations in i from 0.1
to 1.0 Amp and (T_2-T_1) from 2 to 35 K the equation was valid within
3% overall. Measurements were then made from 5 to 350 K using
helium, nitrogen and oil as baths. The current i was chosen to
be as large as possible without ever disturbing the linear profile of the temperature distribution in the wire by more than 10%
of the difference (T_2-T_1) (max. 1 Amp, average 0.7 Amp at higher
temperatures). Thus the profile was near enough to linear that
the small error in the placement of the differential junctions
was irrelevant. If the temperature profiles of the two wires
are parallel no error is introduced.

Equation (2) assumes that all the heat transfer in the wire
is by conduction. A more sophisticated analysis shows that if any
heat is lost from the sample by radiation (or convection) the
observed δt will be too small regardless of the sign of μ. To
determine the extent of the effect of radiation loss a comparison
was made between two arrangements with the apparatus operating at
the highest temperatures. In one case the sample was free to
radiate to its surrounding heat shield. In the second the space
around the sample was packed with an insulating reflective fibre
(Kaowool). The difference was only a few percent so it was concluded that the effect of radiation loss was small.

IV.   RESULTS

The present results are compared with those of the Christian
et al. (1958) scale on the basis of μ/T in Figs. 1 and 2. It is
seen that the only region of serious disagreement is between 20
and 80 K. However, because of the nature of Eq. (1) the effect
on the thermopower carries through to all temperatures above 20 K.
Above 50 K the error in the Christian scale for the thermopower
of lead is about -0.3 μV/K. Another problem is that the region
from 20 to 80 K is of prime importance to those people trying to

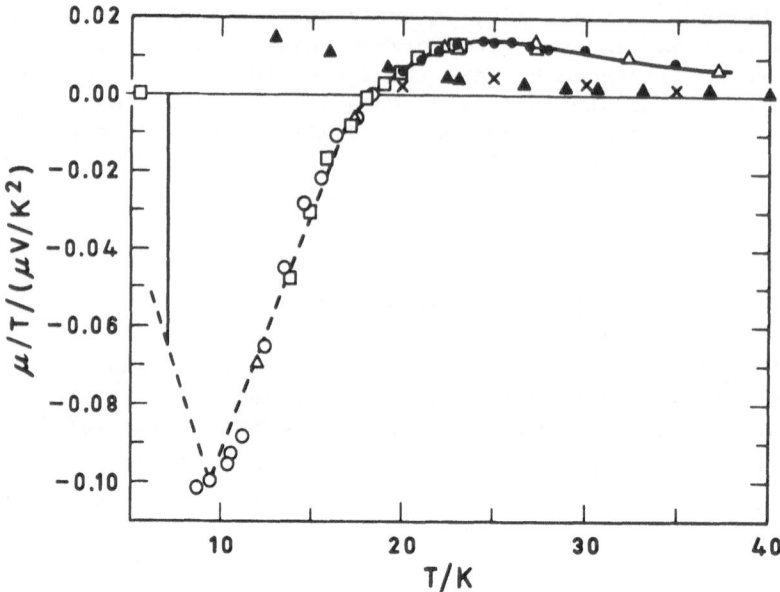

Fig. 1.   Thomson heat as μ/T.
---- from superconducting experiment of Christian et al.[2] for lead.

×   from Borelius'[3] work, as assembled by Christian et al. for lead.
▲   present results for 1.2% gold in silver alloy.
others: present results for lead, various runs.

separate the phonon drag and diffusion components of thermopowers
of, for example, noble metal alloys.

The way to check the accuracy of the method of course is to
measure the Thomson heat of something else, compute its thermopower by means of Eq. (1) and then measure its thermopower against
lead (viz. a thermocouple experiment). If the whole experiment is
valid then the thermopower of the test material calculated from
the Thomson heat measurement should agree with the values given
by the thermocouple experiment, as

$$S_{test} = S_{test,Pb} + S_{Pb}^{(Thomson\ heat\ expt)}. \tag{3}$$

The main point of this test is that the thermocouple experiment is
fundamentally independent of heat losses in the middle portions of
the wire and of the consistency and accuracy of its shape and
thermal conductivity, whereas the Thomson heat calculation depends
critically on these considerations. The better one chooses the
material for the test the more rigorous is the proof. For optimum

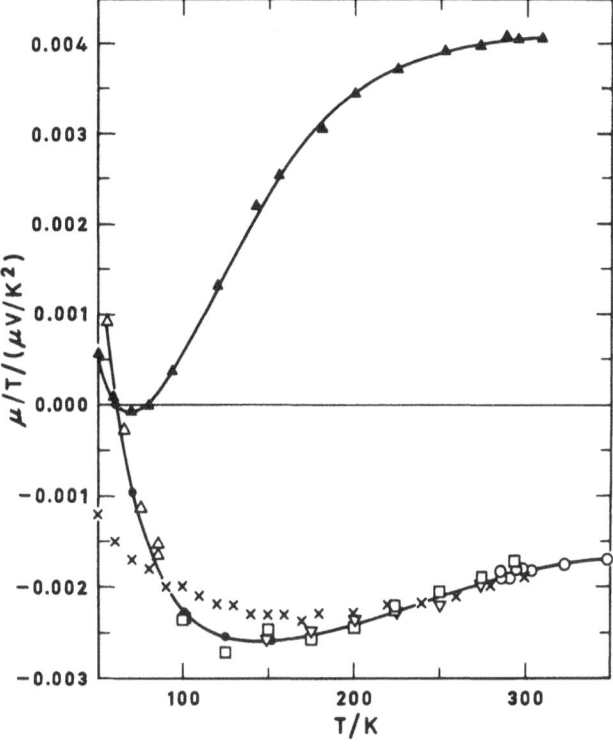

Fig. 2.   Thomson heat as μ/T.   Symbols as for Fig. 1.

accuracy with respect to thermopower the test material should have
a small Thomson heat (so that moderate relative errors have negligible effect on the calculated thermopower) and a small thermopower
(so that the components of Eq. (3) are of similar size). It saves
a lot of work if one already knows the thermal conductivity of the
test material! For optimum accuracy with respect to Thomson heat
the test material should have a Thomson heat exactly opposite to
that of the lead (viz. same magnitude, opposite sign). The
reason for this is that such an arrangement will give similar
observables for similar conditions yet be doubly sensitive to
error because the observed δt will be too small in each case (because of heat loss) yet the errors in Eq. (3) will be additive.
Clearly if the test material were similar to lead one would have
a perfect match yet it would be null and meaningless.

A test material of 1.2% gold in silver was chosen for the
following reasons. It was available. Its thermal conductivity

could be obtained by interpolation into the recent measurements of
Crisp and Rungis (1971). Its thermal conductivity varied very
slowly and uniformly from 20 to 300 K. Its thermopower was expected to be small.

In retrospect it was a very lucky choice indeed. The observed
values of μ/T for the silver gold alloy are also given in Figs. 1
and 2. In the region of most interest its Thomson heat was about
one-tenth that of lead so errors of even 20% are irrelevant to
the calculation of its thermopower. The thermopowers from Eqs.
(1) and (3) agreed over whole range within 0.02 $\mu V/K$ verifying
the present scale of the thermopower of lead to at least this
accuracy. As most of the error probably comes from the determination of the thermal conductivity of the silver alloy, the error
in the thermopower of lead is probably about ± 0.01 $\mu V/K$.

V. CONCLUSIONS

The thermopower of lead is now known from 0 to 350 K to a
precision of about 0.01 $\mu V/K$. From recalculations of the low temperature results of Crisp and Rungis[5] it appears that an alloy of
13% Ag in Au might make a reference material even better than lead,
inasmuch as its thermopower should be less than 0.05 $\mu V/K$ from
20 to 300 K. Neither should it be particularly sensitive to iron
contamination (because of the effect of the 13% silver), or to
silver content. Its thermal conductivity should be low at low
temperature, approaching that of lead above 100 K. However such
a material has the obvious disadvantage of not being universally
available. If we are to learn anything from the history of the
thermopower scale it must surely be that the saving grace is that
even if the old scale was wrong at least everyone used the same
numbers and everybody's corrections are easy to apply.

It remains to be seen if the new scale allows enough simpli-
fication of the interpretation of the thermopowers of other
materials to be worthwhile. In the meantime plans are being made
to extend an absolute scale to at least 1000°C using lead, gold
and platinum as the reference materials and to make simultaneous
thermocouple measurements on copper, tungsten, molybdenum, gold,
and platinum.

REFERENCES

1. R.B. Roberts, Nature 265, 226-27 (1977).
2. J.W. Christian, J.P. Jan, W.P. Pearson and I.M. Templeton, Proc. R. Soc. A 245, 213-221 (1958).
3. G. Borelius, W.H. Keesom, C.H. Johansson, and J.O. Linde, Proc. Acad. Sci. Amst. 35, 10-14 (1932).
4. R.B. Roberts, Phil. Mag. (in press).
5. R.S. Crisp and J. Rungis, Phil. Mag. 22, 217-36 (1970).

EVIDENCE OF AN ANOMALOUS THOMSON EFFECT

P. B. Jacovelli and O. H. Zinke

Department of Physics, University of Arkansas

Fayetteville, Arkansas   72701   USA

Research completed since the title of this paper was submitted indicates that a more correct title would have been "Evidence of an Anomalous Peltier Effect."

The anomalous Peltier effect was discovered, however, as a result of initial attempts by C. E. Canada and myself to measure Thomson coefficients through the use of an AC magnetic Wheatstone Bridge,[1] a device which senses small temperature changes in metal sample through changes in eddy-current resistance in the sample. Through the use of this device we were able to measure Thomson coefficients where the temperature at the point of measurement (T) was not greatly different from the local ambient temperature (T_O), a temperature range heretofore accessible with great difficulty for metals of low resistivity. The results of these experiments will be submitted shortly to Physical Review. Our measured Thomson coefficients showed good agreement with literature values for high resistivity metals such as nickel as can be seen in Fig. 1 where our values are compared to those of the most recent Thomson measurements, which were made by Maxwell, Lloyd and Keller,[2] and an earlier measurement by Borelius.[3] For low resistivity metals such as copper and silver, however, our measured Thomson coefficients showed trends toward infinity as the difference ($T-T_O$) decreased to zero. Possible sources of systematic and computational error did not seem to account for these latter observations. Analysis of our early copper data seemed to indicate that thermal radiation was a factor.

The system was redesigned to produce a single, stable value of T_O, and R. C. Norris and P. B. Jacovelli carried out an extensive series of measurements on a single, silver sample which are shown

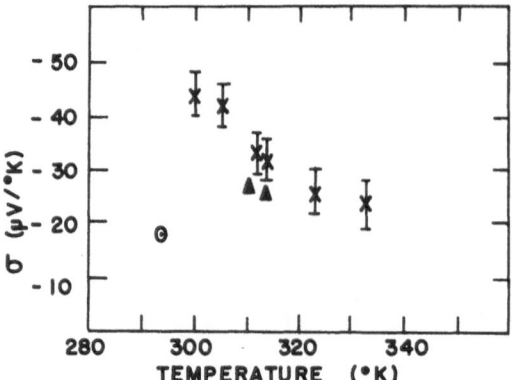

Fig. 1. Measured Thomson Coefficients for Nickel. The closed
triangles are from Maxwell, Lloyd and Keller.[2] The circle with the
dot is from Borelius.[3] The crosses are from Canada and Zinke (to
be submitted).

on Fig. 2. Here, measured Thomson coefficients are plotted against
both positive and negative values of $(T-T_0)$. A trend toward minus
infinity in the measured Thomson coefficients can be seen as negative values of $(T-T_0)$ approach zero and a similar trend to plus
infinity is seen as positive values of $(T-T_0)$ approach zero.

From an exhaustive review of measured Thomson coefficients made
by J. B. Sawyer[4] it was apparent that no previous measurements had
been made for negative values of $(T-T_0)$ or for that matter, for
small values of $(T-T_0)$. Indeed, if $(T-T_0)$ is a variable in measured
Thomson coefficients, our measurements did not seem to contradict
any reported in the literature. While conditions surrounding our
measurements may not be comparable to previous measurements for iron,
a composite of Thomson measurements for iron[4] seems to show an up-
swing similar to ours for silver for positive $(T-T_0)$ as can be seen
from Fig. 3.

Jacovelli and I reasoned that if the effect was real we should
be able to alter the emf of a copper-iron thermocouple by changing
the radiation patterns along the conductors between high-temperature
(T_h) and low-temperature (T_c) reservoirs. Alterations of approximately 50% were observed in a series of experiments one of which
is schematized in Fig. 4. Here, one thermocouple is placed inside
a copper rod which is thermally attached to T_h and T_c, and the other
thermocouple is in ambient air. Both thermocouples are constructed
from one piece of copper and one piece of iron wire arc-welded near
their centers with the weld serving as the junction at T_h and with
the four ends in the reservoir at T_c. This configuration forms

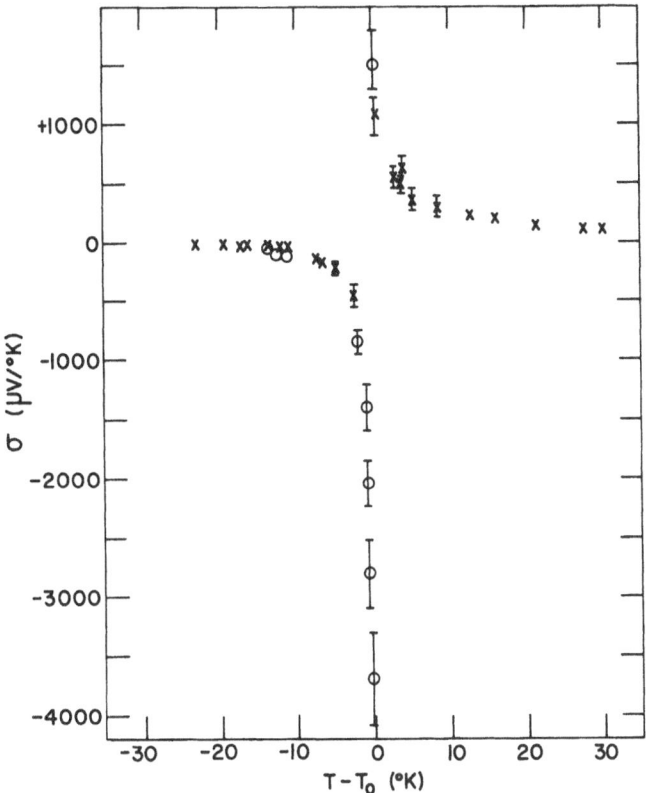

Fig. 2. Measured Thomson Coefficients for Silver. The crosses
represent measurements made by Norris, the open circles represent
measurements made by Jacovelli. The results are to be submitted
for publication. The abscissa represents the difference between
the absolute temperature at point of measurement and the ambient
temperature.

four copper-iron thermocouples, Cu-Fe, $\overline{\text{Cu}}$-$\overline{\text{Fe}}$, $\overline{\text{Cu}}$-Fe, and Cu-$\overline{\text{Fe}}$
where the bar designates a conductor enclosed in the copper rod.
The emfs of these thermocouples are shown on Fig. 5. Displayed on
the upper curve are the Cu-Fe emfs which corresponds to literature
values.[5] On the lowest curve are the $\overline{\text{Cu}}$-$\overline{\text{Fe}}$ emfs, which are about
50% of the upper curve at similar temperatures. In addition we
observed extremely high values of Benedicks emfs for iron, over 600
microvolts for large values of $(T_h - T_c)$ as can be seen on Fig. 6.
There is some question in our minds about the displayed Benedicks
curve for copper. The experiment was repeated with copper-Constantan

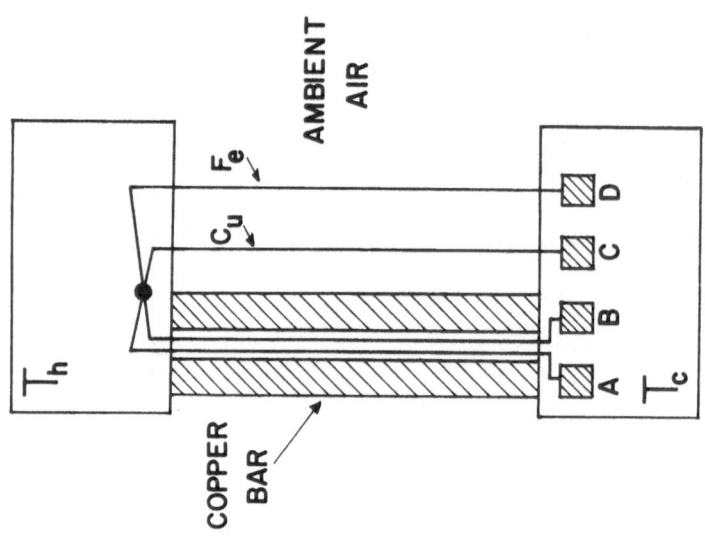

Fig. 4. Schematic of Thermocouple Experiment. The copper and iron wires are single wires arc welded at the T_h reservoir. One thermocouple shown is in air and another is in the center of a copper rod which is thermally joined to T_h and T_c reservoirs.

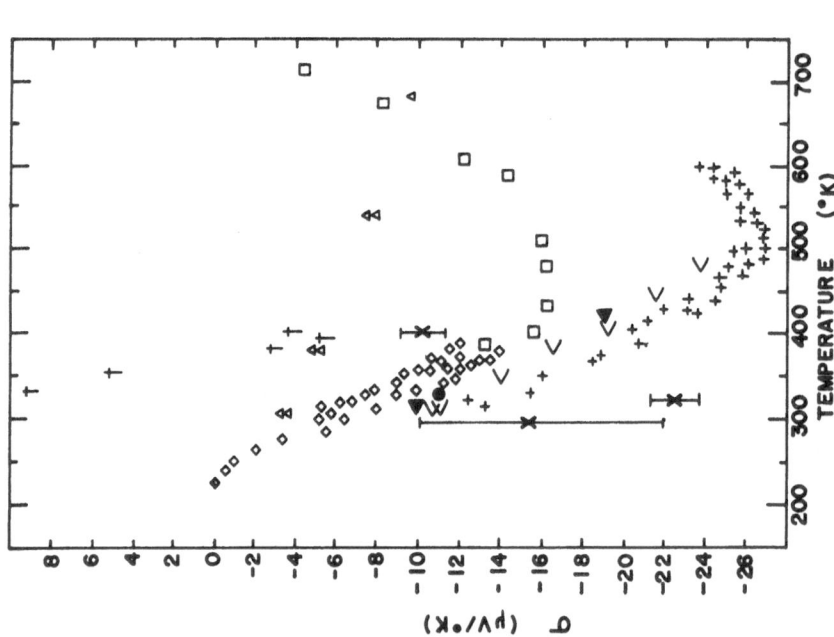

Fig. 3. Measured Thomson Coefficients for Iron. The various symbols represent various experimenters. The key may be found in Ref. 4. The abscissa is absolute temperature.

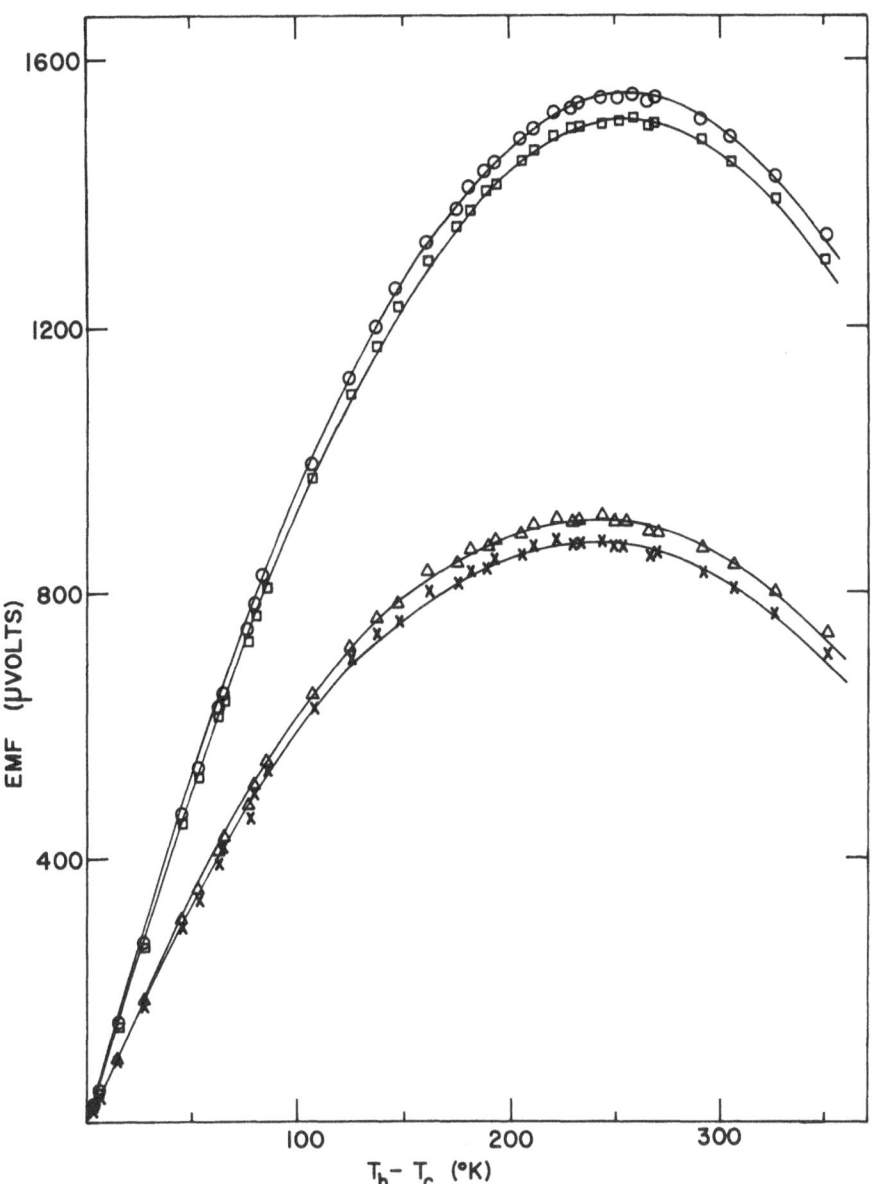

Fig. 5. EMF'S From the Four Thermocouples of Fig. 4. The top curve
is the emf of the Cu-Fe couple, the next is the C̄u-Fe, the next is
the Cu-F̄e, and the lowest curves is the C̄u-F̄e, where the bars indicate that the conductor is in the copper rod and symbols without a
bar indicate that the conductor is in air.

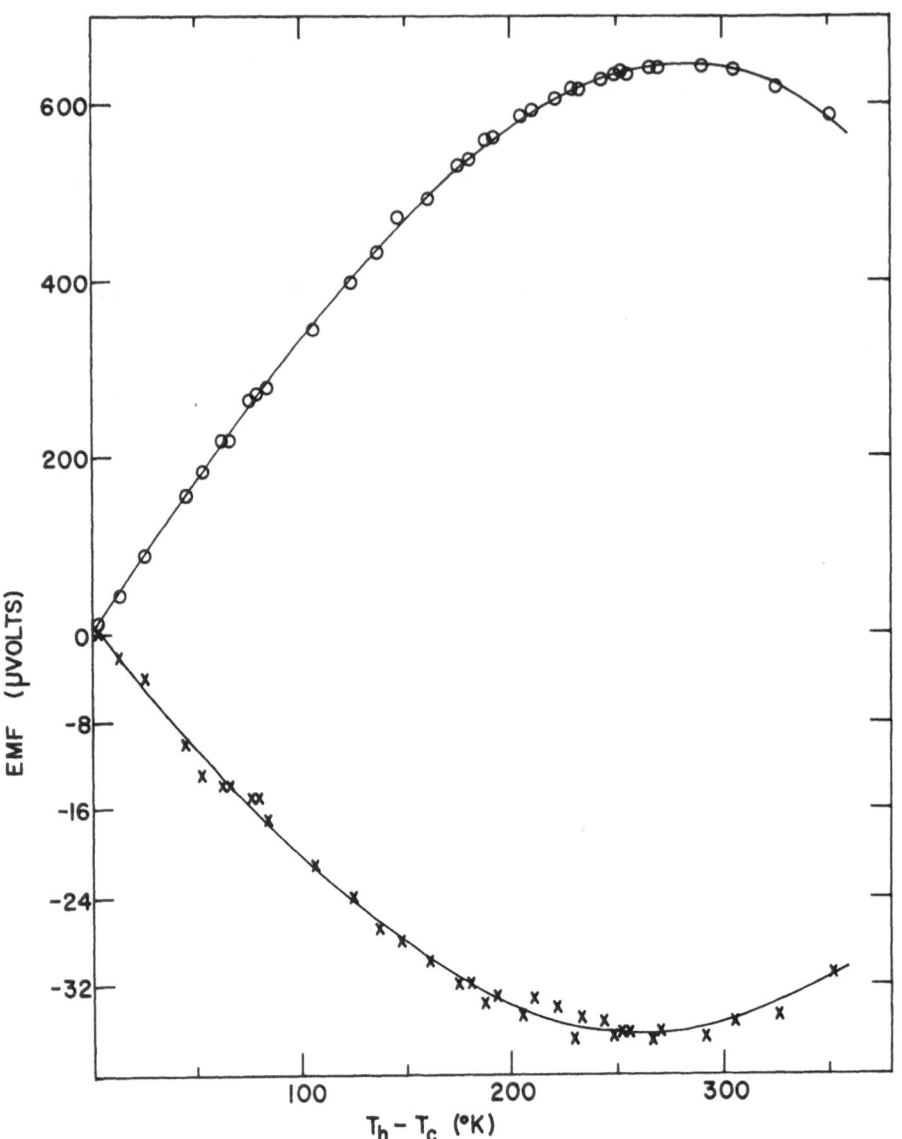

Fig. 6. Benedicks EMF'S From Iron-Iron and Copper-Copper Thermocouples. The upper curve is from the Fe-Fe thermocouple of Fig. 4
and the lower curve is from the Cu-Cu thermocouple of Fig. 4.

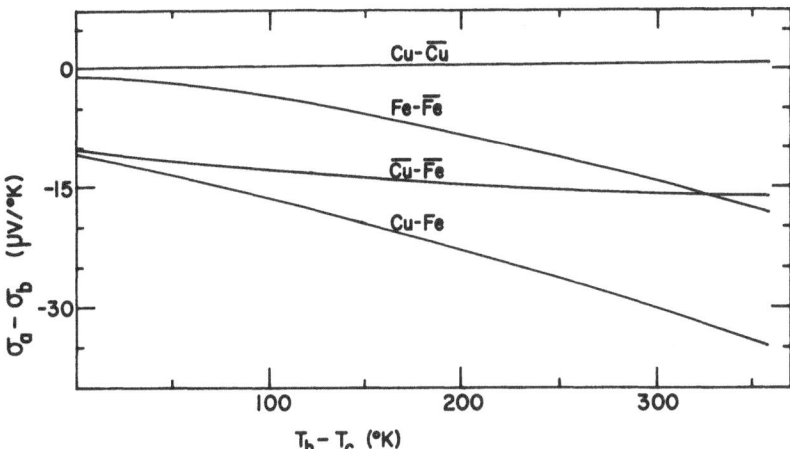

Fig. 7. Calculated Differences in Thomson Coefficients For the Data Presented in Figs. 5 and 6. At the ordinate the top curve is the Cu-Cu thermocouple, the next is the Fe-F̄e, the next the C̄u-F̄e, and the bottom curve is from the Cu-Fe thermocouple.

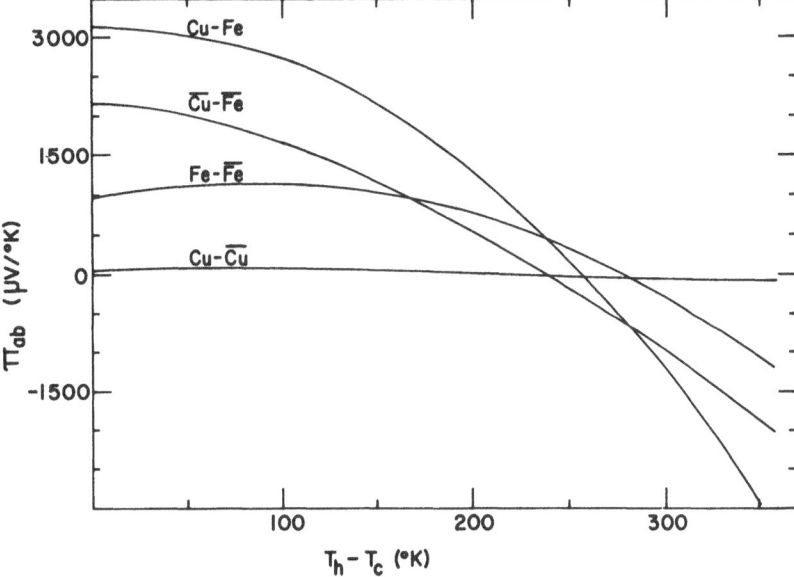

Fig. 8. Calculated Peltier Coefficients For the Data Presented in Figs. 5 and 6. At the ordinate the top curve is from the Cu-Fe thermocouple, the next is from the C̄u-F̄e, the next is from the Fe-F̄e, and the bottom is from the Cu-C̄u thermocouple. Note that the Peltier coefficient of the iron-iron couple is of the order of 1000 microvolts/degree K.

and a different method of assembly of the same apparatus, and the
observed maximum different between enclosed and nonenclosed curves
was less than 10 microvolts which corresponded to the maximum value
of the Benedicks emf for copper.

An analysis of the data of Figs. 5 and 6 based on the thermocouple equations relating the derivatives of the emfs to Peltier and
Thomson coefficients indicates that what we are seeing is not a
Thomson phenomenon. These equations are formulated for the limit
as (T_h-T_c) approaches zero. Fig. 7 displays the calculated values
of differences of Thomson coefficients, and it will be seen that in
the limit as (T_h-T_c) approaches zero the differences tend to zero
for the Benedicks emfs and to the same value for the enclosed and
nonenclosed thermocouples. The indication is that whatever is af-
fecting the output of the thermocouples between the reservoirs is
not a Thomson phenomenon. However, Fig. 8, which displays the
results of the Peltier calculations, shows a clear difference between the Peltier coefficients of enclosed and nonenclosed copper-
iron thermocouples and even shows that the iron-iron thermocouple
has a well-defined Peltier coefficient. The Peltier effect is indeed anomalous in that it obviously does not occur at a junction.

We have other evidence that this anomalous effect exists and
that it is a radiation effect.

REFERENCES

1. O.H. Zinke and P.B. Jacovelli, Rev. Sci. Inst. 36, 916 (1965).
2. G.M. Maxwell, J.M. Lloyd, and O.V. Keller Jr., Rev. Sci. Inst.
    38, 1084 (1967).
3. G. Borelius, Ann. de Phs. 56, 388 (1918) and 63, 845 (1920).
4. J.B. Sawyer, "A Review of Thomson Effect Measurement Tech-
    niques," a thesis submitted in partial fulfillment of the
    requirements for a degree of Master of Science, University of
    Arkansas (1975).
5. A.E. Caswell, "International Critical Tables" Vol. 6, 213
    (McGraw-Hill, New York, New York, 1929).

PURE NOBLE METALS

A.M. Guénault and D.G. Hawksworth*

Physics Department
University of Lancaster
Lancaster, LA1 4YB, United Kingdom

ABSTRACT

It is suggested that the main features of the observed temperature-dependence of S at temperatures below 15 K in Cu, Ag and Au should be interpreted as a change in the dominant electron scattering mechanism. The outstanding result is then that S for the ideally pure metals is strongly negative below about 10 K.

Recent experiments have been made of the low temperature thermoelectric power of pure copper and silver by Rumbo,[1] of pure silver by Ewbank et al.[2] and of pure gold by ourselves (Guénault and Hawksworth,[3] where a fuller account of this work will be found). All of these measurements show a highly non-linear dependence of S on temperature, and in the past attention has been paid to identifying this non-linearity with a phonon drag contribution to S.

The main point of this paper is to point out that any such interpretation must first allow for the cross-over of electron scattering from a 'dirty' regime at the lowest temperatures towards a 'pure' regime at higher temperatures. We can estimate the cross-over temperature T* around which this change occurs from a simple one-band approximation as follows (although in reality a detailed treatment should be of a multiband nature, with different T* values in each band). The thermopower is given by

$$S = (S_o W_o + S_T W_T)/(W_o + W_T)$$

* Present address: Applied Superconductivity Laboratory, University of Wisconsin, Madison, Wisconsin.

where W_O and W_T are respectively the thermal resistances for residual and phonon scattering, and S_O and S_T the characteristic thermopowers for the two scattering mechanisms. The cross-over from S_O to S_T occurs at a temperature T* at which $W_O \simeq W_T$. Using the method of Guénault,[4] one obtains

$$T*^3 = 300 \; \theta_D^2/C.RR$$

where RR is the resistance ratio $\rho(300 \text{ K})/\rho(4.2 \text{ K})$, and C defines the magnitude of W_T. As an example, this gives T* ~ 8 K for our purest oxidised gold sample (RR = 1400), and with this value in mind, the experiments can be interpreted as a switch from S_O (which appears to be small and positive) which dominates at low temperatures, to S_T which becomes important at and above 8 K. This interpretation indicates that S_T itself is positive above 12 K in gold, but becomes strongly negative below this temperature (having a magnitude about -50 nVK^{-1} at 10 K). Purer samples (and hence lower T* values) would be needed for S_T to be studied down to lower temperatures. Similar conclusions about S_T appear to be valid also for Ag and (with slightly less confidence because of residual iron impurity) for Cu.

The observation that S_T is negative at low temperatures is of much interest since:

   (i)   it is in a regime where electron-phonon and phonon-electron coupling are strong, so that a separation into distinct electron diffusion and phonon drag contributions cannot readily be made.
  (ii)   the electron scattering is strongly inelastic, so that the usually simple theoretical approaches to S are not possible.

 (iii)   there is a possible role for electron-electron scattering (e.g., Lawrence[5]) in addition to electron-phonon scattering.

REFERENCES

1. E.R. Rumbo, J. Phys. F: 6, 85 (1976).
2. M. Ewbank, J.L. Imes, W.P. Pratt, P.A. Schroeder and J. Tracy, Phys. Lett. 59A, 316 (1976).
3. A.M. Guénault and D.G. Hawksworth, J. Phys. F: 7, (to be published) (1977).
4. A.M. Guénault, J. Phys. F: 1, L1 (1971).
5. W.F. Lawrence, Phys. Rev. B13, 5316 (1976).

HOMOVALENT NOBLE METAL ALLOYS

A. M. Guénault

Physics Department, University of Lancaster

Lancaster, LA1 4YB, ENGLAND

ABSTRACT

Measurements on dilute homovalent alloys of noble metals show that the magnitudes of the thermopowers at low temperatures can differ markedly from the alloy contributions extracted from room temperature measurements. We discuss possible reasons for this which include effects of relaxation time anisotropy and many body scattering contributions.

During recent years, the Lancaster group have made a systematic study of the thermoelectric properties of dilute homovalent alloys of the noble metals, using oxidation treatments to evaluate and eliminate the contributions to the thermopower of trace iron impurity. A recent paper (Dee and Guénault[1]) includes a compilation of our most reliable data, so that this is not reproduced in detail here.

In this paper we focus on the electron-diffusion contribution $S_{e,o}$ which arises from the electron-impurity scattering. Our results indicate very marked differences between low temperatures where impurity scattering dominates and at room temperature, where phonon scattering dominates and one uses the Nordheim-Gorter rule to estimate $S_{e,o}$ from the difference in thermopower between the dilute alloy and the pure metal. The results are summarised in Table 1, in terms of the usual thermopower parameters x defined from:

$$S_e = \frac{\pi^2}{3} \frac{k}{e} \frac{kT}{E_F} x.$$

TABLE 1

Alloy	x_T	x_o^{NGR}	x_o
Ag Au	− 1.12	+ 1.6	+ 0.8
Ag Cu		?	− ve?
Cu Au	− 1.63	+ 0.3	− 1.2
Cu Ag		+ 2	− 4.4
Au Ag	− 1.47	+ 0.9	− 4.8
Au Cu		− 0.2	− 2.5

Table 1:   Electron diffusion thermopower parameters. In the table, x_T represents the thermopower parameters for the pure metal at room temperature: x_o^{NGR} is from room temperature measurements of the alloys and the Nordheim-Gorter rule; x_o is the measured value at helium temperatures.

The dramatic nature of these differences is readily appreciated from, say, Au Ag alloys, where at room temperature the x_T of − 1.47 for Au is made more positive (i.e. S made more negative) on alloying, whereas the low temperature x_o at − 4.8 is much more negative than x_T. The change on alloying seems to go in the wrong direction.

There are at least two explanations (not exclusive) for this behaviour. Firstly, as discussed in detail by Dee and Guénault[1], one must surely include the influence of relaxation time anisotropy in any discussion of the noble metals. At low temperatures the weighting between necks and bellies (to use a simple picture) is governed by the electron-impurity relaxation times, whereas at room temperature the differential effect of alloying will have neck-belly weightings governed by the electron-phonon relaxation times. We have shown that the results can be completely described using experimental relaxation time anisotropies from dHvA work, if we assume thermopower anisotropies of typically $x_{o,belly} = + 1$, $x_{o,neck} = - 10$, which are not implausible in magnitude or in sign.

However, secondly, these measurements are also candidates for observing in $S_{e,o}$ the type of many-body scattering effects discussed by Nielsen and Taylor[2] and Hasegawa[3]. These virtual-recoil terms, involving absorption and re-emission of a virtual phonon during the electron-impurity scattering process, should be present at low temperatures, but absent at room temperatures where $T \gtrsim \theta_D$. These various new terms include some dependent on host pseudopotentials only (Δx_C, say) and some on the difference between solvent and solute potentials (Δx_B) which should therefore be opposite in sign for say Au Ag and Ag Au. If we neglect relaxation-time anisotropy entirely, the results of Table 1 are roughly consistent with the various Δx_C being -2 to -4 and $|\Delta x_B|$ being up to about 4. Actually these x values seem rather too large to be reconciled with the theory, since $(E_F/k\theta_D)^2$ (m/M), which is the order of magnitude of the expected term if the appropriate pseudopotentials are roughly equal to E_F, is 0.4 to 0.5 only in the three noble metals.

However, this does perhaps point out that there is a good possibility of identifying Nielsen-Taylor effects in $S_{e,o}$ by studying the high and low temperature thermoelectric power in an alloy where relaxation time anisotropy is the same in both limits. K Rb is an interesting possibility here, since one expects the scattering to be almost isotropic in both limits, and recent measurements have been made at Leeds (T. Dosdale, J. S. Dugdale, private communication); however the room-temperature measurements are not easy to do with sufficient accuracy, and the analysis of the data is somewhat inconclusive giving for an 18% Rb in K alloy: $x_0 = -0.6$ (Guénault and MacDonald[4]) $x_0^{NGR} = +0.2 \pm 0.5$. Certainly these values are fairly close together, remembering that x_T is $+3$.

REFERENCES

1. R.H. Dee and A.M. Guénault, J. Phys. F. 7, 153 (1977).
2. P.E. Nielsen and P.L. Taylor, Phys. Rev. B10, 4061 (1974).
3. A. Hasegawa, J. Phys. F. 4, 2164 (1974).
4. A.M. Guénault and D.K.C. MacDonald, Proc. Roy. Soc., A264, 41 (1961).

THE THERMOPOWERS AND RESISTIVITIES OF COPPER ALLOYED WITH HOMOVALENT AND POLYVALENT IMPURITIES AND THE VALIDITY OF THE NORDHEIM-GORTER RELATION

A. Rahim Chowdhry and R. D. Barnard

Department of Pure and Applied Physics

University of Salford, Salford M 5 4WT UK

Although the Nordheim-Gorter rule (NGR) has been used for many years to analyse and predict the diffusion thermopower of dilute alloys, it is surprising that no very detailed examination of its validity has been performed over a wide range of temperatures. Its apparent success in predicting the observed linearity of the diffusion thermopower S_D versus reciprocal of the total electrical resistivity (ρ^{-1}) for a series of dilute alloys of different composition has led to a general acceptance of the rule, but no detailed examination has been made to test that the intercept $S_o{}^{NGR}$, (the characteristic thermopower of the solute) is linear in T as the NGR predicts. However, failure of the NGR was first indicated by Guénault[1] who showed that $S_D{}^{NGR}$ T^{-1} measured at room temperature in noble metal systems was considerably different from the value obtained on the same alloys at very low temperatures. Subsequently Barnard[2] suggested that the original NGR for the diffusion thermopower of an alloy S_D^A given by

$$S_D^A = \frac{\rho_{th} S_{th} + \rho_o S_o}{\rho_o + \rho_{th}} \qquad (1)$$

where ρ_{th} and ρ_o are respectively the thermal and impurity electrical resistivities and S_{th} and S_o the characteristic thermopowers of the solvent and solute, should be modified to

$$S_D^A = \frac{\rho_{th} S_{th} + \rho_o S_o + \Delta S_\Delta}{\rho_o + \rho_{th} + \Delta} \qquad (2)$$

43

thereby including the deviation from Matthiessen's rule (DMR), and introducting S_Δ the characteristic thermopower for DMR. It was shown that provided Δ/ρ_o was independent of concentration C of impurity, (then denoted by k) as is usually the case with dilute alloys in the temperature region where diffusion thermoelectricity is observed, (say T >200K) then S_D^A was still proportional to ρ^{-1} according to

$$S_D^A = \frac{S_o + k\,S_\Delta}{1 + k} + \frac{\rho_{th}}{\rho}\left(S_{th} - \frac{S_o + kS_\Delta}{1 + k}\right) \qquad (3)$$

but now

$$S_o^{NGR} = \frac{S_o + k\,S_\Delta}{1 + k} \quad \text{or} \quad \frac{S_o^{NGR}}{T} = \frac{\left(\dfrac{S_o}{T}\right) + \left(\dfrac{S_\Delta}{T}\right)k(T)}{1 + k(T)} \qquad (4)$$

On the assumption that $S_\Delta \propto T$ equation (4) provides a basis for understanding why $S_o^{NGR} T^{-1}$ and $S_o T^{-1}$ are different and furthermore the variation of $S_o^{NGR} T^{-1}$ with T becomes, for a given system, entirely dependent on k(T). According to the original NGR, $S_o^{NGR} T^{-1}$ is constant at all temperatures for a given system.

In an examination of equation (4) we have chosen alloy systems in which k(T) is independent of T (Ag–Au), k(T) decreases with T, (Cu–Au) and k(T) increases with T (Cu–Zn, Cu–Ga, Cu–Ge). Such temperature dependences of k should, if equation 4 is valid give three distinctly different variations for the temperature dependence of $S_o^{NGR} T^{-1}$.

Figure 1 shows the variation of $S_o^{NGR} T^{-1}$ as a function of T between 200 and 600 K for the above alloy systems in the temperature range where diffusion thermoelectricity only is observed. In all cases S_o^{NGR} was obtained from extrapolation of the linear regions of S_D^A versus ρ^{-1} plots. The values at OK have been taken from Guénault[1,3] and the data on Ag–Au from Crisp and Rungis[4]. There is clearly no agreement with the original NGR, but the results are quite compatible with the predictions of equation 4 provided $S_\Delta T^{-1}$ is negative in each case. Figure 2 shows calculated values of $S_o^{NGR} T^{-1}$ for different values of $S_\Delta T^{-1}$ where good agreement with experiment exists if $S_{\Delta_2} T^{-1}$ is -27 nVK^{-2} (Cu–Au), $-35 nVK^{-2}$ (Cu–Ge) and -40 nVK^{-2} for Ag–Au[2].

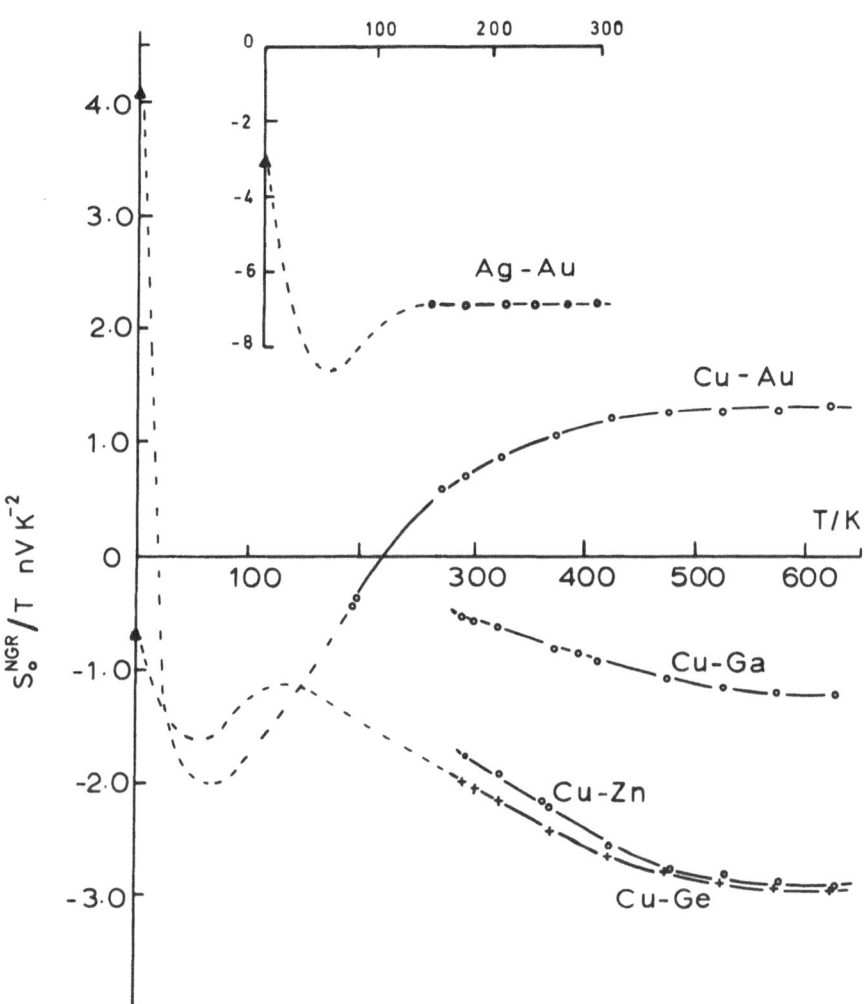

Fig. 1. S_o^{NGR} plotted as a function of temperature for Ag-Au, Cu-Au, Cu-Zn, Cu-Ga and Cu-Ge alloys.

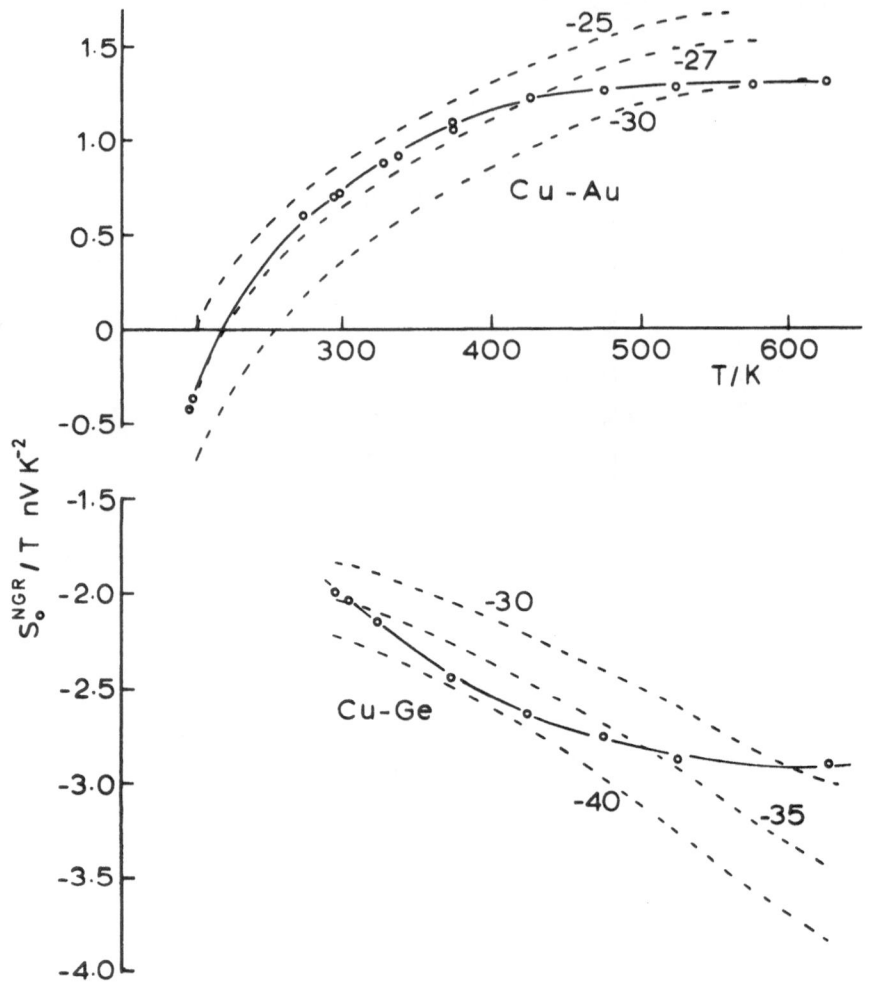

Fig. 2. $S_o^{NGR} T^{-1}$ as a function of T calculated using equation (4)
for different values of $S_\Delta T^{-1}$. The experimental curves for Cu–Au
and Cu–Ge are also shown.

        The direct application of equation (2) when T <200K is not
possible because phonon scattering of electrons becomes increasingly
inelastic and the Wiedemann–Franz law is not obeyed. Strictly
Kohler's formula should be applied here in which the electrical
resistivities of equation (1) or (2) are replaced by thermal resistivities. We can however simply modify equations (1) and (2) to be
applicable in the low temperature region by adopting a procedure

given by Huebener[5] in which ρ_{th} is replaced by $\rho_{th}\left(\dfrac{L_oT}{\rho_{th}\kappa}\right)$

where L_o is the Lorenz number and κ the electronic thermal conductivity of the pure metal. This effectively replaces the electrical resistivity by the thermal resistivity. Since impurity scattering is elastic no modification to ρ_o is required.

In applications at low temperatures, it is the phonon-drag attenuation ΔS_g accompanying alloying that is usually required in which case the relative thermopower of the pure metal and alloy are measured directly i.e.

$$S^A - S^P = \Delta S = \Delta S_g + \Delta S_D \quad .$$

With the modified NGR corrected for change of Lorenz this becomes

$$\Delta S = \Delta S_g + \frac{\rho_o}{\left(\rho_{th}\left(\dfrac{L_oT}{\rho_{th}\kappa}\right) + \rho_o + \Delta\right)}\left[\left(S_o - S_{th}\right) + \frac{\Delta}{\rho_o}\left(S_\Delta - S_{th}\right)\right] \quad (5)$$

while the unmodified form is

$$\Delta S = \Delta S_g + \frac{\rho_o}{\rho_{th}\left(\dfrac{L_oT}{\rho_{th}\kappa}\right) + \rho_o}\left[S_o^{NGR} - S_{th}\right] \quad (6)$$

In figure 3 is shown the calculated variation of ΔS_D for Cu - 0.1 at % Au and Ag - 0.1 at % Au obtained using equations (5) and (6). In equation (6) the values of S_o^{NGR} at 300 K have been used as has been done by previous workers. There is clearly considerable difference in the values of ΔS_D given by the two equations and in both cases the modified NGR produces turning points which occur at the same temperature as the peaks in the total thermopower. Clearly the modified NGR gives entirely different values for ΔS_g from those previously calculated. Indeed in the case of Ag-0.1 at % Au where ample data are available[4], the modified ΔS_D curve quite closely matches the change in the total thermopower observed by Crisp and Rungis in Ag - 0.09 at % Au leaving ΔS_g at 30 K approximately five times smaller than the previous estimate. This result could explain why Crisp and Rungis found that the calculated values of the attenuation of S_g based on Klemens'[6] formula for phonon scattering by point defects (Rayleigh scattering), were about an order of magnitude

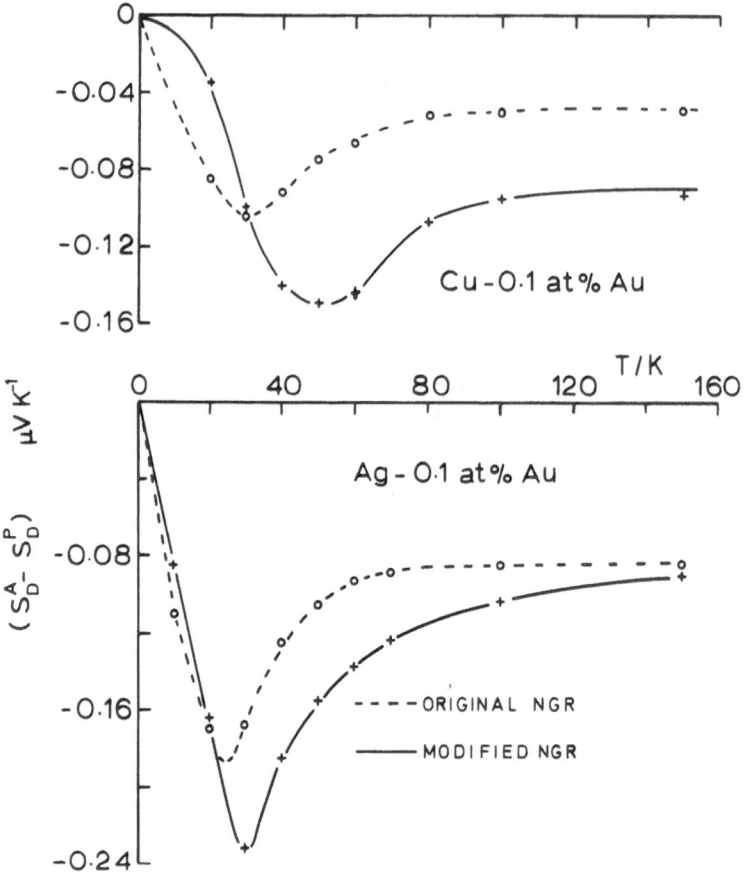

Fig. 3.   The change of diffusion thermopower calculated using the
original and modified NGR.   In each case corrections due to the
change of Lorenz number have been included.

smaller than their estimates of ΔSg.   It should also be pointed out
that the modified NGR, equation (5) is also capable of producing the
two turning points observed in the total ΔS of some $A\ell$ alloys[2], previous explanations having been associated with anisotropy of relaxation times[7] and the Nielsen-Taylor effect[8].

    The alternative form of the NGR in which a single alloy is
measured at a series of temperatures and plots of $S_D^A T^{-1}$ versus
ρ^{-1} made are also compatible with the modified NGR, and the high
temperature results have made possible accurate extrapolation to
obtain S_{th} as a function of impurity concentration.   Figure 4 shows

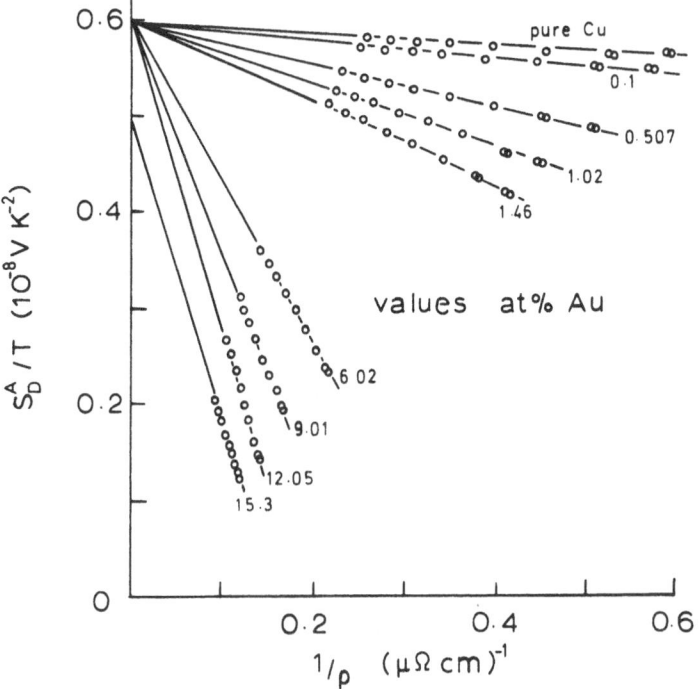

Fig. 4. Alternative NG plot for Cu-Au alloys. The intercept gives $S_{th}T^{-1}$ as a function of concentration.

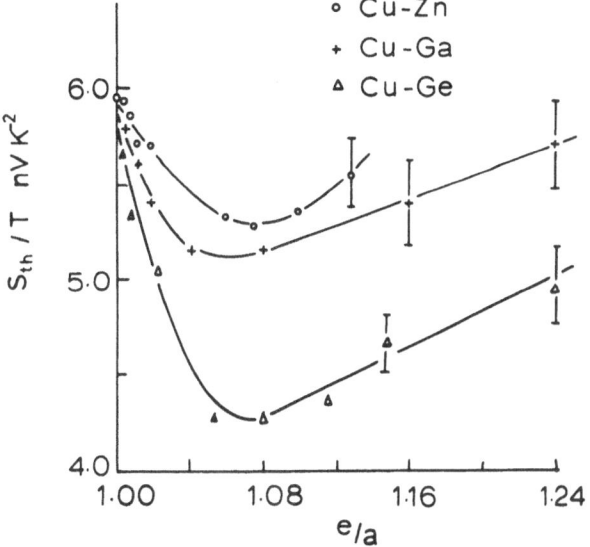

Fig. 5. $S_{th}T^{-1}$ versus electron-atom ratio for Cu-Zn, Cu-Ga and Cu-Ge alloys.

the results for Cu–Au where $S_{th} T^{-1}$ was found to be constant up to 12 at % Au but for Cu–Zn, Cu–Ga and Cu–Ge (figure 5), $S_{th} T^{-1}$ decreased initially reaching a minimum in each system at the electron

atom ratio $\frac{e}{a}$ = 1.08.   The minimum at $\frac{e}{a}$ = 1.08 suggests a connection

with the resurgence of the phonon–drag component in these alloy systems which also occurs at this electron density[9], and suggests it is due to a band structure effect resulting from the swelling of

the Fermi surface as $\frac{e}{a}$ increases and possibly associated with the

proximity of the $\{200\}$ zone faces.

REFERENCES

1.   A.M. Guénault, J. Phys. F 2, 316 (1972).
2.   R.D. Barnard, J. Phys. F 5, 99 (1975).
3.   A.M. Guénault J. Phys. F 4, 256 (1974).
4.   R.S. Crisp and J. Rungis, Phil. Mag. 22, 217 (1970).
5.   R.P. Huebener, Phys. Rev. 135, A1281 (1964).
6.   P.G. Klemens, Proc. Phys. Soc. A 68, 1113 (1955).
7.   R.P. Huebener, Phys. Rev. 171, 634 (1968).
8.   A.W. Dudenhoeffer and R.R. Bourassa, Phys. Rev. B5, 1651 (1972).
9.   R.S. Crisp, W.G. Henry and P.A. Schroeder, Phil. Mag. 10, 553
      (1964).

ELECTRON TRANSPORT IN Ag-Hg ALLOYS

R. Craig and R. S. Crisp

Physics Department

University of Western Australia

ABSTRACT

Measurements are reported of thermopower, thermal conductivity
and electrical resistivity on a series of α-phase Ag-Hg alloys
from 2-300 K. There is a rapid attenuation of the lattice component of thermal conductivity as expected for the large mass difference and a corresponding attenuation of the peak in the thermopower. As in all the other noble metal systems the high temperature
data can be represented by a Nordheim-Görter relation down to at
least 160 K.

EXPERIMENTAL

The measurements reported here are part of a continuing
systematic study of the noble-metal alloy systems. The samples
were supplied by Cambridge Metals Research as 1.0 mm diam. wires
prepared by diluting a master alloy. The lower concentration
alloys were drawn down to a diameter suitable for the thermal
conductivity measurement and all the samples were then annealed
for 24 hours at 600°C in 1/3 atmos. of pure argon in close fitting
pyrex capillary capsules.

A few alloys were also prepared by diffusing mercury into pure
silver wires. Repeated measurements on different pieces of the
same sample for a number of concentrations confirmed that there was
no macroscopic inhomogeneity in the specimens. The concentrations
were inferred from the nominal concentrations and measurements of
the residual resistivity using the resistivity increment data of
Linde (1939).

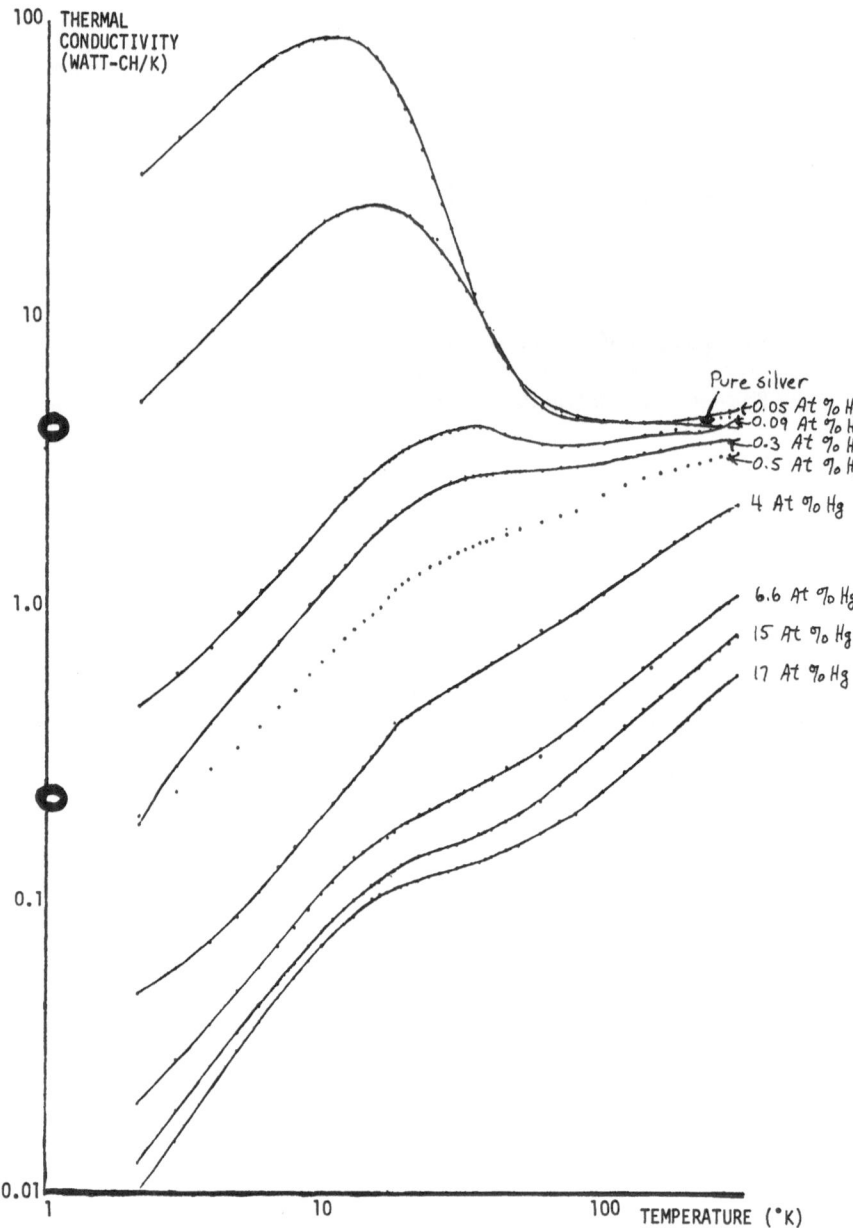

Fig. 1. Measured thermal conductivities as a function of temperature for the Ag-Hg system.

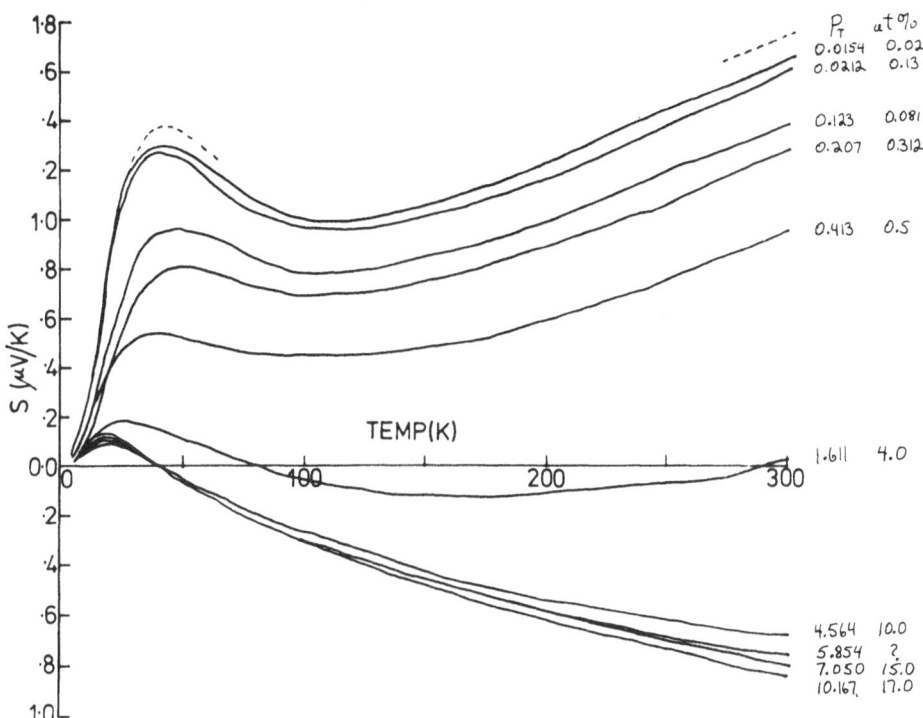

Fig. 2. Measured thermopowers as a function of temperature for the Ag-Hg system.

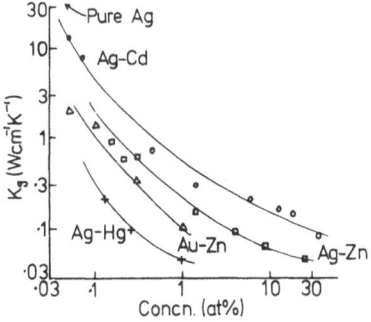

Fig. 3. K_g as a function of concentration for various solutes in Ag.

The total thermal conductivities and thermopowers for all the alloys are shown in Figs. 1 and 2. The lattice thermal conductivity (K_g) has been separated from the electronic component in the usual way and is shown along with K_g for some other solutes in silver in Fig. 3. The large mass difference for the Ag-Hg system explains the very rapid attenuation when Hg is the solute. At 1 at % solute K_g has decreased to a few percent of its value in the pure metal.

The thermopower is typical of a noble metal system and shows a small resurgence of the peak at the highest concentration. The attenuation of the peak appears to saturate at about 4 at % Hg in a very similar fashion to that observed for Ag-Cd and some other series. Nordheim-Görter plots are linear out to surprisingly high concentrations ($\sim$15 at %, electron/atom ratio 1.15) from 300 K down to at least 160 K. NG plots using the thermal resistivities are subject to greater scatter but represent the data equally well at the higher temperatures. The intercept [S_o^{NGR} in the notation of Guénault (1972)] is linear with temperature with an extrapolated value at T = 0 of +0.3 μV/K, very close to that for the solutes Zn and Cd in Ag. Further discussion of these data is included in the paper in these Proceedings by Crisp.

ACKNOWLEDGMENT

This work has been supported by the Physics Department Research Grant Committee through a block grant from the University of Western Australia. R. Craig gratefully acknowledges assistance in the form of a UWA Postgraduate Scholarship.

REFERENCES

1. R.S. Crisp and W.G. Henry, 1977, to be published.
2. A. M. Guénault, J. Phys. F., 2, 316, (1972).
3. J.O. Linde, Ph.D. Thesis, University of Stockholm, (1939).
4. R.B. Roberts and R.S. Crisp, Phil. Mag., in press (1977).
5. R.B. Roberts, Ph.D. Thesis, University of Western Australia (1973).

DISCUSSION

J. F. Goff: As I remember from semiconductors, the phonons involved in phonon drag and the phonons involved in lattice thermal conductivity are not the same phonons, and therefore, there is a chance that introducing point imperfections which reduce the lattice thermal conductivity may not affect phonon drag the same way. How is this in metals?

S. Crisp: There are indeed two sorts of phonon scattering. One is inelastic scattering that gives rise to phonon drag, and these are certainly the phonons that participate in K_g. The other sort of phonon scattering is what we call thermal scattering, and that's something different altogether. That's scattering of an electron wave by the potential due to an atom being displaced. That does not require anisotropy in the phonon distribution. Phonon drag and lattice thermal conductivity imply that the phonon distribution is anisotropic. In thermal scattering, which is due to the same physical phenomenon, the phonons convey no net momentum; it takes place also in an isothermal specimen.

B. R. Coles: I'd like to make a related point. I wonder if it is fair, really, to compare your results with the behavior of the total lattice thermal conductivity. As you point out, one is concerned with the lattice term in the thermal conductivity, and the behavior of the peak as you go from pure silver to the very dilute alloys suggests very strongly that the initial effects of impurities are due to the electron limited lattice thermal conductivity. As you introduce impurity scattering you reduce the effectiveness of electrons in scattering phonons and the initial effect looks every much like it might be due to that. So, I think it is really only a component of the thermal conductivity you should be comparing results with, and you should not expect to find these vast 300 fold changes.

S. Crisp: I thought I made that clear.

F. J. Blatt: One brief comment on the question Dr. Goff raised. The point is that in the case of semiconductors it is only a small fraction of the phonon distribution which can interact with the electrons, but it is the entire phonon distribution which can contribute to the lattice thermal conductivity. In this particular case, that is, in metals, Dr. Crisp is quite right; one might well on this very simple model expect a close parallelism between K_g and S_g. In one case you are concerned with energy transport by the phonons and in the other case you are concerned with momentum transport which is then conferred upon the electrons. And since in the case of phonons at low temperature both are proportional to each other then, no matter what the scattering mechanism -- whether it goes as ω^4 as for Rayleigh scattering, or whatever -- the two phenomena should go hand in hand.

THERMOPOWER IN ALLOYS OF Ag WITH Zn, Ga, Ge AND As AT LOW TEMPERATURES

S. J. Song and R. S. Crisp

Physics Department

University of Western Australia

ABSTRACT

Measurements are reported of the thermopowers of the α-phase alloy systems of Zn, Ga, Ge and As in Ag from about 0.85K to 50K. All the systems show some evidence of Fe contamination in the 5K region. The data have been fitted with

$$S = AT + BT^3 + \frac{CT}{T + T_k}$$

to determine the diffusion, phonon drag and iron impurity terms. The values obtained for the constants and the standard deviation of the fits indicate that other mechanisms are needed for a full explanation.

EXPERIMENTAL

The cryostat utilised a copper He-4 container of about 100 ml capacity pumped by a charcoal sorption pump through a 1 mm orifice to minimise film creep. The vacuum chamber surrounding the pumped container was totally immersed in He-4 which could be pumped down to about 2K to reach the lowest temperatures.

Temperatures were measured with a Cryo-Cal Ge resistor (100Ω at 4.2K) calibrated by NML, Sydney and are believed accurate to better than a few mK. Thermo-emfs against Pb were measured using a super-conducting chopper system giving a resolution of 5×10^{-11} V in a circuit of 0.05Ω. The thermopowers are believed accurate to better than 1nV/K.

The integral method of measurement was used following Crisp, Henry and Schroeder (1964) using the new Roberts (1977) Pb scale to evaluate the alloy thermopowers.

Temperatures down to below 0.85K or higher were achieved, depending on the thermal conductivity of the specimen wire.

The alloy wire specimens are new portions of those used by Crisp and Henry (1964) for characteristic thermopower determinations and later by Crisp and Henry (1977) for measurements of thermopower from about 5 to 210K. Annealing procedures were the same as described previously and the residual resistivities quoted here are those determined in the earlier work.

DISCUSSION

Representative plots of Thermopower (S) versus temperature (T) for the four systems are shown in Figs. 1-3. The negative thermopower in the most dilute alloys is due to the residual Fe impurity (< 1 ppm by analysis) and is quenched in the usual way at higher concentrations. Guénault (1967) has made measurements down to 0.3K on a few examples from Ag alloy systems. His results were fitted with

$$S = AT + BT^3 + \frac{CT}{T + T_K}$$

and the values of B obtained demonstrated the enhancement of phonon drag in very dilute alloys with some solutes due to the anisotropy of relaxation time for impurity scattering between neck and belly areas of the Fermi surface.

The present series of results have been fitted to the same function and the coefficients A, B and C so determined are given in Table 1.

The coefficient $A = \frac{\pi^2 k^2}{3e} \left(\frac{d \ln \sigma(\epsilon)}{d\epsilon}\right)_{\epsilon_F}$ should be constant or at least vary smoothly with concentration for a given solute; this very evidently is not the case here. We note the A values are negative corresponding to impurity scattering on the convex free electron like belly regions of the Fermi surface and conclude that these contributions must dominate in this temperature range.

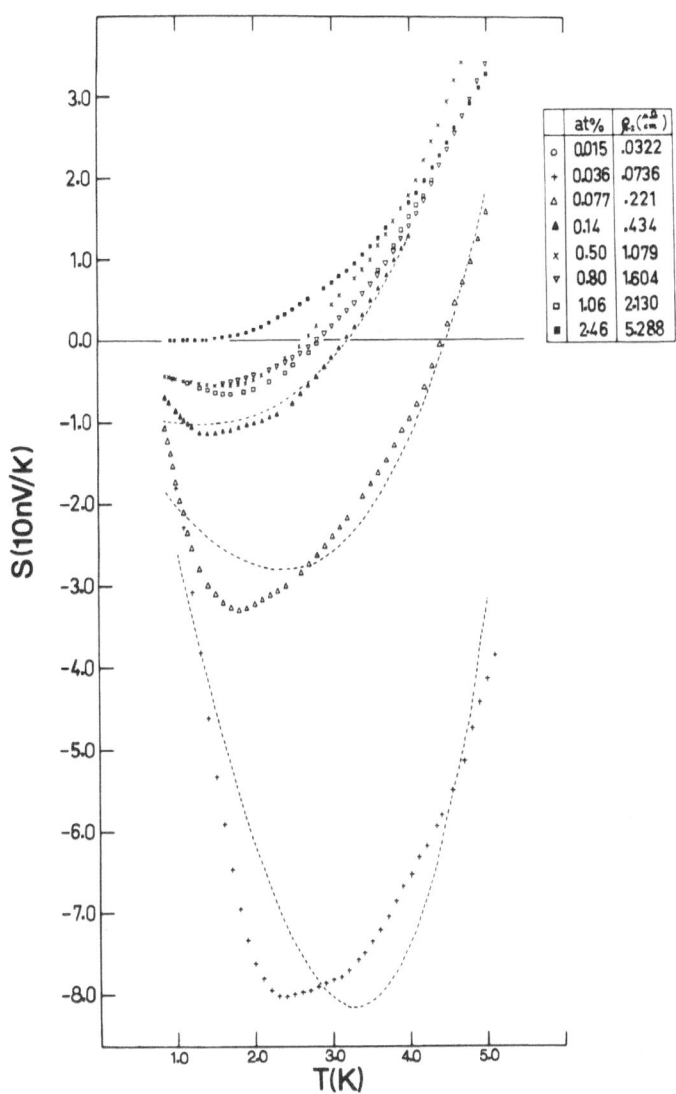

Fig. 1.   Thermopower in the system Ag-Ga, dotted lines are fitted
curves.

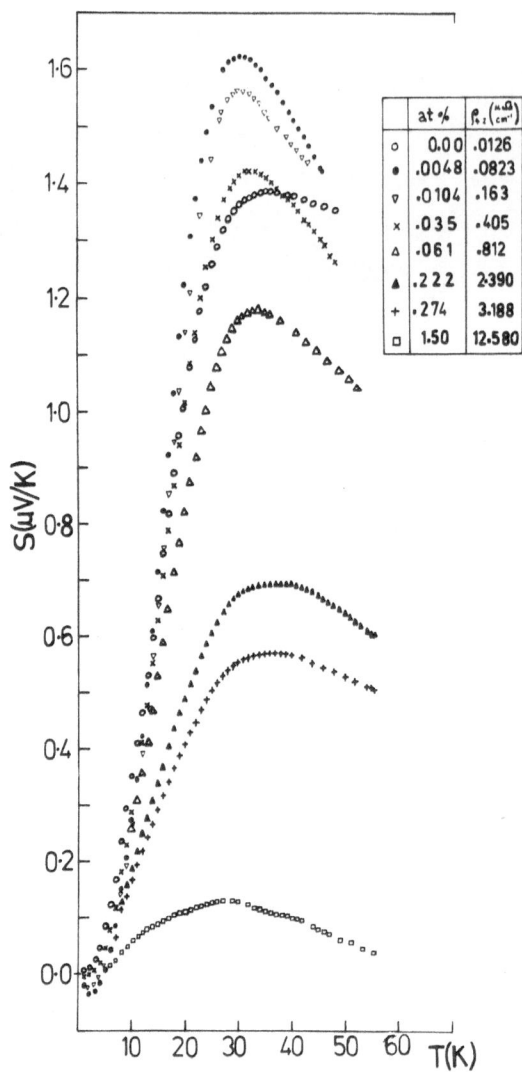

Fig. 2.   Thermopower in the system Ag–As, up to 50K.

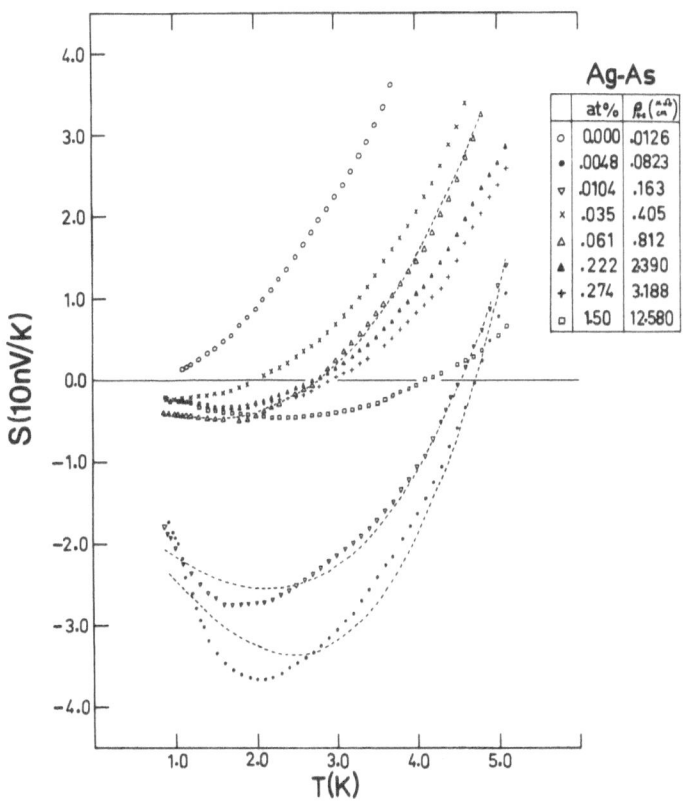

Fig. 3. Thermopower in the system Ag-As, dotted lines are fitted curves.

An alternative evaluation of the parameter A is via a Nordheim-Gorter plot. These have been made at 1K (see Fig. 4 for example) and the values are indicated in Table 1. Guénault (1972) has pointed out that such a procedure does not give simply the characteristic thermopower for a particular solute; what we are observing here is S_O of his equation (4) while the A above is S_O/T. B, the coefficient of the phonon drag term varies as expected, being reduced as concentration increases and at the lowest concentrations increases to a value greater than that for pure Ag as observed by Guénault (1967). The "phonon drag enhancement" phenomenon is observed in all the noble metal alloy systems and is most marked for the least charged impurities where belly electrons are scattered preferentially with respect to neck electrons.

Table 1. Experimentally determined parameters for pure Ag and the systems Ag-Zn, Ga, Ge, As.

Solute (at %)	Resistivity ($\mu\Omega$ cm)	A (nV K^{-2})	B (nV K^{-4})	C (nV K^{-1})	Std. dev. of fit (nV)	S_{NGR} (1K) (nV K^{-1})
		SILVER − ZINC				
0.35	0.176	−6.7 ± 0.9	0.70 ± .03	−15.0 ± 1.6	2.7	
0.56	0.288	−15.3 ± 2.1	0.85 ± .07	−58.8 ± 3.7	6.6	
1.13	0.555	−.04 ± 0.5	0.48 ± .02	−13.0 ± 0.9	1.5	
1.62	0.823	+1.25 ± 0.4	0.40 ± .01	−12.7 ± 0.7	1.1	
2.60	1.338	+0.88 ± 0.18	0.374 ± .006	−8.6 ± 0.4	0.5	−4.64 ± 1.12
3.0	1.574	−0.45 ± 0.24	0.296 ± .007	−20.0 ± 0.5	0.6	
4.6	2.365	+0.62 ± 0.11	0.221 ± .004	−6.1 ± 0.2	0.3	
5.5	2.743	+1.11 ± 0.17	0.277 ± .006	−9.6 ± 0.3	0.5	
		SILVER − GALLIUM				
0.015	0.0322	−91.2 ± 6.8	2.33 ± .02	+22.0 ± 12.8	19.5	
0.036	0.0736	−46.8 ± 2.9	1.44 ± .09	+22.2 ± 5.3	8.3	
0.077	0.221	−11.1 ± 1.2	.678 ± .04	−11.3 ± 2.3	3.8	
0.14	0.434	−1.65 ± 0.41	0.45 ± .01	−10.1 ± 0.8	1.2	
0.50	1.079	−0.76 ± 0.27	0.415 ± .009	−8.0 ± 0.5	0.7	−1.42 ± 1.45
0.80	1.604	−1.04 ± 0.12	0.359 ± .004	−5.3 ± 0.2	0.3	
1.06	2.130	−0.93 ± 0.36	0.353 ± .01	−5.6 ± 0.7	1.0	
2.46	5.288	+0.48 ± 0.11	0.253 ± .004	−1.4 ± 0.2	0.3	
		SILVER − GERMANIUM				
.023	0.109	−12.1 ± 1.2	0.710 ± .04	−14.2 ± 2.2	3.5	
.080	0.367	−4.92 ± 0.41	0.502 ± .01	−4.9 ± 0.8	1.1	
0.15	0.785	−2.48 ± 0.27	0.422 ± .009	−6.7 ± 0.5	0.7	
0.30	1.567	−0.38 ± 0.21	0.319 ± .006	−6.5 ± 0.4	0.6	
0.54	2.572	−2.21 ± 0.14	0.313 ± .005	−1.9 ± 0.3	0.4	−2.35 ± 1.45
0.73	3.775	−2.71 ± 0.25	0.322 ± .007	−1.2 ± 0.5	0.5	
0.99	4.946	−0.34 ± 0.15	0.234 ± .005	−3.2 ± 0.3	0.4	
2.0	9.893	−0.75 ± 0.11	0.187 ± .003	−.4 ± 0.2	0.3	
		SILVER − ARSENIC				
0.0048	.0823	−11.9 ± 1.0	0.682 ± .03	−15.0 ± 2.0	3.0	
0.0104	.163	−6.10 ± 0.44	0.485 ± .01	−18.2 ± 0.8	1.3	
0.035	.405	−1.31 ± 0.28	0.425 ± .008	−1.1 ± 0.6	0.7	
0.061	.812	−2.06 ± 0.28	0.414 ± .009	−3.2 ± 0.5	0.8	−2.51 ± 0.58
0.222	2.390	−1.68 ± 0.09	0.295 ± .003	−1.7 ± 0.2	0.2	
0.274	3.188	−2.60 ± 0.15	0.295 ± .005	−0.2 ± 0.3	0.4	
1.5	12.580	−2.05 ± 0.08	0.138 ± .002	−1.5 ± 0.2	0.2	
		PURE − SILVER				
	.0123	+3.27 ± .41	.544 ± .012	−2.81 ± .8	1.0	

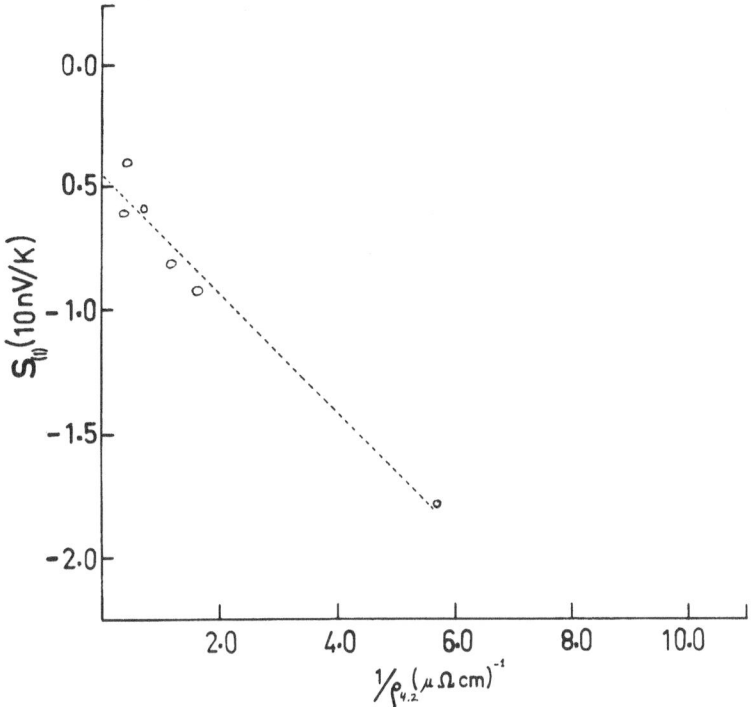

Fig. 4. Nordheim-Gorter plot for system Ag-Zn at 1K.

The coefficient C is proportional to the Fe concentration and varies in a fairly haphazard fashion. This is consistent with the usual determinations of Fe concentration by analysis or other means. The numerical values correspond to Fe concentrations of 1.0 ppm or less.

The values for A and the standard deviations of the fits suggest that this function does not fully describe the situation in this temperature range where the BT^3 term dominates. This work is continuing and the measurements will soon be extended to much lower temperatures in a He-3 cryostat under construction. Further discussion appears in these Proceedings in the paper by Crisp.

ACKNOWLEDGEMENTS

This research has been supported by the Australian Research
Grants Committee and by the Physics Department Research Committee
through a block grant from the University of Western Australia
Research Grants Committee.   S. J. Song gratefully acknowledges
receipt of a Commonwealth Postgraduate Award.

REFERENCES

1. R.S. Crisp, W.G. Henry and P.A. Schroeder, Phil. Mag. 10, 553,
    (1964).
2. R.S. Crisp, and W.G. Henry, Phil. Mag. 11, 841, (1965).
3. R.S. Crisp and W.G. Henry, (1977), to be published.
4. A.M. Guénault, Phil. Mag. 133, 17, (1967).
5. A.M. Guénault, J. Phys. F, 2, 316, (1972).
6. R.B. Roberts, Nature, 265, 226, (1977).

THE DIFFUSION THERMOPOWER ON THE TWO-BAND MODEL--SEPARATION OF DIFFUSION AND PHONON DRAG COMPONENTS IN NOBLE METAL ALLOY SYSTEMS

R. S. Crisp

Physics Department

University of Western Australia

ABSTRACT

The recent redetermination of the Pb scale of thermopower by Roberts (1977) has shown significant errors in the previously accepted scale above 20 K. In the light of this, earlier measurements on noble metal alloy systems have been recomputed and combined with some more recent studies and some general comments regarding phonon drag and diffusion thermopower can be made.

II. EXPERIMENTAL RESULTS

The following general comments can be made regarding the behaviour of the α-phase noble metal alloy systems.

(i). The peak which is observed between about $\frac{\theta}{7}$ and $\frac{\theta}{5}$ is attenuated with increasing solute concentration; the attenuation saturates at about 5 at % solute and the peak persists to the highest concentrations and in some cases resurges.

At very low concentrations (~ 0.05 at .%) the peak may be enhanced to a value greater than that in the pure metal. (See Figures 1 and 2.)

(ii). At higher temperatures ($> \frac{2\theta}{3}$) the data can be very well represented by a modified Nordheim-Gorter relation (a plot of S against ρ_ℓ/ρ at constant T is linear) and the intercepts at $\rho_\ell/\rho = 0$ from such plots are linear in T. This holds out to surprisingly

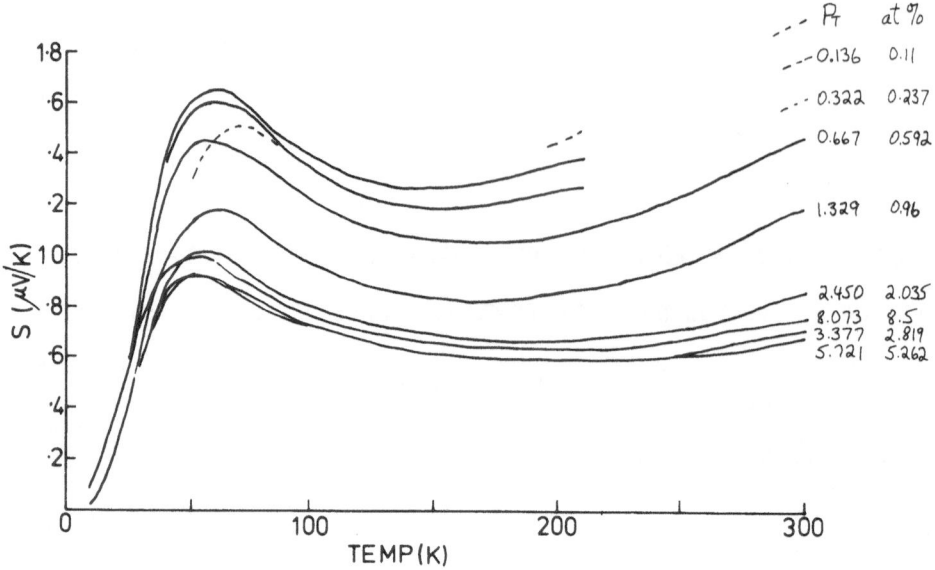

Fig. 1.   Thermopower in the system Cu-Ga showing the persistence of
the peak to high concentrations and its resurgence at 8.5 at %.

high concentrations (e/a = 1.15 for Ag Hg) and is observed when
thermal rather than electrical resistivities are used.   (See Figures
3 and 4.)

(iii).   The peak shows a dependence on T^3 at low temperatures
and $1/T$ at high temperatures as expected for phonon drag, however,
such dependences are obviously affected by the choice of background
subtraction of the diffusion term.

(iv).   The lattice thermal conductivities (K_g) which have been
measured for some systems are very strongly attenuated with increasing solute concentration and mass difference.   Typically 1 at % of
solute reduces K_g to less than 1% of its value in the pure metal.
(See Figure 3, Craig and Crisp, these proceedings.)

III.   DISCUSSION

(i).   The Peak.   The phonon drag peak is believed to be due to a
delicate balance between positive (neck) and negative (belly) contributions both of which are attenuated by solute addition in the same
fashion as K_g.   If the peak observed at a few percent solute is indeed phonon drag then in the pure metal the observed peak must
represent the almost exact balance between two contributions of

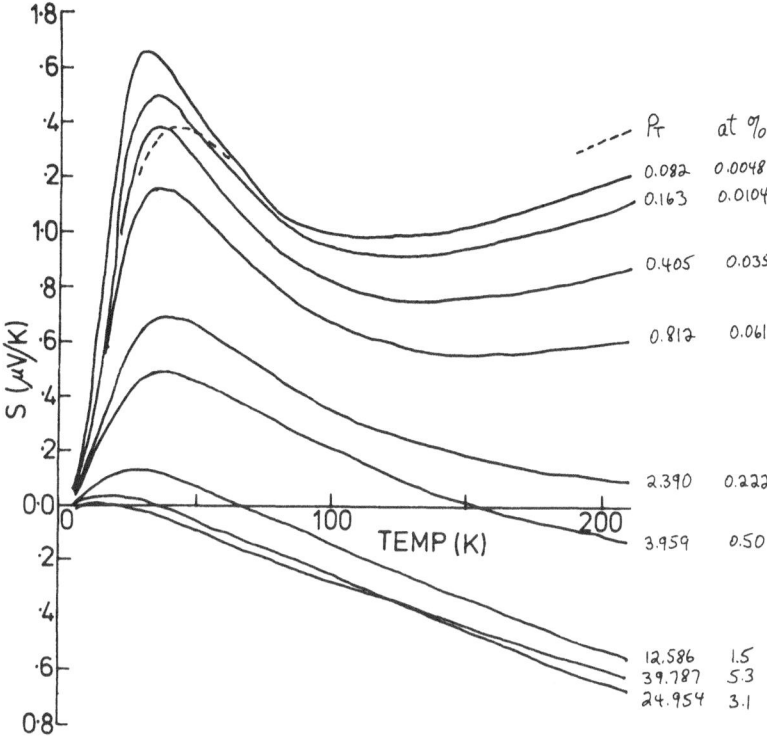

Fig. 2.   Thermopower in the system Ag–As showing the enhancement of
the peak at the lowest concentrations (pure Ag shown dotted).

magnitude some hundreds of μV/K. While this model explains satis-
factorily the "enhancement" observed at fairly small concentrations
it is difficult to explain the peak observed out to the highest concentrations. Whatever band structure changes are invoked, K_g has
become vanishingly small.

(ii). The Diffusion Thermopower. The high temperature data
is remarkably well represented by a N.G. relation and this can be
anticipated from a two (or more) band model. The NG intercept
(S_0^{NGR} in the notation of Guénault (1972)) is a function of impurity
and phonon scattering of neck and belly electrons and of the anisotropy parameters. While the two band model is a gross oversimpli-
fication and neglects for example interband scattering, one is
tempted to use it to predict the behaviour of the diffusion thermopower at lower temperatures.

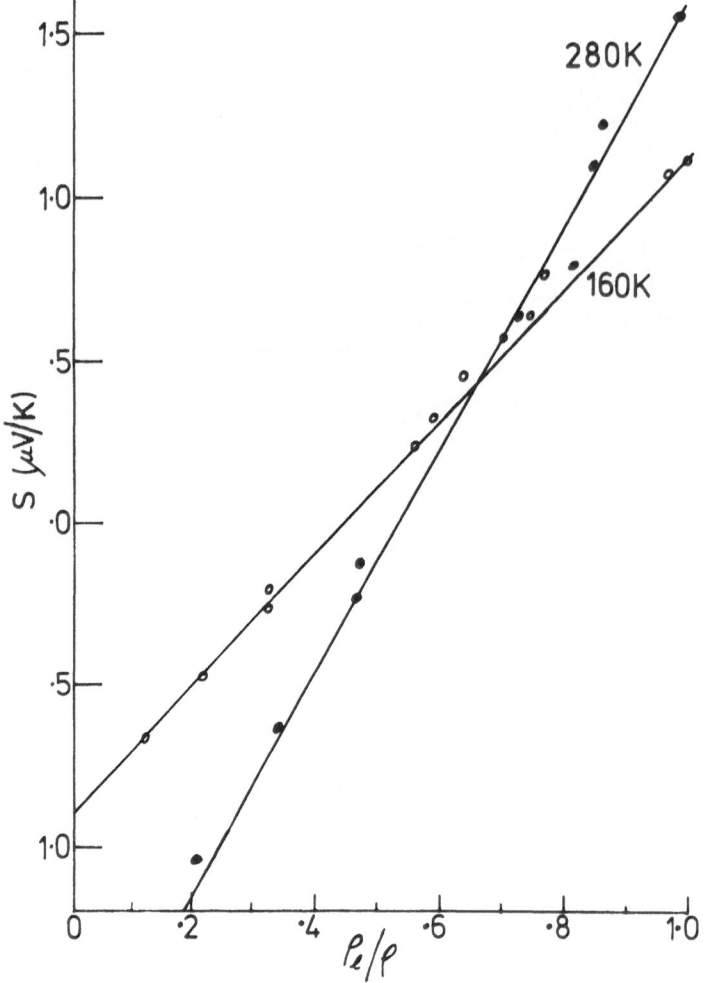

Fig. 3. Nordheim-Gorter plot for Ag-Au at 280 K and 160 K.

(iii). <u>Extrapolation to Lower Temperatures</u>. Least squares fits
have been made to the thermopower data as a function of temperature
above about 150 K. The four characteristic thermopowers (linear in
T) were used as adjustable parameters, the four resistivity components were found from the measured residual and total resistivities
and anisotropy parameters taken from Dugdale and Basinski (1967).
Good fits are obtained; however, with four adjustable parameters and
only slowly varying data the determinations are poor and in fact the

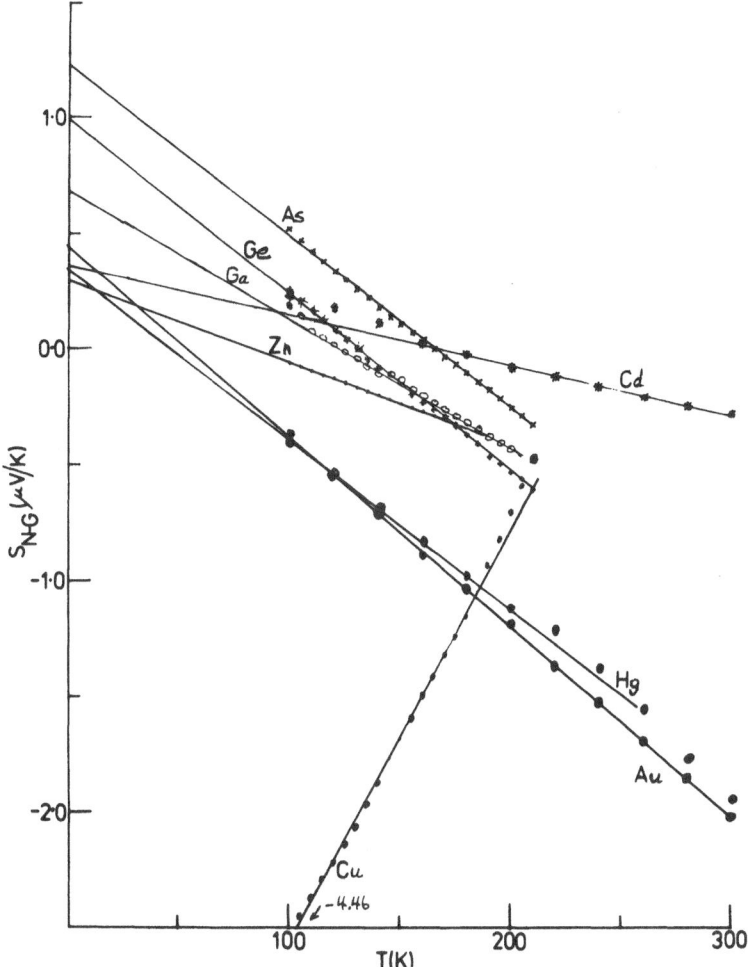

Fig. 4. Variation of Nordheim-Gorter intercept with temperature for various solutes in Ag.

characteristic thermopowers found are not constant within any particular system.

One can go a step further and assume that by say 5 at % solute
in Ag-Hg there is no phonon drag and the observed thermopower is
all diffusion. The data can then be fitted over the whole temperature range down to about 5 K. Good fits are obtained provided rather
more extreme values are used for the anisotropy parameters

$$(\sigma_N/\sigma_B)_{imp} = 0.05; \quad (\sigma_N/\sigma_B)_{ph} = 0.05 \text{ at } 5 \text{ K}$$

$$\text{and } (\sigma_N/\sigma_B)_{ph} = 0.15 \text{ at } 300 \text{ K}$$

Using the parameters determined at this concentration one can now calculate the diffusion term at higher and lower concentrations.

At higher concentrations the model has the property of a "resurging" peak while at lower concentrations the separated phonon drag term varies rapidly with concentration as predicted from the lattice conductivity. The high temperature fit to the low concentration data is poor however; the model has been pushed beyond its limit. It had been hoped to fix the impurity scattering parameters from precise low temperature (below 4 K) measurements (see Song and Crisp, these Proceedings), however, until new measurements are performed down to 0.3 K this has not been achieved.

IV.   CONCLUSION

If one accepts the postulate that beyond a percent or so of solute there is negligible phonon drag then the peak remaining in the thermopower must be due to the diffusion thermopower or some other mechanism (e.g. "phony" phonon drag). It has certainly been demonstrated that the two-band model, which recognises the basic features of the noble metal Fermi surfaces, has the required properties to fit the observations. In particular there is a difficulty in the 4–10 K region where lattice scattering is still small and the observed thermopowers vary very rapidly with themperature. However it is just in this region that one might expect a rapid variation in the anisotropy parameters or in the characteristic thermopowers themselves. Either would be sufficient to obtain reasonable fits to the model. This work is proceeding.

V.   ACKNOWLEDGMENTS

The thermopower project is supported by the Australian Research Grants Committee and by the University of Western Australia Research Grants Committee.

VI.   REFERENCES

1. J. S. Dugdale and Z. Basinski, Phys. Rev. <u>157</u> 552 (1967).
2. A.M. Guénault, J. Phys. F <u>2</u> 316 (1972).
3. R.B. Roberts, Nature <u>265</u> 226 (1977).

PRESENT STATE OF EXPERIMENTAL KNOWLEDGE ON THERMOPOWER OF METALS AT

HIGH TEMPERATURES - ABOVE 77 K

M. V. Vedernikov and A. T. Burkov

A. F. Ioffe Physico-Technical Institute 194021

Leningrad, USSR

ABSTRACT

The present state of knowledge of thermoelectric properties of metals is characterized by an absence of general empirical relationships. It is rather difficult to give theoretical interpretation to a number of observed features of thermopower. In order to solve these problems a systematic study of thermopowers of all metals is necessary. In this paper the high temperature thermoelectric properties of all transition, rare earth and normal metals are considered together for the first time. The basic part of this paper is an analysis of the thermopower of solid polycrystalline metals in the magnetically disordered state. It is shown that the thermopower behaviour at high temperatures is often complicated and disagrees with predictions of simple models. The close similarity of thermopowers is revealed for metals which belong to the same group of the Periodic System. Both thermopower change at the structural transformations and the thermopower anisotropy for single crystals are discussed. The data on thermopowers of metals reviewed here can serve as a reliable base for theoretical interpretation. Last, a connection between the high temperature thermopower of metals and their electronic structure is discussed.*

I. INTRODUCTION

This paper deals only with the experimental knowledge of TEP of metals, namely, with its completeness, reliability, classification, as

---

*The absolute thermopower is hereinafter referred to as TEP or S.

well as with conclusions that may be drawn from this analysis. The
lower boundary of the temperature range considered, 77 K, is defined
not only by specific features of measurements above and below the
nitrogen boiling point. For most metals, the region above 77 K is
high temperature in the physical sense since it is the region where
the typically low-temperature factors do not noticeably affect TEP.
The history of investigation of the thermoelectric properties of
metals in this temperature region, particularly near room temperature,
dates back 150 years. These investigations proceeded in parallel with
studies of electrical resistance which helped to establish already
in the 19th century the following basic empirical relations: the
electrical conductivity of metals is higher than that of any other
substances; resistivity increases with increasing temperature; resistivity has a linear temperature dependence; resistivity follows
the Wiedemann-Franz law and Matthiessen's rule. These relations
served as a basis not only for physical interpretation of the electrical properties but also for the development of the electron theory
of metals as well. In contrast to this, for TEP no general empirical
relationships of its behaviour have apparently been revealed up to
now. This is the first characteristic aspect of our current experimental knowledge on TEP that we have to point out. On the other hand,
it is well known that some essential features of the thermoelectric
properties of metals cannot be at present explained theoretically.
It is natural to assume that if one succeeded in experimentally revealing common features in the behaviour of TEP, this would facilitate its physical understanding.

Let us try to portray in more detail the current state of our
empirical knowledge of TEP of metals. In the absence of universally
accepted relations, the literature naturally contains various statements, estimates and concepts of a more or less general nature. First
of all, there exists an opinion that TEP is a poorly reproducible
property which is extremely sensitive to weak, hard-to-control factors.
One often attempts to describe the behaviour of TEP for the whole
class of metals by constructing a "thermoelectric series" in which
TEP of elementary metals are arranged in increasing order of magnitude at a given (usually room) temperature. A small TEP of the order
of a few microvolts per degree at room temperature is considered
typical for metals. It is said that at high temperatures the behaviour of TEP is trivially linear without any singular features.
Some of these ideas had apparently appeared due to the approximation
of nearly free electrons (NFE), in accordance with which TEP can be
written as

$$S = \frac{\pi^2 k^2}{3e} \frac{1}{E_F} T \tag{1}$$

Here k is the Boltzmann constant, E_F the Fermi energy, e the carrier
charge which should be taken with its sign. Taking into account that

the Fermi energy typical for metals is about 5 eV, we obtain for the case of one conduction band the TEP shown in Fig. 1. This simple simple picture has the following meaning: the sign of TEP is defined by that of the current carriers and coincides with the sign of the Hall constant; the magnitude of TEP at room temperature is of the order of 1 μV.K^{-1}; the temperature dependence of TEP is linear. To what extent all these ideas agree with reality could be learned by comparison with a sufficiently complete experimental picture. By the "sufficiently complete experimental picture" we understand, for the high temperature range, a simultaneous consideration of TEP of all or almost all metals in as wide a temperature range as possible. This we intend to do in our report. As far as we know, such an analysis has not been carried out before. This may apparently be attributed to the fact that before the 1930's one had studied, as a rule, the integral thermo-e.m.f. of thermocouples, which masks the true thermoelectric properties of a metal. Only the introduction of the concept of absolute differential thermopower, the adoption of the technique of differential thermopower's measurement by two thermocouples to specimens of arbitrary shape and gradual introduction of all metals of the Periodic Table into laboratory practice (including recently the broad rare-earth metals group) provided the present possibility for such an analysis to be performed.

If we want this analysis to cover all metals, we shall have to restrict ourselves to considering polycrystalline metals in the magnetically disordered state. There are still only scarce data on TEP of single crystals at high temperatures. Only a small fraction

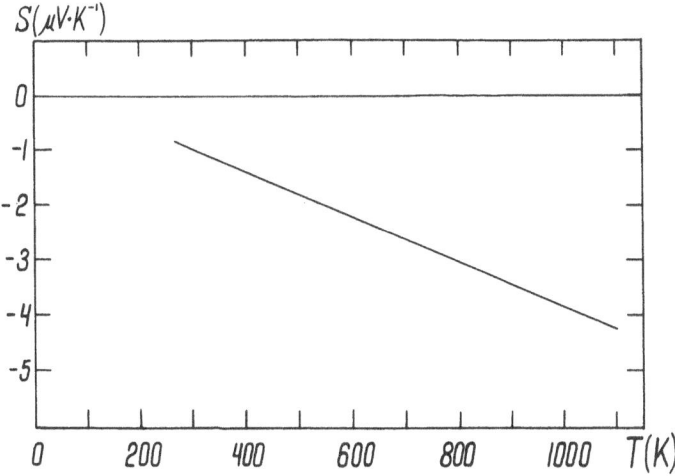

Fig. 1. TEP versus temperature for metal in the nearly free electron approximation.

of metals can exist in a magnetically ordered state which affects
strongly the TEP. We will give below examples of behaviour of TEP
of magnetic metals which should be studied separately.

II. GENERAL EMPIRICAL ANALYSIS OF THE THERMOPOWERS OF
POLYCRYSTALLINE METALS

Transition metals of the d-series and the rare-earth metals,
their number being about 40, constitute more than one half of all
elemental metals. Their TEP has been analyzed in detail by the
present authors in the temperature ranges 80-1800 K[1] and 80-1000 K,[2]
respectively. Information on other metals was taken from reviews[3,4]
and a number of more recent works, references to which are given below. The method of analysis used is illustrated by the study of
transition metals.[1] The available data were collected and compared,
several errors were revealed, and TEP of some metals were measured
for the first time. Of particular importance was a substantial extension of the temperature range covered toward higher temperatures
for many metals. As a result, it was found that if one considers
metals in the magnetically disordered state, the temperature dependences of TEP above 77 K for transition metals of any group of the
Periodic Table turn out to be strongly similar. This similarity is
so high that TEP of all metals of any group can be described in a
first approximation by one common curve. As an illustruation, Fig. 2
shows the properties of chromium, molybdenum and tungsten, which are
group VI metals. If one considers chromium above the Néel point, all
of them are seen to reveal a positive TEP with a nonlinear temperature

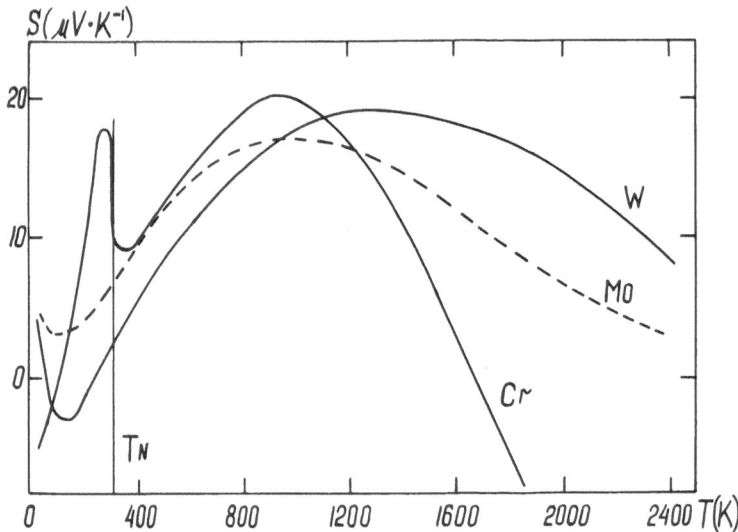

Fig. 2. TEP versus temperature for transition metals of group VI.[1,5]
T_N is the Néel point for Cr.

dependence. The maximum lies at about 1000 K, the magnitude of TEP at maximum being about 20 $\mu V \cdot K^{-1}$. At the highest temperatures, as shown by the case of Cr,[5] TEP should become negative. Presented in Fig. 6 is a generalized TEP for metals of the group VI which has all these features. In the same way generalized TEP are constructed for transition metals of all other groups (Figs. 3-8). Note that of all the transition metals, only data for Tc were not available. Recently, its thermo-e.m.f. has been measured at high temperatures[6] and turned out to agree with our generalized relation for the VII group. To substantiate the need for a separate interpretation of TEP of magnetically ordered metals, we refer besides antiferromagnetic Cr, to ferromagnetic Co which has the highest magnetic ordering temperature (Fig. 9). In this case the effect of magnetic ordering on TEP is very strong which is seen from comparison of the observed TEP in the ferromagnetic region with its hypothetical magnitude for nonmagnetic cobalt shown with a dashed line in the figure. This line was obtained by extrapolating the temperature dependence in the paramagnetic region, in accordance with the behaviour of the generalized curve for group VIII-2 in Fig. 8. The TEP of all d-magnetics can be found in ref. 1. Data on rare-earth magnetics support the conclusion on the strong effect of magnetic properties of metals on TEP and the validity of the idea for a separate study of this region.[2]

TEP of all rare-earth metals (REM) was studied above room temperature up to 1000 K for the first time in ref. 2. Taking into account the close connection between TEP and the position of metals in the Periodic Table, one should expect the metals in this broad group to be very similar in their thermoelectric properties. Data on

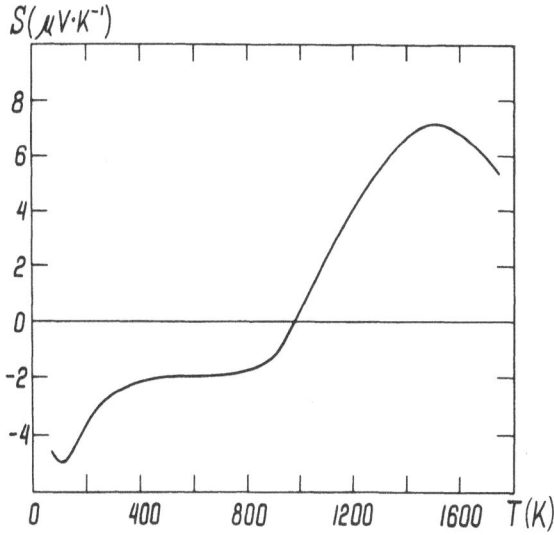

Fig. 3. Generalized TEP versus temperature for transition metals of group III Sc and Y.

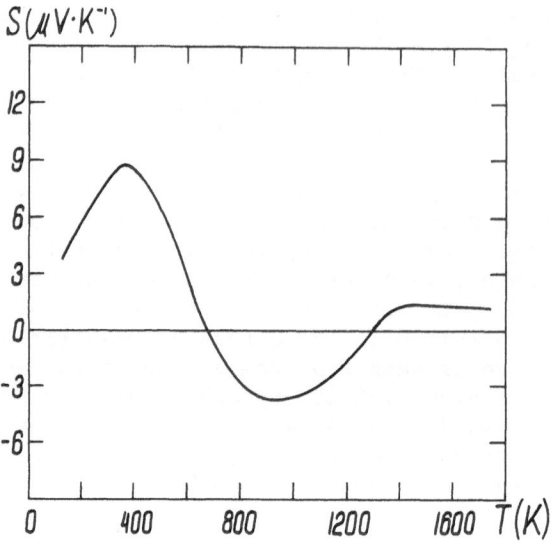

Fig. 4.   Generalized TEP versus temperature for transition metals of group IV Ti, Zr and Hf.

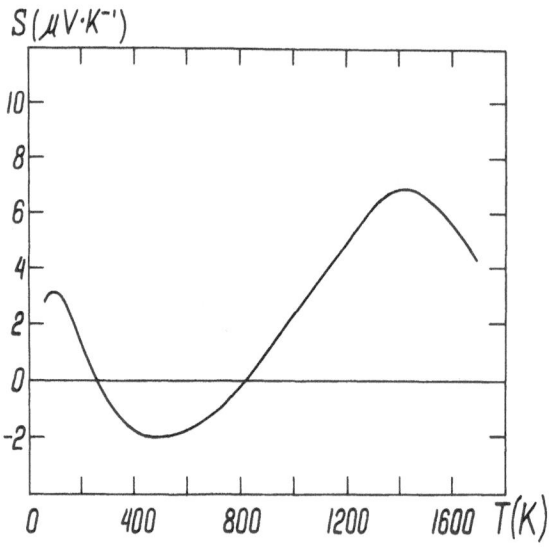

Fig. 5.   Generalized TEP versus temperature for transition metals of group V: V, Nb and Ta.

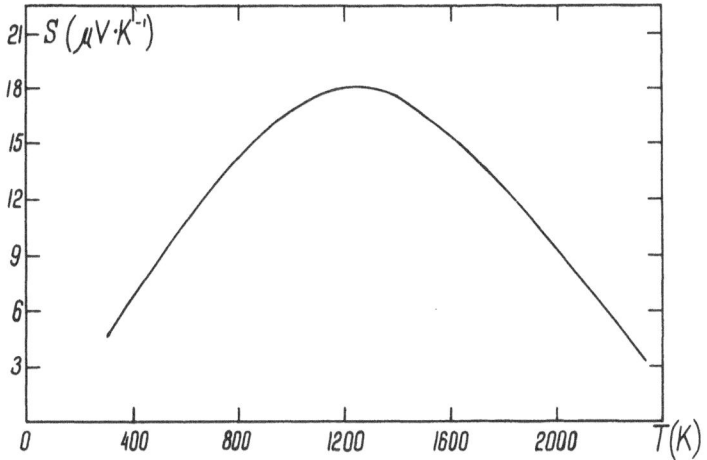

Fig. 6. Generalized TEP versus temperature for transition metals of group VI: Cr, Mo and W.

Eu and Yb which are anomalous REMs disagreed with this, however; indeed, their TEP at room temperature exceeds by a factor of 10 the TEP of other REMs. As shown in ref. 2, TEP of all normal REMs are close to each other at high temperatures (some examples can be seen in Fig. 10). The temperature dependence of REM TEP is a convex curve with a maximum at T=(0.5±0.1) T_M, where T_M is the melting point. The

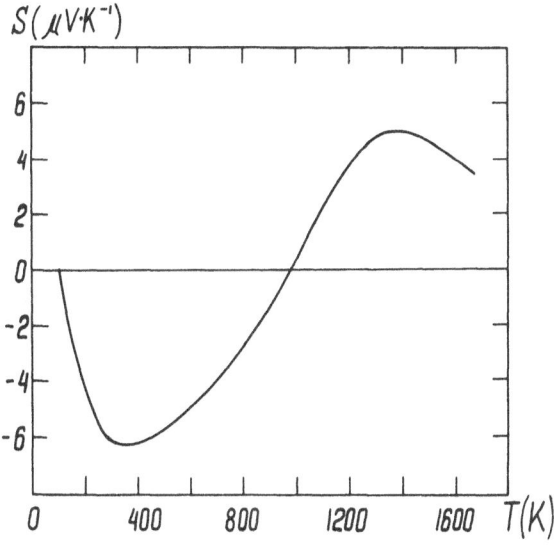

Fig. 7. Generalized TEP versus temperature for transition metals of group VII Mn, Tc and Re.

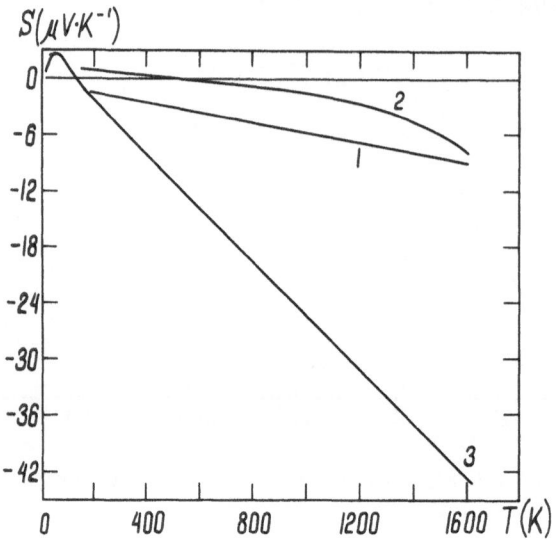

Fig. 8. Generalized TEP versus temperature for transition metals of group VIII. 1, metals of subgroup VIII-I Ru and Os. 2, metals of subgroup VIII-2 Co, Rh, and Ir. 3, metals of subgroup VIII-3 Ni, Pd, and Pt.

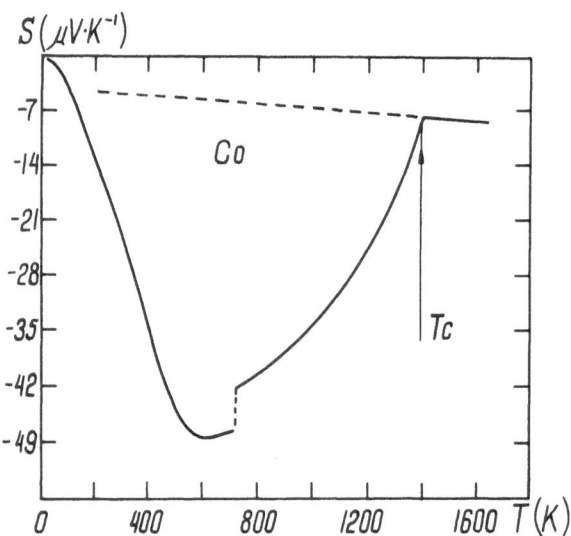

Fig. 9. TEP versus temperature for Co (full curve).[1] T_C is the Curie point.

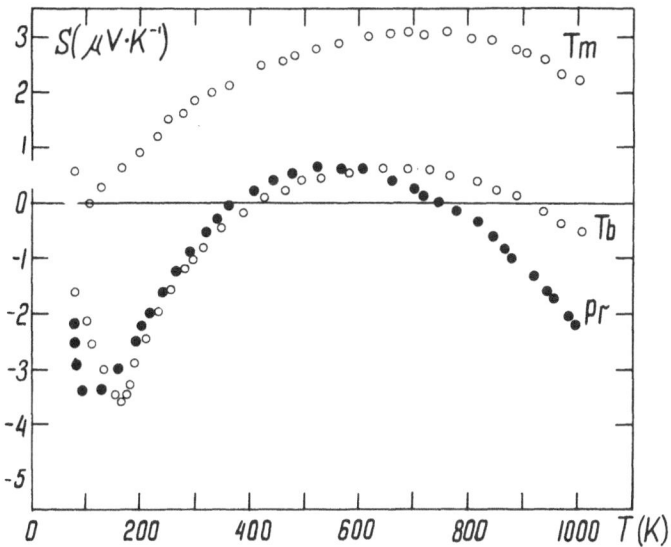

Fig. 10.   TEP versus temperature for rare earth metals Pr, Tb, and Tm.[2]

magnitude of TEP does not exceed a few microvolts per degree throughout the temperature range studied, being usually positive at maximum.
Such TEP can be described by a generalized curve similar to curve 2
in Fig. 11. A study of Eu and Yb at high temperatures revealed a
qualitative similarity with normal REMs (Fig. 12). It turns out that
their TEP are likewise described by curves with maxima lying near
$T=0.5T_M$ (curve 1 in Fig. 11). Thus the temperature dependences for
normal and anomalous REMs are qualitatively similar, but quantitatively different. This supports the conclusion on the similarity
of TEP for metals belonging to the same group of the Periodic System.
For completeness, one should consider the available data for normal
metals. Data for alkali metals of group Ia seem to be reliable.
The temperature dependences of ref. 3 are confirmed by more recent
data of Pearson and Dugdale (ref. 4). For the first time, we do
not observe here a common temperature dependence pattern (Fig. 13).
Na, K and Rb in their TEP resemble very much metals in the NFE
approximation. TEP of Li is positive, that of Cs occupying an intermediate position. TEP of noble metals of group Ib are studied well
and can be portrayed by a generalized curve (Fig. 14). The data used
here are those of ref. 7. TEP of the group II metals are known insufficiently well. The situation in group IIa where the data are scarce
is particularly poor. The upper temperature limit is too low, and
the data are not always reliable. Nevertheless, one common feature
is evident, in that all metals have TEP curve with a positive slope.
A generalized curve for this case is shown roughly in Fig. 15.

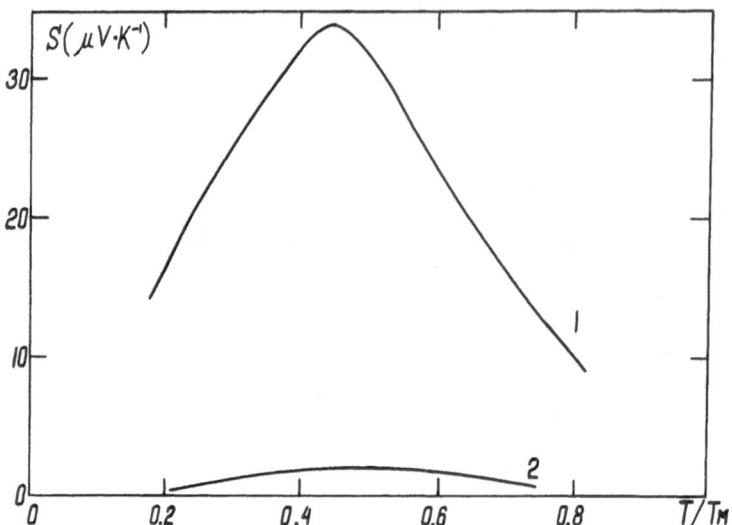

Fig. 11. Generalized TEP versus relative temperature for rare earth metals. T_M is melting temperature. 1, curve for anomalous metals Eu and Yb. 2, curve for normal REMs. (2).

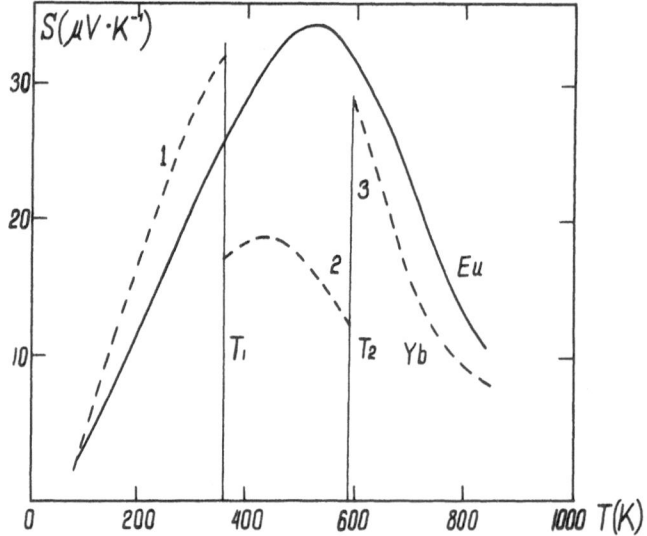

Fig. 12. TEP versus temperature for rare earth metals Eu and Yb. T_1 and T_2 are temperatures of Yb structural transformations (2).

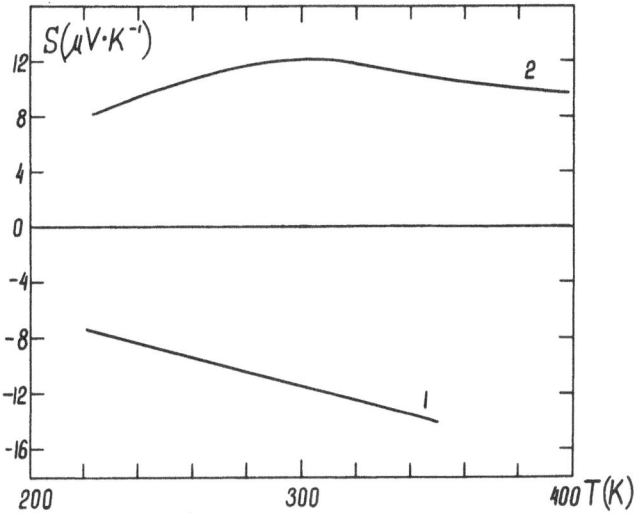

Fig. 13. TEP versus temperature for alkali metals of subgroup Ia.
1, generalized curve for Na, K, and Rb. 2, Li (3).

The magnitudes of TEP and the temperature ranges of measurement
for different metals deviate strongly from this graph. The sources
employed here are as follows: Be;[3] Mg;[8] Ca;[9,10] Sr;[10] Ba;[11] Zn,
Cd;[12] Hg,[3]. Among metals of group IIIa only Al was studied thorough-
ly, its TEP being close to the prediction of the NFE approximation.
Ga, In and Te were studied inadequately. It is obvious, however,
that their TEP bear some similarity with one another while differing
from that of Al (Fig. 16). The references are available in (3,4).
TEP of group IVa metals are close to the generalized dependence shown

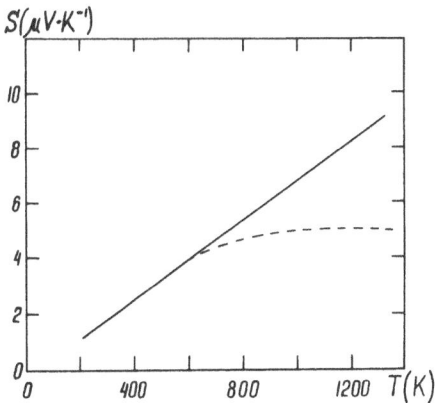

Fig. 14. Generalized TEP versus temperature for noble metals of
subgroup Ib (full curve).

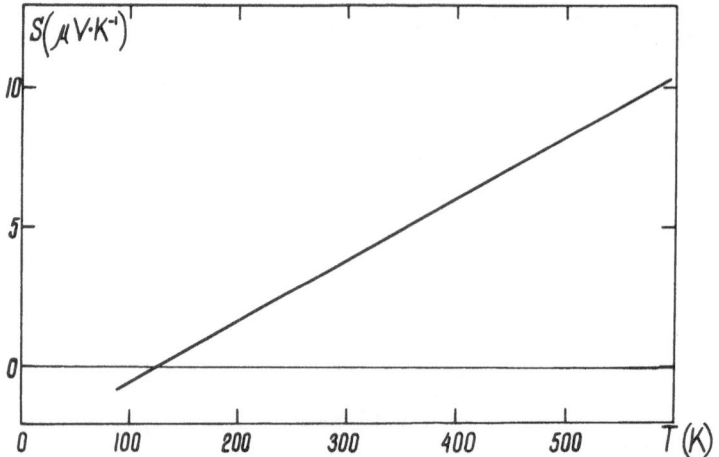

Fig. 15. Generalized TEP versus temperature for normal metals of group II.

in Fig. 17 [see refs. (3, 13), as well as refs. in (4)]. Finally, information on TEP of actinide metals Th, U, Np and Pu can be found in (1, 4). Thus the conclusion on the close similarity in the behaviour of TEP at high temperatures for metals belonging to the same group of the Periodic System turns out to be of a general nature. This is the major result of our analysis. Deviations pointed out for groups Ia, IIIa as well as the classification of actinides will be discussed below.

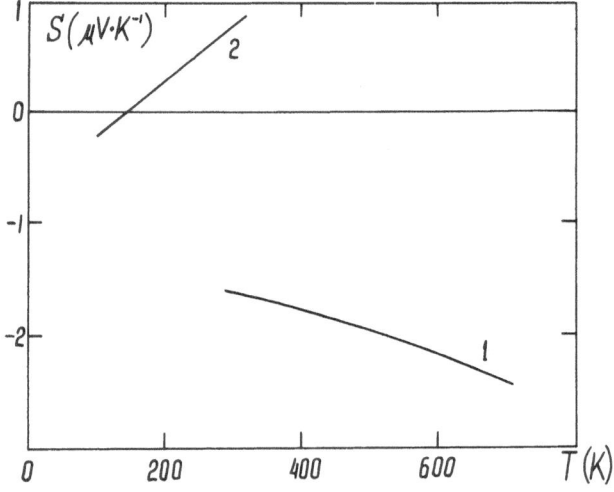

Fig. 16. TEP versus temperature for normal metals of subgroup IIIa. 1, Al (3, 4). 2, generalized curve for Ga, In, Tl.

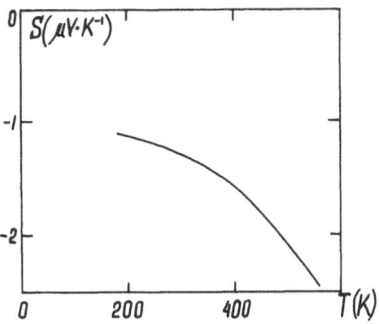

Fig. 17. Generalized TEP versus temperature for normal metals of
subgroup IVa Sn and Pb.

A consideration of the generalized TEP shows the behaviour of
TEP at high temperatures to be, generally speaking, quite complicated. The temperature dependences are in many cases nonlinear and
have extrema and inflections. TEP may frequently reverse its sign,
acquire either anomalously high or anomalously low values. It is
due to this complexity of behaviour that we can clearly reveal the
above mentioned similarity of the curves by matching of their main
features. We see also that in many cases TEP of a real metals does
not have either one or several or even all the features which would
follow from the NFE approximation. Starting from this pure empirical
fact, we can say that in a general case this approximation is not
valid for a description of the high-temperature TEP of metals.

In the high-temperature range, TEP of many metals reverses its
sign. The temperature at which sign reversal takes place does not
frequently coincide with a phase transition, TEP at this point varying in a continuous way. For such TEP the sign at a given temperature does not represent a fundamental characteristic. Therefore, a
construction of a "thermoelectric series" would not have a physical
meaning. It may be interesting for purposes of applications only.

Summing up, one may say that at present the only reasonable
way to characterize the high temperature TEP of metals lies in considering the temperature dependence over as wide temperature range
as possible. Using for the purpose of interpretation only the value
of TEP at one given temperature of within a narrow temperature region
may result in wrong conclusions.

The accuracy of TEP measurements by the differential technique,
i.e. by means of two thermocouples at a small temperature difference,
has been analyzed earlier.[1,14] The accuracy is poor for several
reasons. It is difficult to match precisely the points on the

specimen between which the potential difference is measured with
those between which one measures temperature difference. With
specimens of a small size, this may result in a noticeable error in
determining true temperature differences. Also, the absolute thermopower of the reference metal is usually not known with high accuracy.
The specific feature of the absolute thermopower lies in its being
not measured directly but rather found as a difference between the
measured and tabulated values. As a result, errors in measurements
of TEP by the differential techniques may frequently reach 1-2
$\mu V \cdot K^{-1}$. At a small value of the absolute thermopower of interest
this may even involve an error in sign. This is, undoubtedly, one
of the reasons for the belief that TEP is poorly reproducible. However, our experience shows that if some precautionary measures are
taken, TEP measurements can be made with good reproducibility within
the above random error.

The effect of metal purity on its high-temperature TEP was
esimated.[1,2] Specimens of different degrees of purity were studied
in some cases. One apparently may conclude that results obtained
for metals of 99.8-99.9 at.% purity or higher above room temperature
do not differ from one another (within the above mentioned random
error).

III.   EFFECT OF CRYSTAL STRUCTURE ON TEP

An interesting feature of the above general consideration is
that we did not have to consider the crystal structure of the metals
of interest. The transition metals of some groups have different
crystal structures. Among REMs there are metals with different
lattices, too. However, the temperature dependences of their TEP
usually turn out to be similar.[1,2] The same can be said about some
normal metals. We did not carry out a special study of TEP change
at polymorphic transformations. However, most frequently the value
of TEP at the temperature of a structural transition is observed to
change discontinuously, whereas the temperature dependence remains
unchanged. This is shown schematically in Fig. 18. If one shifts
part 1 of the curve vertically to 1', this part and part 2 will form
one curve without any singularity at the point of transformation.
One ordinarily does not observe any noticeable change in the nature
of the temperature dependence similar to curves 2' and 2''. An illustrative example is given by Yb exhibiting two structural transformations at nearby temperatures. As seen from Fig. 12, TEP of Yb
is very close to that of Eu, differing from it only within the
region T_1-T_2. If, however, part 2 is raised up, it will connect in
a natural way with parts 1 and 3. The curves for Yb and Eu are very
similar now. We may see another example of invariability of TEP
temperature dependence in the case of Co at about 700 K (Fig. 9).

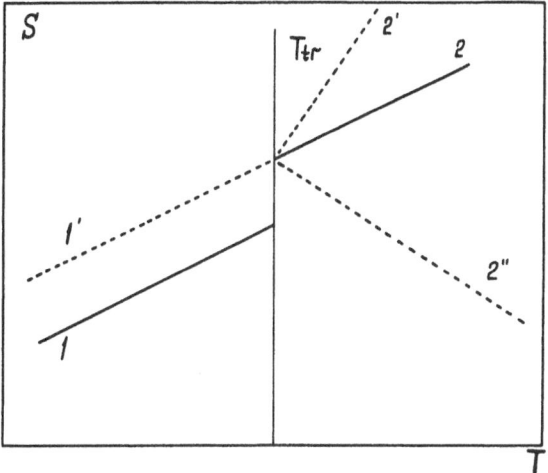

Fig. 18. Typical change of TEP at structural transformation (full curves). T_{tr} is the temperature of transformation.

IV. ANISOTROPY OF TEP

The observations described in the preceding section can be summarized as follows: differences in the crystal structure (for metals of one group or for the same metal in different polymorphic states) can affect markedly the value of thermo-e.m.f. while influencing only slightly its temperature dependence at high temperatures. Then, as regards anisotropy, one may suggest that TEP of an anisotropic noncubic single crystal will exhibit the same temperature dependence for different directions. It would be interesting to check this suggestion. There are available at present some data on TEP of noncubic single crystals, however only for a too narrow temperature region below 300 K. As far as we know, only TEP of single crystal Zn was measured by Linder up to the melting point.[3] Therefore, we have embarked on a study of single crystals at high temperatures. Preliminary results are shown in Figs. 19 and 20. TEP of Re and Er disagree with our suggestion. According to Linder, the TEP of Zn behaves the same way. However TEP of Gd, Dy and Ho along the principal directions turned out similar (data for Dy is not presented here). At this stage, it is too early to draw a general conclusion, and more experimental data should be accumulated. Note only the following difference between the metals studied. Re, Er and Zn do not have polymorphic modifications and retain hexagonal structure up to the melting point. All metals revealing similarities in TEP along different axes (Gd, Dy, Ho) have polymorphic transformations at temperatures higher than those covered in the measurements (they transform from hexagonal to b.c.c. lattice).

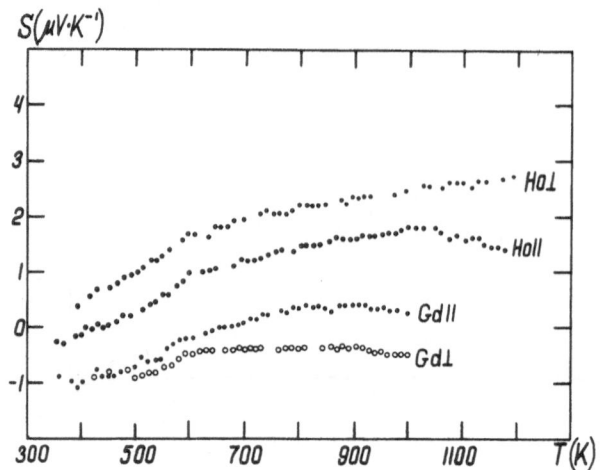

Fig. 19. TEP versus temperature for single crystals of Gd and Ho. TEP are measured along directions parallel and perpendicular to the hexagonal axis.

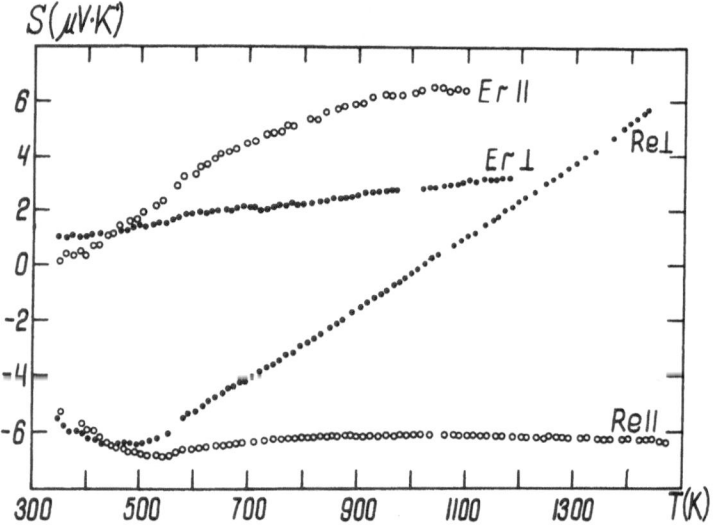

Fig. 20. TEP versus temperature for single crystals of Er and Re. TEP are measured along directions parallel and perpendicular to the hexagonal axis.

V. CONCLUSION

Data on the high-temperature TEP of metals collected and discussed here are sufficiently reliable, although additional experimental study is needed, particularly on group IIa and IIIa metals.
We succeeded in revealing a relation between the behaviour of TEP
in the wide range of high temperatures and the position of a metal
in the Periodical System. This is a clear indication that we have
here a true property of the metal. The position of an elemental
metal in the Periodic System is determined by its electronic structure. Hence, the actual type of TEP temperature dependence is
closely associated with electronic structure. Some other evidence
also points in favor of such an opinion. First of all, we note that
the temperature dependences for normal metals are simpler than those
for transition metals and REMs. This agrees with the energy spectrum of normal metals being simpler. However we do not want to say
that for the interpretation of thermoelectric properties of normal
metals, one may restrict oneself to the NFE approximation. The
second observation applies to actinide metals. The TEP temperature
dependences of Th, U, Np and Pu turn out rather similar to those of
transition metals of IV, VI, VII AND VIII-I groups, respectively.
Here we must take into account that the number of electrons in un-
filled shells is the same for Th and group IV metals, U and group VI
metals and so on. In our opinion, the above permits one to formu-
late the major problem for theoretical interpretation of TEP. The
theory should find the reason (or reasons) resulting in a complicated
temperature dependence of TEP at high temperatures. It should, first
of all, reveal the direct effect of electronic structure on TEP. We
mean, for example the part of TEP calculated by Ziman.[7] There is
available at present vast and reliable material on deviations of
the Fermi surfaces from spherical shape in most metals. One should
always keep this in mind when trying to interpret TEP. Naturally,
we do not want to underestimate the role of current carrier scattering. However its calculation is more complicated since it involves
a larger number of arbitrary parameters.

In our report at this Conference on the calculation of TEP of
group I metals it is shown that the anomalous positive sign of TEP
of the noble metals and Li can partially be attributed to their
energy spectrum being different from that of free electrons. There
are indications of deviations from the free electron model for the
group IIIa metals Ga, In, Tl while the behaviour of Al agrees with
it (see, e.g., ref [15]). Thus the behaviour of actinides and metals
of groups I and IIIa do not disagree with our conclusion of a direct
connection between TEP and electronic structure. We believe our
empirical observations support the well known opinion of theorists
about thermopower being the most sensitive probe of the electronic
structure of metals among other transport properties. The complicated behaviour of TEP demonstrated in the present paper undoubtedly

carried much valuable information on fine features of the electron
energy spectrum. This makes further development of the theory of
thermopower still more urgent.

The analysis of temperature dependences leads to interesting
observations if we also take into consideration TEP of liquid metals.
The physical properties of liquid metals are well explained on
the basis of the NFE approximation. It is known at the same time
that TEP of almost all liquid metals have negative temperature derivatives as do free electrons. The noble metals Cu and Ag behave
in the same way.[16] But in the solid state they have an anomalous
positive TEP (Fig. 14). Therefore, there exists a break of TEP curve
at melting. Using Fig. 18 one can represent such a behaviour by
means of the curves 1 and 2" provided $T_{tr} = T_M$ (melting temperature). Thus the change of electronic structure at melting is followed by sharp change of TEP temperature dependence. Let us turn
now to the third metal of this group, Au. Its TEP in the middle
temperature region is similar to that of Cu and Ag but it loses its
positive slope when heated (Fig. 14, dashed line). One can say
that TEP of solid Au at sufficiently high temperature becomes like
TEP of the metal in NFE approximation if its temperature dependence
is considered. When melting, the TEP temperature dependence of Au
does not change, differing in that from Cu and Ag.[16] Summing up all
these observations it is possible to suggest that in the case of Au
the transformation of a complicated energy spectrum into a simpler
one takes place (or begins) not at the melting point, as for Cu and
Ag, but still in the solid state. If we consider the generalized
TEP of different groups of metals once more, we shall notice for a
number of cases at high temperatures the same peculiarity that Au has:
TEP curve achieves maximum and then has negative slope. One can
think in such cases TEP reflects also a change in electronic structure. It is clear that examination of thermoelectric properties only,
is insufficient to make any final conclusion on electronic structure.
But, perhaps we have here an example where TEP is more sensitive
to the changes of electronic structure than other properties of the
metal. It confirms the necessity of wider and more thorough experimental and theoretical investigations of TEP at high temperatures.
It would be useful to study TEP of all metals in solid state up to
the melting point. TEP of liquid metals in the wider temperature
range are of significant interest too.

REFERENCES

1. M.V. Vedernikov, Advan. Phys. 18, 337 (1969).
2. M.V. Vedernikov, A.T. Burkov, V.G. Dvunitkin, and N.I. Moreva,
    J. Less-Common Metals 52, 221 (1977).
3. J. Nystrom, in "Landolt-Bornstein, Zahlenwerte und Funktionen",
    edited by K.H. Hellwege (Springer, Berlin, 1959), vol. 2,
    part 6, p. 929.

4. R.P. Hubener, in "Solid State Physics", edited by H. Ehrenreich, F. Seitz, and D. Turnbull (Academic Press, N.Y. and London, 1972), vol. 27, p. 63.

5. A.D. Stewart and J.M. Anderson, Phys. Stat. Solidi B 45, K89 (1971).

6. V.I. Spitsyn, V.E. Zinov'ev, P.V. Gel'd, and O.A. Balachovskiy, Dokl. Ak. Nauk SSSR 221, 145 (1975) - in Russian.

7. J.M. Ziman, Advan. Phys. 10, 1 (1961).

8. F.J. Blatt, Helv. Phys. Acta 41, 693 (1968).

9. J.G. Cook, M.J. Laubitz, and M.P. Van der Meer, Can. J. Phys. 53, 486 (1975).

10. J.G. Cook and M.P. Van der Meer, J. Phys. F: Metal Phys. 3, L130 (1973).

11. J.G. Cook and M.J. Laubitz, Can. J. Phys. 54, 928 (1976).

12. O.K. Kuvandikov, A.V. Cheremushkina, and R.P. Vasil'eva, Fiz. Metal. Metalloved. 34, 867 (1972) - in Russian.

13. J.G. Cook, M.J. Laubitz, and M.P. Van der Meer, J. Appl. Phys. 45, 510 (1974).

14. M.V. Vedernikov, Prib. Techn. Eksper. No. 5, 209 (1975) - in Russian.

15. B. Bosacchi and R.P. Huebener, J. Phys. F: Metal Phys. 1, L27 (1971).

16. T. Ricker and G. Schaumann, Phys. Kondens. Mater. 5, 31 (1966).

DISCUSSION

J.F. Goff: I'd like to point out that if you consider the thermo-electric power as a ratio of two integrals, the amount of energy in the integrals is very, very large - about plus or minus 10 to 15 times kT - because essentially you are balancing off contributions from above and below the Fermi energy. So you are looking at small differences. If you take as a reasonable value an energy range at 1000° of about 2 electron volts, this means that in the transition metals you are encompassing a large portion of the band structure of that metal. So the peculiar shape of the thermopower, it seems to me, may come about from the fact that the high energy and low energy portions of the contributing integral have different effects in different temperature ranges which causes, perhaps, in the case of iron, a sinusoidal shape of the thermopower. There seems to be a strong correlation in this case with band structure. I'd like to comment, along this line, that the peculiar effect in the thermo-power of one of the metals - ytterbium, I think it was, that had the transition - might come about from the fact that in a phase change you affect different portions of the Fermi surface in differ-ent ways. The portion contributing to the conductivity could change by a small amount, as it does quite often in phase change, while the portion in the numerator might not change very much. This

would cause the thermopower to simply increase or decrease, depending on the way the change went, without changing the overall shape
of the thermoelectric power itself.

THERMOELECTRICITY IN TRANSITION METALS

D. Greig

Physics Department
University of Leeds
Leeds LS2 9JT, U.K.

At any temperature thermoelectric effects in transition metals
are typically an order of magnitude greater than in "simple" metals,
so that at least one branch of any commercial thermocouple consists
of a transition metal or alloy. It was because of this large magnitude that many of the earliest studies of thermoelectricity were
made on transition metals,* with the change in slope of the temperature dependence of S at the Curie point first observed in Ni
more than a century ago.[1] Very briefly the values of S reflect
large values of $d\ln\sigma/d\varepsilon$ which, in turn, are due to the strong energy
dependence of the density of states $N(\varepsilon)$ close to the Fermi energy,
ε_F.

The study of transport properties in transition metals is also
complicated by phenomena that are either not present or are not
observed in "simpler" metals, and much of the present work on transition elements is centred around these effects. The most notable
of these arise from various aspects of magnetism (magnons, spin
fluctuations (paramagnons), anomalies at the Curie and Néel temperatures), but there is also a continuing interest in electron-electron
effects and the general question of s-d scattering. A survey of
this work up to 1974 is given in "Thermoelectric Power of Metals",[3]
and the present article is a review of the direction that the work

---

* The largest (negative) thermopower at room temperature is observed
  in the semi-metal Bi, and one of the earliest thermocouples used
  by Seebeck was Bi-Cu. Apart from this, however, Seebeck's series
  (in which he gave elements studied in order of their emf's) began
  Bi, Ni, Co, Pd, Pt, ...[2]

has taken since then.   The article does not, however, include any
reference to phase transitions as this is a subject of so great
importance that it will be reviewed separately later in the conference.

1.   PURE METALS

Very broadly, recent work on pure metals consists of accurate
measurements on the purest materials at high and low temperatures.
In the former category we may instance work by Laubitz and his
colleagues on Pd[4] and Ni[5], with the results summarised in Fig. 1.
Although the general trend of these results is already well-known,
the authors point out (i) the close similarity of the results on
Ni and Pd in the paramagnetic range and (ii) the concave nature of

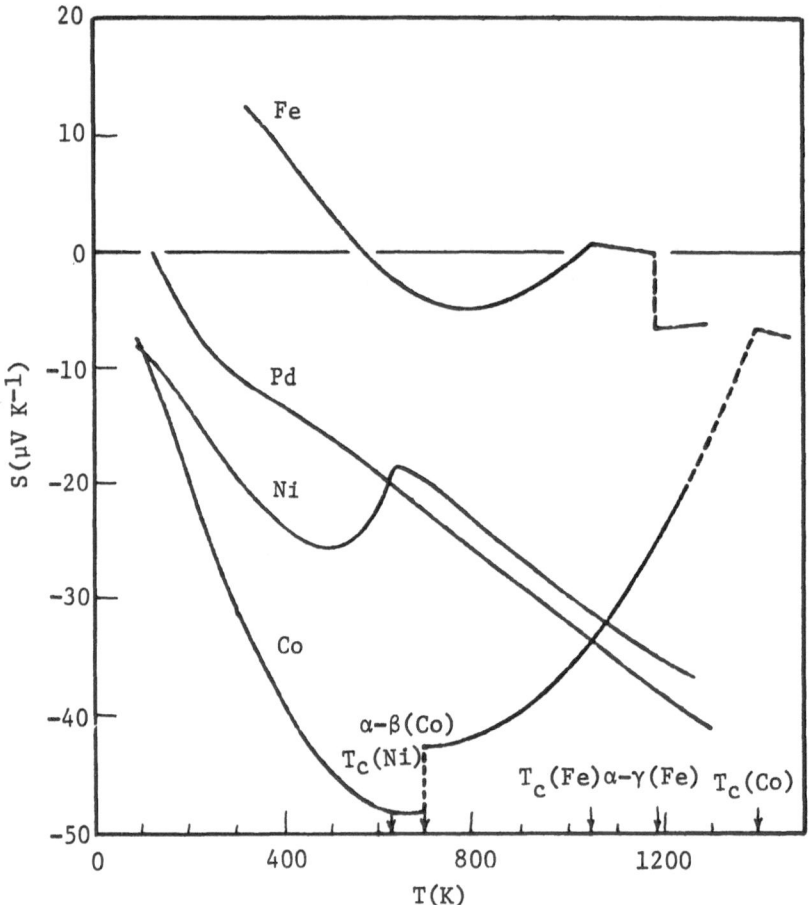

Fig. 1.   Temperature variation of the absolute thermoelectric power
of Ni, Fe, Pd and Co.   From Laubitz et al refn. 5.

the curves for the ferromagnetic range.

These new data were, in fact, part of a wider study that included measurements of electrical and thermal conductivities, σ and κ respectively, with the objective of providing a simple coherent model to explain all the observations on any particular metal. The argument is essentially to use a simple model of conduction electrons scattered only by phonons with κ, σ, and S related through the transport equations by means of an adjustable function $X(\varepsilon)$. This function, which is represented by the authors as a Taylor series, $X(\varepsilon) = \sum_n a_n (\varepsilon - \varepsilon_F)^n$, is related to the relaxation time, τ, by $X(\varepsilon) = f(T) \int v\tau dS$, where v is the electron velocity, dS an element of a surface of constant energy ε, and f(T) a simple specified function containing temperature and parameters of the lattice such as the atomic mass M and θ_D. Unfortunately, in the case of paramagnetic Ni, it is found impossible to obtain a self-consistent model giving σ, κ, and S for one particular form of $X(\varepsilon)$. The authors therefore conclude that at least one other scattering mechanism, possibly electron-electron, is required. Furthermore, they note that the peak in the "experimental" variation of $1/X(\varepsilon)$ with energy is much broader than expected. They argue that, as this energy dependence comes almost entirely from the energy dependence of τ, it should follow from the Mott model that $1/X(\varepsilon) \alpha 1/\tau \alpha N(\varepsilon)$. In fact the peak in the energy dependence of $X(\varepsilon)$ is almost twice as broad in energy as the density of states peak in all band structure calculations. This clearly throws serious doubt about any simple application of the Mott model for numerical estimates of themopower at high temperatures. The authors intend, however, to investigate whether or not the addition of electron-electron scattering process can remove the discrepancy.

As a final comment it should be mentioned that Shimizu and Sakoh[6] calculated the transport properties of paramagnetic bcc Fe by using transport integrals together with the calculated band structure and density of states. Their calculation gives quite good agreement with experiment in the very limited temperature range above the $\gamma - \delta$ transition (1667K to 1810K), but, of course, does not attempt to reproduce the sharp change in S at the transition itself.

Turning to low measurements we note particularly a series of papers on W beginning with measurements by Trodahl[7] at temperatures between 2K and 9K. The most significant feature of the data is a positive peak in S at $\sim$ 6K with the temperature dependence at lower temperatures containing both terms in T and higher powers of T.

Trodahl suggests that the curvature in S may be due to a combination of a negative electron-impurity thermopower S_d^i combined with a positive electron-electron diffusion term, S_d^{ee}. Although both S_d^i and S_d^{ee} are proportional to T, the observed curvature in S arises when they are weighted with the corresponding thermal resistivities.

The second set of measurements on W over approximately the same temperature range was reported by Garland and Van Harlingen[8] who used more accurate experimental techniques on higher purity specimens. One of these crystals had a residual resistance ratio (RRR) as high as 77,300. The temperature dependence of S was found to be roughly the same as that observed by Trodahl; that is, positive at the lowest temperatures with a peak somewhere between 4K and 6K for the various samples. However these authors do not analyse their thermopower results in detail but instead concentrate attention on the thermoelectric ratio G. They find that, in all specimens for T < 4K, the temperature dependence of G is accurately given by

$$G = AT^{-1/2} + BT.$$

At the very lowest temperatures where electron scattering is dominated by impurity processes we should expect G to be independent of temperature. This is clearly shown at temperatures below ∿ 0.5K in a recent series of experiments by Stone et al.[9] whose results are reproduced in Figure 2. In the temperature range between about

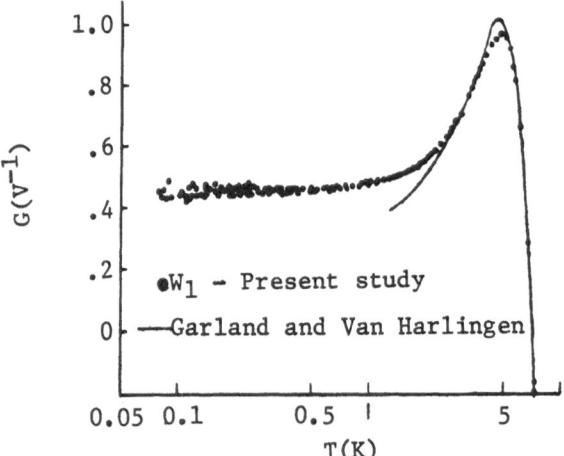

Fig. 2. The temperature variation of the thermoelectric ratio G for a tungsten sample W1 (RRR = 44,000), and for a sample of comparable purity (RRR = 63,000) as measured by Garland and Van Harlingen.[8] From Stone et al. refn. 9.

1 and 4K some additional scattering mechanism must therefore exist, but Garland and Van Harlingen argue that it <u>cannot</u> be electron-electron. When multiple scattering processes are present the appropriate G will be weighted -- for the same reason as in the Nordheim-Gorter rule -- by the ratio of the corresponding electrical resistivity to the total resistivity. Consequently, at these temperatures the electron-electron component of G, G_{ee}, is observed as $G_{ee}\rho_{ee}/\rho_{Total} \sim G_{ee}\rho_{ee}/\rho_o$ where ρ_{ee} and ρ_o are the electron-electron and impurity resistivities respectively. As $\rho_{ee} \propto T^2$ we see that G_{ee} should also vary as T^2 which is quite unlike the temperature dependence actually observed. This they have to explain on the basis of different electron phonon contributions to G from different sheets of the Fermi surface.

Returning to the recent experiments below 1K referred to above, we note that it is only in the purest of 3 tungsten samples used by Stone et al.[9] that G becomes independent of T. For the other two (RRR = 34,000 and 22,000) G changes sign and falls rapidly with decreasing temperature, showing no evidence, even at 45mK, of reverting to a constant value as would be expected from thermodynamic arguments. The authors attribute this anomaly to the presence of Fe impurities, and this seems to correlate with the presence of a very shallow resistance minimum in one of their samples.

2. DILUTE ALLOYS

2.1 Non-magnetic metals

Although most work on dilute alloys of transition metals has been undertaken in materials in which some aspect of magnetism is involved, two papers on non-magnetic alloys are of general interest. In the first Uray[10] has shown the value of a Nordheim-Gorter plot in determining the type of impurity present in nominally pure metallic wires. As an example he has studied W containing up to $\sim$ 150 ppm of one of the solutes Fe, Ni, Pt, Al, Ge or Mo, and has shown that the differential thermopower at 310K between the alloy and the pure starting material is quite different for each of the six cases (Fig. 3). It is therefore evident that such a measurement is an extremely useful means of chemical analysis. Naturally when several impurities are present simultaneously the investigation is more difficult. However, even then Uray has shown that, by studying Nordheim-Gorter plots made before and after evaporating some of the impurity, it is possible to deduce whether or not more than one solute is present.

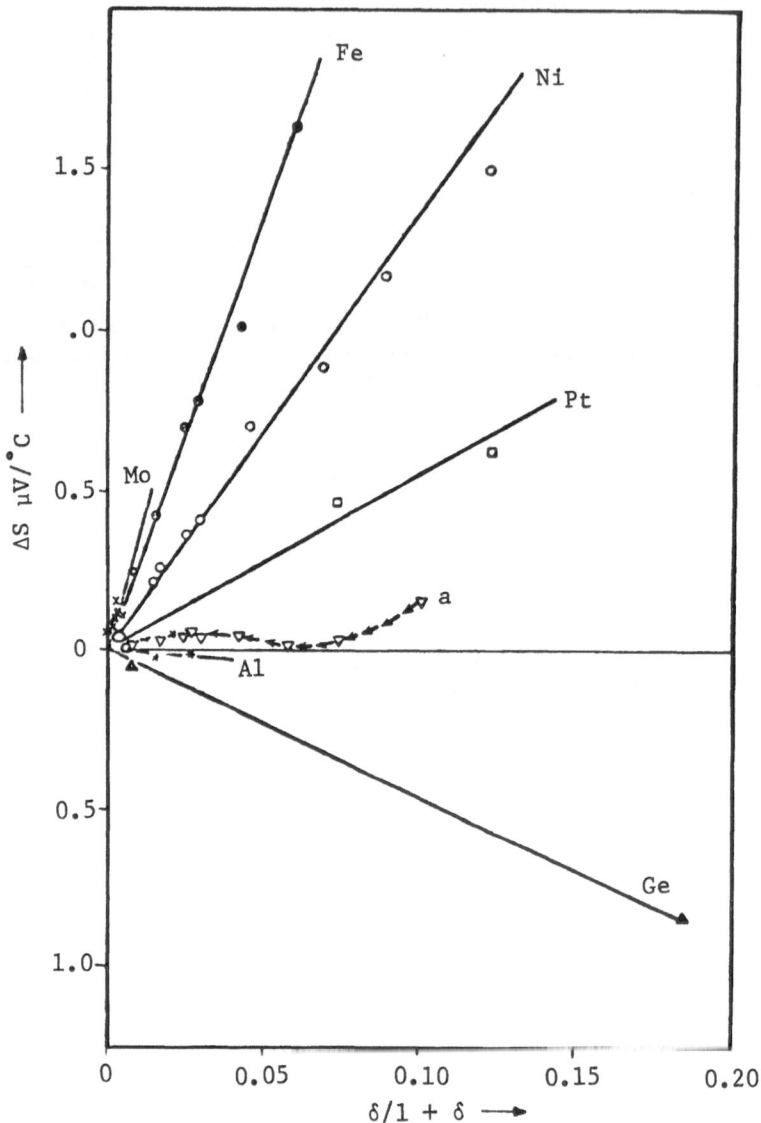

Fig. 3. The Nordheim-Gorter diagram for different alloys of tungsten after annealing first at 2000K and then at 2800K for 15 min.
ΔS is the difference in thermopower between the alloy wire and a
wire of pure tungsten, while δ is the ratio of the excess resistivity
of the alloy to the total resistivity of the pure metal at a particular T (in this case 310K). The annealing curve for the pure metal
(marked by a) was first measured "as drawn" and then after 15 min.
annearling at 1000, 1200, 1400 ... 2800K. From Uray, refn. 10.

The second general point of quite different nature is contained in a paper by Muir and Zuckermann[11] who have shown that, for
s-d scattering with $\tau \propto 1/N_d(\varepsilon)$, the expansion of $d\ln\sigma/d\varepsilon$ leads to
a term in T^3 in the diffusion thermopower. The magnitude of this
term is related to the electronic specific heat, and is also connected with the coefficient of T^2 in the temperature dependent
modification of residual resistivity arising through the same
approximations.[12] In the one system in which these expressions
have been examined numerically - Pd + 5% Rh - the values of the
coefficients obtained in this way are unfortunately very different
from those measured by experiment. However, it must be emphasised
that since, in this alloy, ε_F lies very close to the peak of the
density of states curve, $(dN_d/d\varepsilon)_{\varepsilon=\varepsilon_F} \sim 0$ and this is possibly not
the best system with which to test the calculation. Furthermore,
the T^3 term is naturally competing with phonon-drag effects -- not
to mention 'phoney' phonon-drag -- and the detection of this com-
of S_d may always remain elusive.

2.2 Ferromagnetic metals

As regards measurements on ferromagnetic alloys much of the
recent work has been undertaken by Beylin et al.[14,15] who have
investigated the effect of no fewer than 13 different types of
solute atom in Ni. In their earlier paper[14] in which they studied
the effect of adding In and the transition elements, Ti, Zr, Ta,
Cr, Mo, V, W, and Re, they found that, in each case, S was positive
over practically the whole temperature range from 4K to 400K.*
They have interpreted their results in terms of a model proposed
by Korenblit and Lazarenko[17] who showed that, in the presence of a
combination of inelastic scattering and elastic impurity scattering that is opposite for opposite directions of spin, S_d should be
proportional to $[\tau_o(\uparrow) - \tau_o(\downarrow)]$ where $\tau_o(\uparrow)$ and $\tau_o(\downarrow)$ are the
impurity relaxation times for spin-$\uparrow$ and spin-$\downarrow$ electron respec-
tively. In comparing these results with measurements on Ni-Co,
Ni-Fe, and Ni-Cu where S is predominantly negative, the signs of
S correlate well with known values of $[\tau_o(\uparrow) - \tau_o(\downarrow)]$. Nevertheless two points should be noted. (i) The present results do not
scale with solute concentration as predicted by Korenblit and
Lazarenko and (ii) the measurements can be explained equally well

---

* They did not however study Ni-Mn where rather more complicated
  effects are observed.[16]

by a 'characteristic' thermopower for each solute without invoking this additional scattering mechanism. This is certainly an area for further study.

In their later paper Beylin et al.[15] have studied the high-temperature thermopower (77-1300K) of Ni containing Al, Ga, In, and Nb; that is apart from Nb, elements from two columns to the right of Ni in the periodic table. These measurements should be compared and contrasted with their earlier measurements[18] over the same temperature range of Ni alloyed with W, Re, and Hf. It is found that, <u>for the most part</u>, alloying with the transition metals leads to large positive values of S in the paramagnetic region with S reaching a maximum and then decreasing slightly with rising T, whereas adding the non-transition metals makes S more negative than in the pure metal with S falling less quickly than T. However, it must be emphasised that for <u>Ni</u>-In the trend is in the opposite direction.

The authors have obtained an expression for the electron-phonon diffusion thermopower by expanding the transport integrals. It is

$$S_d^{phonon} = \frac{a\tau'T}{b + c\tau''T^2} \tag{1}$$

where $\tau' = (d\tau/d\varepsilon)_{\varepsilon=\varepsilon_F}$ and $\tau'' = (d^2\tau/d\varepsilon^2)_{\varepsilon=\varepsilon_F}$, and where a, b, and c are constants. (The full expression also contains the T^3 term referred to in the last section,[11] but the present authors consider this negligible.) It is then argued that the experimental results can be qualitatively explained by (1) simply on the basis of $\tau \propto 1/N_d(\varepsilon)$ together with a rigid band model. In view of the quite different results in <u>Ni</u>-Al and <u>Ni</u>-In, this view might be altogether too naive.

One further paper that is of interest as it embraces work in both this and the previous section is that concerned with measurements between 4.2K and 300K on Pd-Fe alloys at both ends of the composition range.[19] The results are particularly interesting for the Pd-rich material (Fig. 4), since even the most dilute alloy containing 0.5% Fe is ferromagnetic at helium temperatures. We see that, for each of these alloys, the variation of S with T passes through a negative minimum in the ferromagnetic range although the temperature of this minimum is in all cases below 30K. For the more concentrated alloys, this is far below the Curie temperature.

The authors do not discuss the origin of this minimum but we may speculate that it could be due to the large diffusion thermo-

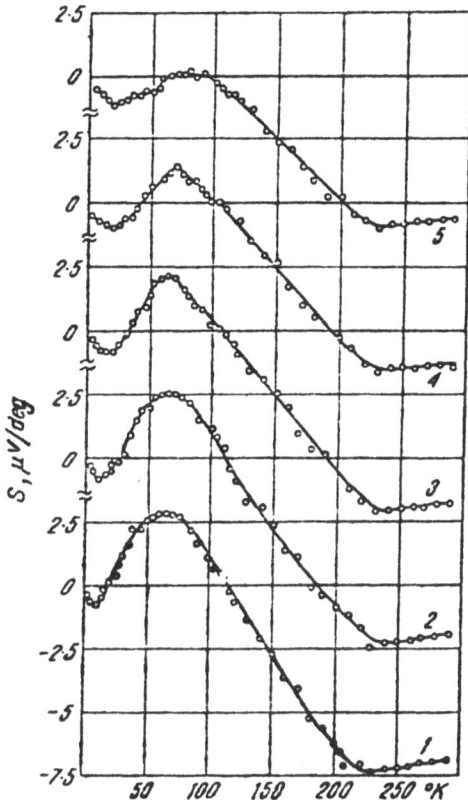

Fig. 4. Temperature variation of absolute thermoelectric power of
dilute Pd-Fe alloys. (1) Pd + 0.5% Fe; (2) Pd + 1% Fe; (3) Pd +
1.5% Fe; (4) Pd + 2% Fe; (5) Pd + 3% Fe.   From Abramova et al.,
refn. 19.

power predicted for ferromagnetic metals by Kasuya[20] when asymmetric
elastic scattering co-exists with inelastic scattering at magnetic
ions.

2.3 Nearly magnetic metals

One of the most interesting -- and controversial -- developments in the study of thermopower is in that class of dilute alloys
that are not quite ferromagnetic; that is in materials such as
Pd-Ni in which the magnetic excitations are spin-fluctuations. A

useful survey on this subject has been written recently by Kaiser[21] based on experimental measurements on the three systems Pd-Ni,[22] Ir-Fe,[23] and Rh-Fe.[24]   The characteristic behaviour of the thermo-power of these 3 systems may be summarized as follows:

(1) A large peak occurs in all 3 systems at a temperature, T_p, which is of the order of the spin-fluctuation temperature, T_{sf}, deduced from resistivity measurements.

(2) In each of the 3 systems T_p is essentially independent of solute concentration.

(3) The sign of the peak is negative for Pd-Ni and Rh-Fe, but positive for Ir-Fe.

(4) The magnitude of the peak increases with concentration in Pd-Ni (Fig. 5) and Ir-Fe, but decreases in the case of Rh-Fe.

(5) At the lowest temperatures the experimental values of S are proportional to T.

On the basis of this summary Kaiser argues that any explana-tion of the peaks in terms of an enhanced diffusion power by

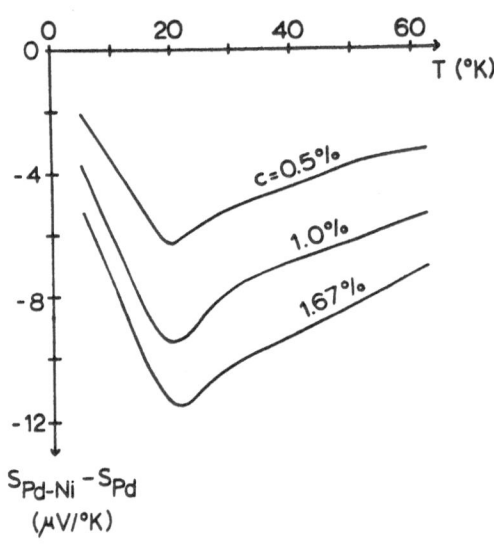

Fig. 5. Temperature variation of the difference of the thermopower of Pd-Ni and that of pure Pd, where C is the concentration of Ni. From Kaiser, refn. 21.

electron-paramagnon scattering is most unlikely. For one thing
there is no simple mechanism for the large asymmetric variation
of τ across the Fermi surface required to produce the required
enhancement of S_d, and furthermore, the variation of peak height
with solute concentration implies that higher order (Kondo) effects
can also be discounted. It is for these reasons that Kaiser does
not make use of an earlier calculation by Fischer[25] of the electron-
paramagnon diffusion thermopower.*

Instead, he has proposed that the origin of the peaks is
paramagnon drag. This mechanism, labelled S_{sf}, has just the required properties to account for the experimental observations on
a qualitative basis:

(1) S_{sf} can be either positive or negative depending on whether or
not Umklapp processes are present.
(2) The net rate at which spin-fluctuations are absorbed by electrons is proportional to the concentration of solute. (For
Rh-Fe there are suggestions that the process is complicated by
interactions between Fe atoms.[24])
(3) At high temperatures the paramagnon spectrum becomes blurred
leading to a maximum effect at $\sim T_{sf}$.
(4) At low temperatures S_{sf} is proportional to the spin-fluctuation
specific heat which, in turn, is proportional to T.

The general agreement with experiment is clearly impressive
and a detailed calculation of paramagnon drag is awaited with great
interest.

3. SUMMARY

The interpretation of the diffusion thermopower of transition
metals continues to be centred around the Mott model, according to
which conduction is by s-electrons that are scattered into d-states,
with a relaxation time that is therefore proportional to $1/N_d(\varepsilon)$.

There is no doubt that, taken very broadly, this model gives
qualitative agreement with experimental results, in that the high
temperature thermopower of elements to the left hand end of the
transition series is largely positive, while that of the Group VIII
elements is largely negative. (For a survey of this point see
reference 3.) These general remarks also apply to the sign of the

---

* This theory is, in any case, restricted to hosts with very weak
  exchange enhancement and is not really relevant to alloys like
  Pd-Ni.

thermopower of dilute alloys.[15] Not surprisingly, however, highly
detailed theoretical investigations are less convincing. Following
their success with the calculation of S in Mo, W, Pd, and Pt[26]
Shimizu and co-workers[6] have turned their attention to Fe, but the
model is then satisfactory only over a very limited temperature
range. When the calculation is made in great detail to correlate
several transport properties, the results are certainly disappoint-
ing.[5] Some account has also been taken of higher order terms in
the temperature dependence of S_d,[11] but as the major term in this
expansion is proportional to T^3, there is clearly a problem to
separate this from the measurements unambiguously. Nevertheless,
in spite of these difficulties, most authors continue to interpret
their results on the basis of the Mott model, and there have been
no attempts to discuss alternatives.

As regards the search for effects that are 'special' to transition metals, we must conclude that evidence for electron-electron
scattering is still non-existent. The suggestion that the maximum
in S at $\sim$ 6K originates from this has now been discounted,[8] and
although new anomalies have now been observed below 1K in W[9] (and
possibly Pt[27]) these do not seem to originate in electron-electron
scattering either. The only possibility of isolating electron-
electron effects seems to be in measurements at high temperatures.[4,5]

For phenomena that are magnetic in origin, however, many interesting results have been obtained and it is their interpretation
rather than their existence that is controversial. In dilute
nickel alloys, for example, the sign of S correlated with the sign
of $[\tau_o(\uparrow) - \tau_o(\downarrow)]$ and the measurements then interpreted on the
basis of a model combining elastic and inelastic scattering.[17] However, the results can be equally explained by a change in $d\ln\sigma/d\varepsilon$
in the presence of impurities, and this is an area for further
study. So too is the interpretation of the low temperature peaks
in dilute ferromagnetic alloys such as are observed in Pd-Fe.[19]

Finally, on the subject of magnon drag, we should mention that
no advances have been made in the last 3 years, and that the only
clear evidence for its existence in metals or alloys is still in
the field variation of the Peltier coefficient in concentrated
nickel alloys.[28] The evidence in favour of paramagnon drag,[21] on
the other hand, in systems such as Pd-Ni is quite encouraging.

ACKNOWLEDGMENTS

I should like to thank Professor J. S. Dugdale for reading and commenting on the first draft of this manuscript.

REFERENCES

1. P.G. Tait, Proc. Royal Soc. Edinburgh, 8, 182 (1872)
2. J.Yarwood, Electricity and Magnetism, (University Tutorial Press, London, 1973) p. 552.
3. F.J. Blatt, P.A. Schroeder, C.L. Foiles and D. Greig, Thermo-electric Power of Metals (Plenum Press, New York, 1976), p.135.
4. M.J. Laubitz and T.Matsumura, Canad. J. Phys. 50, 196 (1972).
5. M.J. Laubitz, T.Matsumura, and P.J. Kelly, Canad. J. Phys. 54, 92 (1976).
6. M.Shimizu and M.Sakoh, J. Phys. Soc. Japan 36, 1000 (1974)
7. H.J. Trodahl, J. Phys. F: Metal Physics 3, 1972 (1973)
8. J.C. Garland and D.J. Van Harlingen, Phys. Rev. B10, 4825 (1974)
9. E.L. Stone, M.D. Ewbank, W.P. Pratt, Jr., and J. Bass, Physics Letts. 58A, 239 (1976)
10. L. Uray, Acta. Tech. Acad. Sci. Hungary 78, 435 (1974).
11. W.B. Muir and M.J. Zuckermann, Canad. J. Phys. 53, 1777 (1975).
12. See, for example, D.Greig and J.A. Rowlands, J. Phys. F: Metal Physics 4, 232 (1974).
13. A. Cafro, F.T. Hedgcock, W.B. Muir, and M.J. Zuckermann, Low Temperature Physics - LT14, Ed. M. Krusius and M. Vuorio (North Holland/Elsevier, Amsterdam and New York, 1975), 3 282.
14. V.M. Beylin, T.I. Zeynalov, I.L. Rogel'berg and V.A. Cherenkov, Phys. Metals and Metallog. 38(6), 182 (1974).
15. V.M. Beylin, T.I. Zeynalov, I.L. Rogel'berg and Z.I. Ali-Zade, Phys. Metal and Metallog. 39(5), 63 (1975).
16. D.Greig and J.A. Rowlands, Low Temperature Physics - LT14, Ed. M. Krusius and M. Vuorio (North Holland/Elsevier, Amsterdam and New York, 1975), 3, 286.
17. I.Ya. Korenblit and Yu.P. Lazarenko, Soviet Phys.-JETP 33, 837 (1971).
18. Z.I. Ali-Zade, V.M. Beylin, T.I. Zeynalov and I.L. Rogel'berg, Phys. Metal and Metallog. 36(1), 184 (1973).
19. L.I. Abramova, G.V. Fedorov, and N.V. Volkenshteyn, Phys. Metal and Metallog. 36(3), 43 (1973).
20. T.Kasuya, Prog. Theor. Phys. 22, 227 (1959).
21. A.B. Kaiser, Magnetism and Magnetic Materials - 1975, Ed. J.J. Becker, G.H. Lander, and J.J. Rhyne (American Inst. of Physics, New York, 1976) 29, 364.
22. C.L. Foiles and A.I. Schindler, Phys. Letters 26A, 154 (1968). (See also, V.A. Matveev, G.V. Fedorov, and N.V. Volkenshteyn, Phys. Metal and Metallog. 42 (1976)).

23. J.S. Touger and M.P. Sarachik, Solid State Commun. <u>17</u>, 1389
     (1975).
24. J.E. Graebner, J.J. Rubin, R.J. Schutz, F.S.L. Hsu and W.A.
     Reed, <u>Magnetism and Magnetic Materials - 1974</u>, Ed. C.D.
     Graham, Jr., G.H. Lander, and J.J. Rhyne (American Inst. of
     Physics, New York, 1975), <u>24</u>, 445.
25. K. Fischer, J.Low Temp. Physics <u>17</u>, 87 (1974).
26. T. Aisaka and M. Shimizu, J. Phys. Soc. Japan, 646 (1970).
27. C. Uher, C.W. Lee and J. Bass, submitted to Physics Letters
     (1977).
28. G.N. Grannermann and L. Berger, Phys. Rev. B <u>13</u>, 2072 (1976).

DISCUSSION

K.H. Fischer: I would like to make one short comment on your last
remarks about paramagnon drag with respect to the theory of Kaiser.
Kaiser uses a model where he has a d-band and a conduction band, and
the conduction electrons give rise to the transport properties,
while the d-electrons are responsible for the magnetic properties.
The problem is, how to couple these two bands. This is done in the
theory of Kaiser by exchange coupling. Now it is known from the
Kondo effect, for instance, that if you have only exchange coupling
then you have symmetry between electron-electron and electron-hole
scattering. This symmetry makes the relaxation time symmetric with
respect to the Fermi energy, and the thermopower exactly vanishes;
more correctly, the thermopower contribution due to this effect.
The same thing happens in the model of Kaiser. That means his model
is unable to explain the large thermopower without invoking other
mechanisms. What would have to be done is to take into account mixing between the two bands. If you do that I'm sure you would get a
big effect in the thermopower and you would not need to invoke magnon
or paramagnon drag effects. I shall describe a somewhat different
model in my talk which I think gives perfect agreement for the data
on iridium-iron and rhodium-iron, based on a theory of localized
spin fluctuations.

D. Greig: Thank you very much. I knew that this would be a controversial point. May I ask, does your model agree with all three
systems?

K.H. Fischer: I think one should distinguish between the palladium-
nickel case and the rhodium-iron and iridium-iron case. The physical difference is that in the case of rhodium-iron and iridium-iron
you may imagine fairly localized paramagnons about the impurities,
which are here the iron atoms. This is different in the case of
palladium-nickel, where we think the paramagnons are much more extended and, therefore, the theory simply cannot be applied to this
case.

B.R. Coles: This is a comment on what Denis has said and what
Konrad added. The first point is that Konrad is absolutely right
that even in the ordinary Kondo system, if you treat the coupling
to the d-electrons as a pure exchange, a pure Kondo Hamiltonian,
then you throw away the fact that you have projected out your effective J from the mixing term, and things go wrong. The same thing
is true also with the f and d-bands of uranium or nickel. You also
must recognize that the coupling is a hybridization coupling and
not a simple exchange coupling. The other point I'd like to make
is that I am very attracted by the idea of a paramagnon treatment
of the Schindler-Foiles peak, but the treatment of the peak as
representative of a local spin fluctuation temperature, I think, cannot be right, since any local spin fluctuation temperature must
extrapolate in a simple mean field approximation toward zero as you
approach the critical concentration for ferromagnetism. So whatever
the temperature of the peak is, I don't think it can be regarded as
a local spin fluctuation temperature and it is more likely to be
something associated with the paramagnon drag effect, taken over the
whole system. On a slightly different point, I would like to emphasize very strongly that, of course, the Mott model is bound to
break down in niobium and tungsten, in the middle of the transition
series, simply because it is no longer valid to treat a group of
carriers as quasi s-like and the group of states into which they are
scattered as quasi d-like; we can see that going wrong even as we
go back from palladium to rhodium. However, it turns out by strange
coincidence that in pure niobium, if Yamashita and Asano's analysis
of the resistance is correct, some sort of treatment of d-like and
s-like parts of a Fermi surface is still possible, although the
tight binding functions are all built up from the atomic d-functions,
and one can then talk about rather distinct parts of the Fermi surface.

D. Greig: Yes, that is a good point. I think all the cases I
was giving, where people have been trying to really squeeze this
model in detail, concerned nickel and palladium and alloys based on
these metals.

C.L. Foiles: I'd just like to emphasize Brian's comment about local
spin fluctuation temperature. We have more data on palladium-
nickel. We have made measurements up to 3 atomic% although we have
not done our concentration intervals fine enough to find where
things stop exactly. But certainly between 1.6%, which was the last
concentration in Figure 5, and 3% the peak still does not shift
although the 3 atomic% alloy is ferromagnetic at 12K; the magnitude
of the peak shrinks only very slightly. By the time you've gone
from 1.6% to 3% it has diminished by 10 or 15%. So nothing drastic
is happening to the peak although the magnetic properties are changing very drastically.

D. Greig: You haven't tried putting a magnetic field on, I take it.

C.L. Foiles: Only because our present rig does not work at 18°K.
We are reactivating one that will, and I hope within 3 months to
give you an answer to that.

J. Bass: Lest your remarks leave the impression that there was
only one tungsten sample for which G behaved simply by becoming
constant at low temperatures, let me note that we now have additional data and the pattern, I think, is fairly clear. As tungsten
is zone-refined beyond the first, second, and even third zone-pass,
the behavior of G becomes more nearly constant. Three out of four
three-or-more-pass zone-refined samples now have constant G's at
low temperatures, and the fourth displays only a small anomaly. So
there is no question, I think, that some "complex" impurity is being
taken out by the zone refining, leaving only simpler impurities.
One other comment, regarding the reason why as you raise the temperature G rises above its low temperature constant value. G is S/LT
where S is the thermopower, L is the Lorenz ratio and T is the
absolute temperature. If S is linear in T and if $L = L_0$, then the
temperature cancels out, and G is independent of temperature. But
now as you raise the temperature, L begins to drop rapidly below L_0
in these very high purity materials, which causes G to rise. Thus
the rise in G as T rises above 1K is not primarily a thermoelectric
effect at all. The primary reason for this rise is just the deviation of L from L_0 as phonon scattering begins to become important
relative to impurity scattering. At still higher temperatures, G
peaks out and then becomes negative, presumably because of a thermoelectric effect.

D. Greig: You are saying that the detailed interpretation given by
Garland and van Harlingen is unnecessary.

J. Bass: That's correct.

J. Garland: I think that Jack is correct. In light of his new
data the detailed explanation which van Harlingen and I proposed
probably is not true. On the other hand, I would point out that
there is also a low temperature peak in the thermopower which we
observed in tungsten, and so I'm not sure I would agree that the
low temperature peak observed in G is simply a function of the
decrease of the Lorenz number.

J. Bass: I tried to be careful not to use the word "simply"; I
said primarily. I don't want to rule out the other possibility,
but if one takes your own Lorenz numbers for tungsten samples of
comparable resistance ratio, then one gets 80% to 120% of the peaks
we see just by taking G_0 and dividing by the appropriate L. So I
think it is a major component; I do not want to argue that it is
all. It could be that there is some thermoelectric portion.

ON THE THERMOELECTRICITY IN URANIUM INTERMETALLICS

H.J. van Daal, K.H.J. Buschow and P.B. van Aken

Philips Research Laboratories

Eindhoven, The Netherlands

ABSTRACT

The thermoelectric power in intermetallic compounds $UNi_{5-x}Cu_x$ of the cubic $AuBe_5$ structure type and in UX_3 (X=Al, Ga, In, Si, Ge, Sn) of the cubic Cu_3Au structure type has been measured between 5 and 300K. The interpretation of the thermopower in $UNi_{5-x}Cu_x$ is given in terms of an s-d band model for the range $0 \leqslant x < 4$ and of mixed U^{4+}/U^{3+} valency in the range $4 \leqslant x \leqslant 5$. It is suggested that the thermopower in UX_3 compounds is dominated by paramagnon scattering.

Reports on the thermopower of uranium intermetallics are still scarce. The present paper shows that these compounds can have a large thermopower varying remarkably with temperature and composition. The interpretation of the thermopower and other electronic properties is still controversial, one of the points of dispute being the localization of the uranium 5f-electrons. In the metal uranium, where the nearest U-U distances in the various phases are relatively small (< 3Å), the spatial extent of the 5f-electrons leads to hybridization with 6d and 7s electrons and thus to narrow band formation[1]. In UX (X=S, P) compounds of the NaCl structure type with intermediate nearest U-U distances ($\lesssim$ 4Å), the 5f bandwidth is not smaller than in γ-U due to overlap and hybridization of 5f with anion p states[2].

The present work is concerned with two systems of uranium
intermetallic compounds which we classify differently with respect
to 5f localization. In the first system, pseudo-binary $UNi_{5-x}Cu_x$
compounds of the cubic $AuBe_5$ structure type, the nearest U-U
distances are large, ranging from 4.8 to 5.0Å. The electronic
properties of these intermetallics have been described in terms
of predominantly localized 5f-electrons[3]. In the second system,
compounds UX_3 (X=Al, Ga, In, Si, Ge, Sn) of the cubic Cu_3Au
structure type, the nearest U-U distances are in an intermediate
range from 4.0 to 4.6 Å. The properties of these compounds have
been described in terms of predominantly hybridized 5f-6d narrow
bands.[4,5] This paper discusses the thermopower of these intermetallics within the framework of data pertaining to lattice
constant, magnetic susceptibility, specific heat and electrical
resistivity. Data for the thermoelectricity in $UNi_{5-x}Cu_x$ have
already been presented and were briefly discussed in ref. 3. The
data for UX_3 have not yet been published.

The thermopower S of compounds $UNi_{5-x}Cu_x$ at temperatures
between 5 and 300 K with x in the range of $0 \leqslant x \leqslant 3$ is presented in
Fig. 1. It is seen that S(T) reaches large positive values for
x=0, for a slight increase of x (x=0.1 and 0.5) switches rapidly
to comparably large negative values, then with a further increase
of x changes gradually from negative back to positive values,
ending at x = 3.0 in a behaviour roughly similar to that of x = 0.
The behaviour at very low temperatures, up to T in the range 30 to
100K, is approximately linear in T, except for UNi_5. At higher
temperatures, up to the range 150 to 250K, deviations from the
low-T linear behaviour are proportional to T^3. The sign of the
curvature of S(T) is invariably opposite to that of the low-T
slope. The thermopower as described above is clearly dominated
by the diffusion term, phonon drag being dominant only in UNi_5 at
low temperatures. In UNi_5, phonon drag leads to a negative peak
at about 60K $(\approx \Theta_D/5)$. In the mixed compounds, evidently phonon
drag is suppressed by the alloying effect[6]. The diffusion thermopower can be described in a two-band model where conduction (s)
electrons in a broad band are scattered predominantly to a narrow
band at the Fermi level. This narrow band has been established
from the average value of the electronic specific heat coefficients
(γ) for $0 \leqslant x < 4$ to be the Ni 3d-band. This nearly filled band does
not contribute significantly to the magnetic properties of the
compounds. The only contribution to a temperature independent
Van Vleck paramagnetism stems from the uranium atoms which are

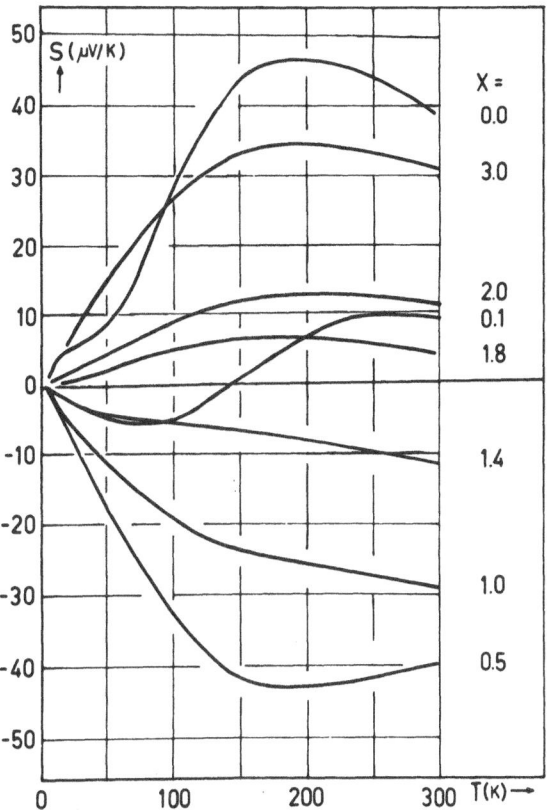

Fig. 1.   The thermopower of compounds $UNi_{5-x}Cu_x$, with $0<x<3$.

tetravalent ($5f^2$) with a non-magnetic ground state separated due
to crystal field splitting by a large energy distance from the
lowest magnetic state. An indication of the fine structure in
the density of states of the Ni 3d-band at the Fermi level $N(E_F)$
can be obtained from the measured γ-values as a function of x
(see Fig. 2). The γ-values have been converted into values of
$N(E_F)$, while the x-values have been regauged in approximate energy
values on the assumption that substitution of one Cu for one Ni
results in an increase of one electron per formula unit. In
first order approximation $S = \pi^2/3. \ k/|e|.kT.\nu_1$, where $\nu_i = [1/N(E).\partial^i N(E)/\partial E^i]_{E_{F_o}}$. The sign of the magnitude of the low-T
term in S, linear in T, determined by ν_1, is for the compounds
with $0\leq x\leq 3$ in quite satisfactory agreement with the $N(E_F)$ vs.

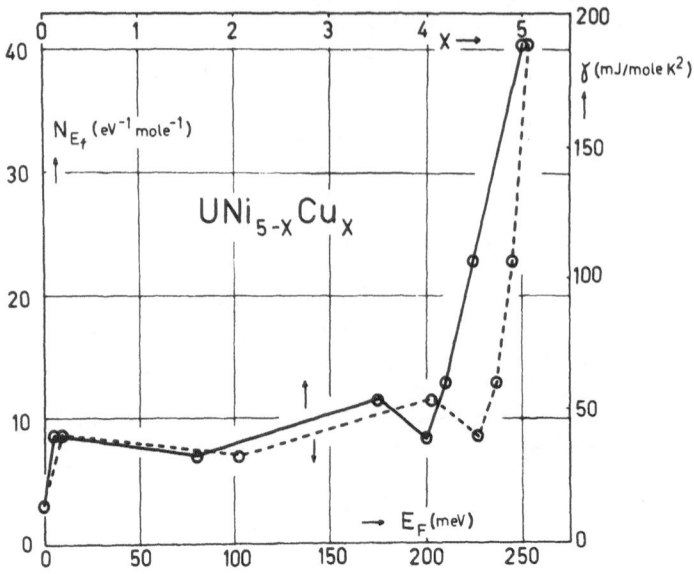

Fig. 2. The density of states $N(E_F)$ and the electronic specific heat coefficient   of compounds $UNi_{5-x}Cu_x$ as a function of x and of the Fermi energy E_F.

E_F curve established from γ data. In a second order approximation, all data, up to temperatures in the range 150 to 250K, can consistently be described by $S = \pm \pi^2/3.|k/e|. \quad T/T_1\{1-A(T/T_F)^2\}$, where $T_1=|v_1|^{-1}$, T_F is the effective degeneracy temperature: $T_F^2 = (6/\pi^2 k^2)|v_1^2 - v_2|^{-1}$, which is of the order of 1000 K, and A is a constant of the order of 10.

The thermopower of compounds $UNi_{5-x}Cu_x$ with $4\lesssim x\lesssim 5$ is quite different from that in compounds with $0\lesssim x\lesssim 3$ (see Fig. 3). Compounds with $3\lesssim x\lesssim 4$ are considered to belong to a transition region. The thermopower in UCu_5 varies from high positive to high negative values in a small temperature range around 11K, while in all compounds (except x=4.0) large negative peaks occur at temperatures T_s with $15K<T_s<35K$. We have suggested that when x comes into

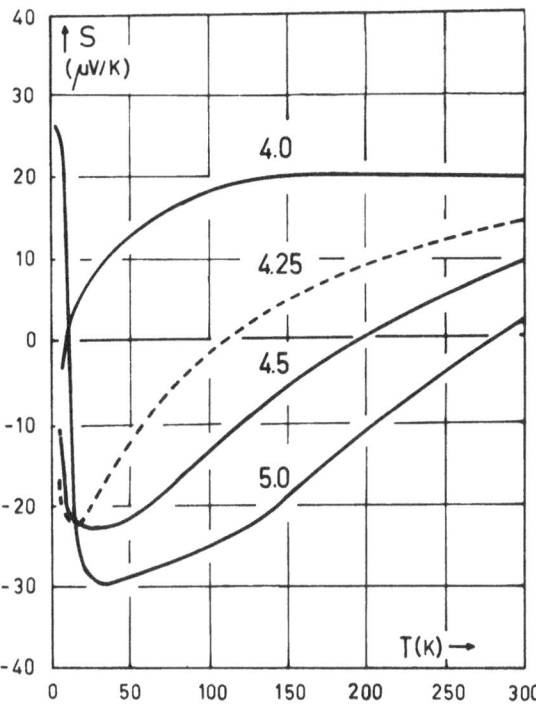

Fig. 3.   The thermopower of compounds $UNi_{5-x}Cu_x$, with $4<x<5$.

the range 4 to 5, the uranium atoms, driven by the electron
increase, change from a U^{4+} to a mixed U^{4+}-U^{3+} state. This suggestion has been based on various experimental facts. The lattice
constant shows an extra increase for $x>3$ additional to the
extrapolated linear increase valid for x up to 3. This is thought
to be directly due to the partial decrease in valency of the
uranium atoms. Furthermore, the coefficient γ tends to very
high values for $x>4$ (see Fig. 2) and the resistivity (ρ) shows
a Kondo-like behaviour with a Kondo temperature $T_K \approx 100K$. The
latter facts are in conformity with the theory of interconfiguration fluctuations. In the mixed valency state the minimum
interconfiguration excitation energy (E_{exc}) approaches a value
of the order of the width Δ of a virtual bound state. This state
becomes situated close to the Fermi level and gives occasion to
very high values of $N(E_F)$.[7,8] In this situation a Kondo effect

results with $T_K \approx T_F \exp(-E_{exc}/\Delta)$. The large negative peaks
observed in S(T) are typical of a Kondo effect with $T_S \leqslant T_K$ in the
in the presence of a positive impurity potential[9]. The transition
from non-magnetic UNi_5 to (antiferro) magnetic UCu_5 arises from
the appearance of U^{3+} ions which have a magnetic ground state for
any crystal field. The specific heat of UCu_5 shows a λ-type peak
at 15.2K marking the Néel temperature T_N. The marked variation of
S around T=11K, just as the peak found in ρ at about the same
temperature, i.e. clearly below T_N, point to spin density wave
(Brillouin-superzone) effects (see also ref. 10).

The thermopower of UX_3 compounds is shown in Fig. 4. S has
positive large values ranging at 300K from about 10 to 60μV/K.
Most of the curves, when disregarding the various types of

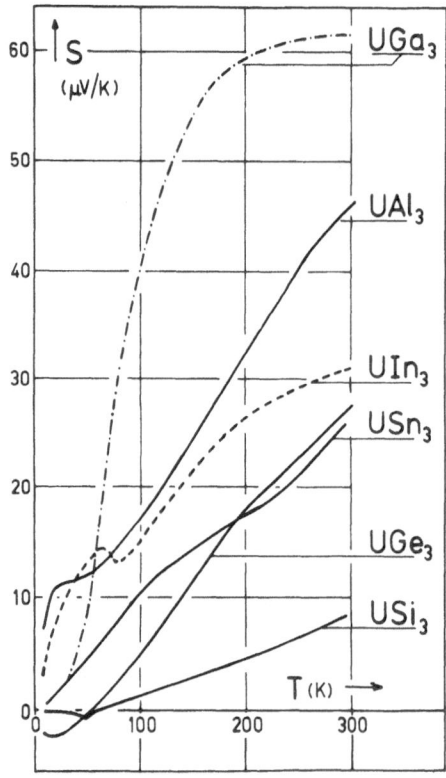

Fig. 4.   The thermopower of compounds UX_3 (X=Al, Ga, In, Si, Ge, Sn).

superstructure, have a small positive curvature in a large temperature range. In UGa_3 and UIn_3 only, S(T) tends clearly towards
saturation at higher temperatures. All compounds are nearly magnetic,[4,5] except UIn_3. The latter compound is an antiferromagnet
as shown by neutron diffraction at 4.2K.[11] A Néel temperature
of about 100K follows from susceptibility data. The onset at
80K of a shoulder on the S(T) curve of UIn_3 is another measure of
T_N. Phonon drag would be expected to give a differently shaped
peak at lower temperatures ($\lesssim \Theta_D/5 \approx 30K$). We will exclude UIn_3 from
the further discussion on nearly magnetic compounds. The nearly
magnetic character of the other compounds stems from narrow
hybridized 5f-6d bands. Spin fluctuations (paramagnons) in these
narrow band states, with a characteristic temperature T_{sf}, determine susceptibility, resistivity and in USn_3 possibly even low-
temperature specific heat behaviour. The thermopower too is
dominated by paramagnon scattering. A theory of the diffusion
thermopower due to paramagnons has been developed recently[12]. In
the theoretical model, electrons in a broad parabolic band, with
a Fermi temperature T_{Fc}, are scattered by paramagnons in an also
parabolic but narrow and nearly magnetic band, with a Fermi temperature T_{Fi}. The model calculations lead to results qualitatively
much the same as measured in these compounds, except for the
saturation effect in UGa_3, if account is taken of Umklapp scattering processes and allowance is made for distortions of the broad
band Fermi surface from the spherical shape. In the theory, the
magnitude of S appears to be independent of the spin-enhancement
factor (D) of the susceptibility. For a comparison between theory
and experiment we have taken $T_{Fi} = DT_{sf}$, where T_{sf} has been determined from our ρ data and D from our susceptibility and γ data.
T_{Fi} ranges from 160K (USn_3) to 1200K (UGe_3). Quantitative agreement between theory and experiment is obtained for ratios T_{Fc}/T_{Fi}
in the range of 5 to 30. Although the average ratio is of the
right order of magnitude, the variations in this ratio suggest that
the specific band structures should be taken into account. A
theory of the effect of paramagnon-drag on the thermopower does
not yet seem to exist. It would be worthwhile to investigate
whether the broad hump around 100K in the S vs. T curve of USn_3
could be due to this effect.

REFERENCES

1. A.J. Freeman and D.D. Koelling, "The actinides, electronic
    structure and related properties", Ed. A.J. Freeman and
    J.B. Darby Jr., Academic Press 1974. Vol I, p. 51.
2. H.L. Davis, see ref. 1, Vol. II, p.1.
3. H.J. van Daal, K.H.J. Buschow, P.B. van Aken and M.H. van
    Maaren, Phys. Rev. Letters 34, 1457 (1975).
4. K.H.J. Buschow and H.J. van Daal, AIP Conf. Proc. No. 5,
    1972, "Magnetism and Magnetic Materials-1971, eds. C.D.
    Graham Jr. and J.J. Rhyne, p. 1464.
5. M.H. van Maaren, H.J. van Daal and K.H.J. Buschow, Solid
    State Communications 14, 145 (1974).
6. R.D. Barnard, "Thermoelectricity in metals and alloys"
    Taylor and Francis ltd. London, 1972, p. 130.
7. L.L. Hirst, Phys. Kondens. Mater. 11, 255 (1970).
8. M.B. Maple and D. Wohlleben, Phys. Rev. Lett. 27, 511 (1971).
9. M.D. Daybell, in "Magnetism", edited by H. Suhl, Academic New
    York, 1973, Vol V, p. 121.
10. M.B. Brodsky and N.J. Bridger, in "Magnetism and Magnetic
    Materials", 1973, eds. C.D. Graham Jr. and J.J. Rhyne, AIP
    Conf. Proc. No. 18 (AIP, New York, 1973), p. 357.
11. A. Murasik, J. Leciejewicz, S. Legenza and A. Misiuk,
    Phys. Stat. Sol. (a) 20, 395 (1973).
12. B. Coqblin, A.A. Gomés, J.R. Inglesias – Sicardi and R.
    Jullien, Proc. 2nd conf. electr. structure actinides, ed.
    J. Mulak, W. Suski and R. Troć, Wroclaw, Poland, 1976, p. 223;
    J.R. Iglesias-Sicardi, thesis, université Paris-Sud, Centre
    d'Orsay, 1977.

DISCUSSION

B.R. Coles: One of the interesting questions about valence fluctuations in intermediate valency systems is the argument that if
magnetic order takes place, does it force a charge ordering? So I
would like to ask whether in UCu5 you are restricting your charges
to specific sites, and can that be seen in the distribution of
magnetic moments?

H.J. van Daal: I don't know if I can answer your question. We
think that what you see in UCu5 are spin density wave effects, not
charge wave effects. How do you see these things? One of the
phenomena is that the resistivity shows a peak below the Néel temperature, marked by a λ-type peak in the specific heat, just as in
rare earth elements like dysprosium, erbium, etc. In the thermoelectric power too you see a large effect at the temperature of the
resistivity peak. We think that the antiferromagnetism is of the
commensurable spin-density wave type. But when you suggest that

you should see effects of charge density waves, I ask, how can I decide if I have to consider charge density waves from these measurements?

B.R. Coles: The thing you would have to know is whether the moments are only restricted to certain uranium sites, or if some uranium sites have different moments than others.

THE THERMOPOWER OF METALLIC BINARY CONTINUOUS SOLID SOLUTIONS

M. V. Vedernikov, V. G. Dvunitkin, and A. Zhumagulov

A. F. Ioffe Physico-Technical Institute

194021, Leningrad, USSR

ABSTRACT

The thermopower composition dependences at 293 K are measured for 10 series of continuous solid solutions (CSS): Cr-V, Mo-Nb Mo-V, Nb-V, Cr-Mo, Ti-Zr, Ti-Hf, Hf-Zr, Sc-Zr, Sc-Hf. The data for 15 series of CSS are analysed together for the first time. It is found that thermopower-composition dependences are different when CSS are formed by two isoelectronic or two nonisoelectronic metals. In the latter case there is a singularity on the thermopower-composition isotherm. The investigation of temperature dependence of thermopower is carried out for two series of CSS.

The properties of disordered systems are of great interest now. It is natural to try to understand the properties of the simplest disordered systems at first. The substitution binary solid solutions, whose structures are closest to the elementary metals are the simplest systems among the metallic alloys. The series of continuous solid solutions (CSS) are of particular interest because in their analysis we can use our knowledge of electronic structure of both metals. However, the thermoelectric properties of CSS are studied poorly. Indeed, only few series of CSS were studied experimentally. The attempts of the theoretical interpretation were made for alloys Ag-Au and Pd-Ag. The behaviour of these alloys is substantially different (Fig. 1). The thermopower of Ag-Au alloys changes markedly only at the ends of the diagram, while in the middle part it changes weakly. The thermopower of Pd-Ag alloys changes considerably within the whole composition range 0-100%. The behaviour of alloys Cr-V looks much more complicated (Fig. 4). The presence of

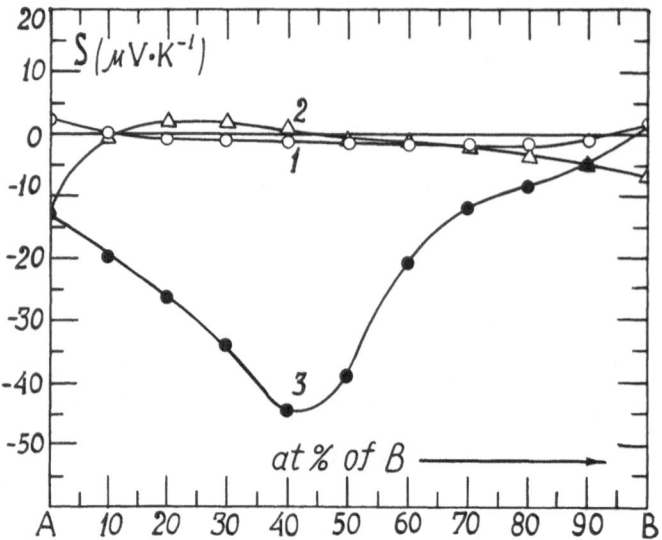

Fig. 1. Absolute thermopower versus composition for series of CSS
at 293 K: 1. Au-Ag; 2. Pd-Pt; 3. Pd-Ag (A:Au,Pd; B:Ag,Pt). According to the literature data.

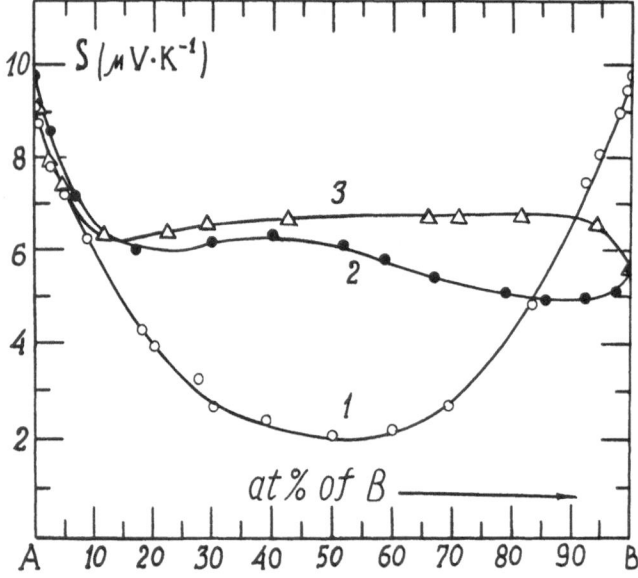

Fig. 2. Absolute thermopower versus composition for series of CSS
at 293 K: I. Ti-Zr (A: Ti); 2. Zr-Hf (A: Zr); 3. Ti-Hf (A: Ti).
Present authors.

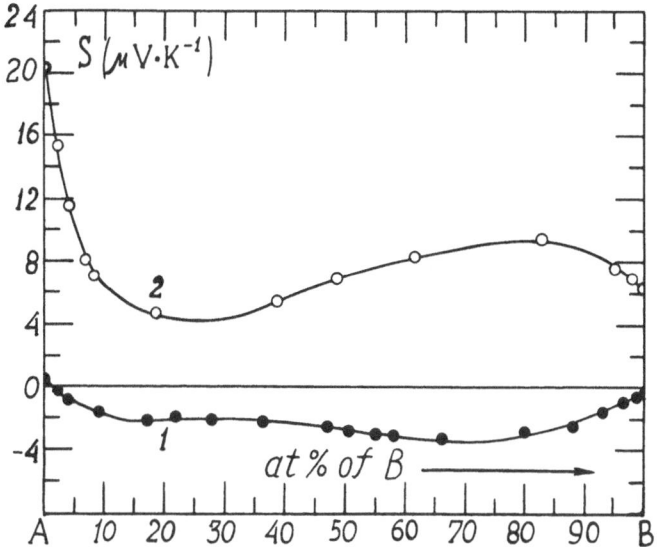

Fig. 3. Absolute thermopower versus composition for series of CSS at 293 K: 1. V–Nb; 2. Cr–Mo (A: V, Cr; B: Nb, Mo). Present authors.

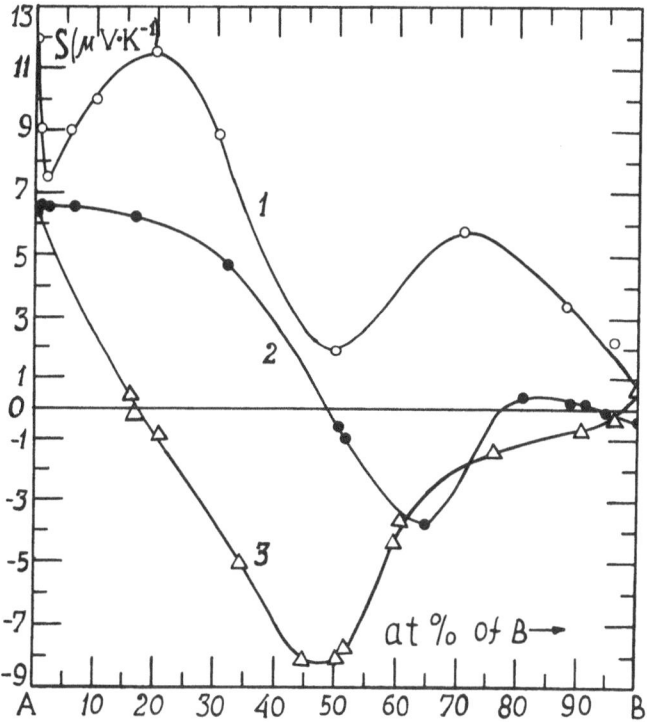

Fig. 4. Absolute thermopower versus composition for series of CSS at 293 K: 1. Cr-V; 2. Mo-Nb; 3. Mo-V (A: Cr, Mo; B: B, Nb). Present authors.

a singularity (minimum) on this curve was unexpected for CSS. We
don't know of any general empirical or theoretical regularities for
thermopowers of CSS. We suggested that the general features of CSS
thermopowers can be found by consideration of a substantially wider
set of CSS series than before. For this purpose we analysed the
data on the thermopower-composition dependences 15 series of CSS.
The data for five series were taken from the literature and the
other ten series were studied by us. Our results are presented in
Fig. 2-5. The literature data for the alloys Au-Cu are close to
curve 1 in Fig. 1; the data for the alloys Pd-Au are close to curve
3; curve 2 for alloys Pd-Pt is qualitatively like curve 1. Among
the alloys examined, the alloys Ti-Zr, Zr-Hf, Ti-Hf (Fig. 2) and
V-Nb, Cr-Mo (Fig. 3) have dependences like curves 1 and 2 of Fig. 1.
The thermopower of these alloys changes more or less strongly at the
ends of the diagram, but in the wide middle range of compositions it
changes weakly and doesn't have any sharp peculiarities. The strong
change of thermopower of the alloys Cr-Mo near Cr is undoubtedly
connected with magnetic order. In other cases (alloys Cr-V, Mo-Ng,
Mo-V: Fig. 4; Sc-Zr, Sc-Hf: Fig. 5) there exist strong and complicated thermopower-composition dependences in the whole composition
range. In the middle part of the diagram it is possible to mark out
a singularity (extremal point) which would seem to divide the diagram in two parts. Although for scandium alloys, the marking of the
extremal point is not so evident, supplementary research confirms
the presence of two parts of the diagram. First it is seen from
the behaviour of resistivity. The resistivity of alloys Ti-Zr
(Fig. 8) and the like changes versus composition symmetrically and
doesn't allow to mark out any part of the diagram. The resistivity

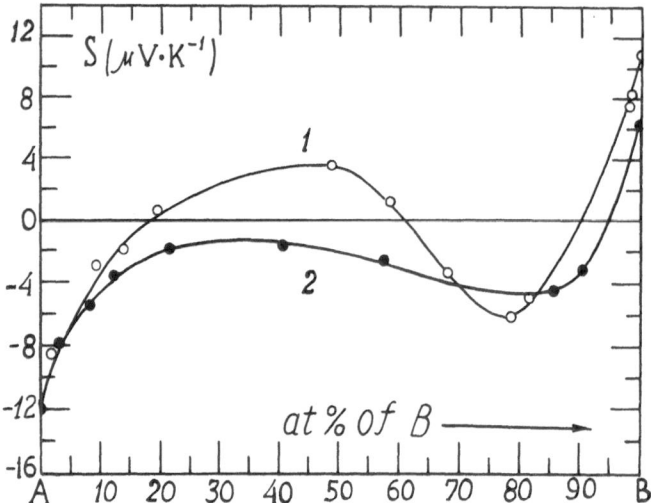

Fig. 5. Absolute thermopower versus composition for series of CSS
at 293 K: 1. Sc-Zr; 2. Sc-Hf (A: Sc; B: Zr, Hf). Present authors.

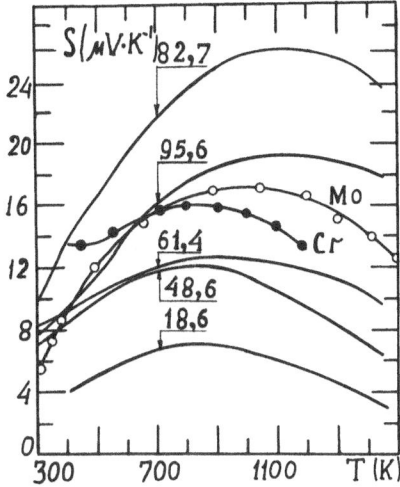

Fig. 6. Absolute thermopower versus temperature for alloys Cr-Mo. Figures show the content of Mo in atomic per cent. Present authors.

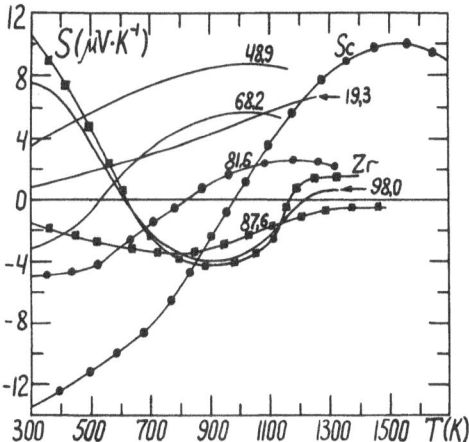

Fig. 7. Absolute thermopower versus temperature for alloys Sc-Zr. Figures show the content of Zr in atomic per cent. Present authors.

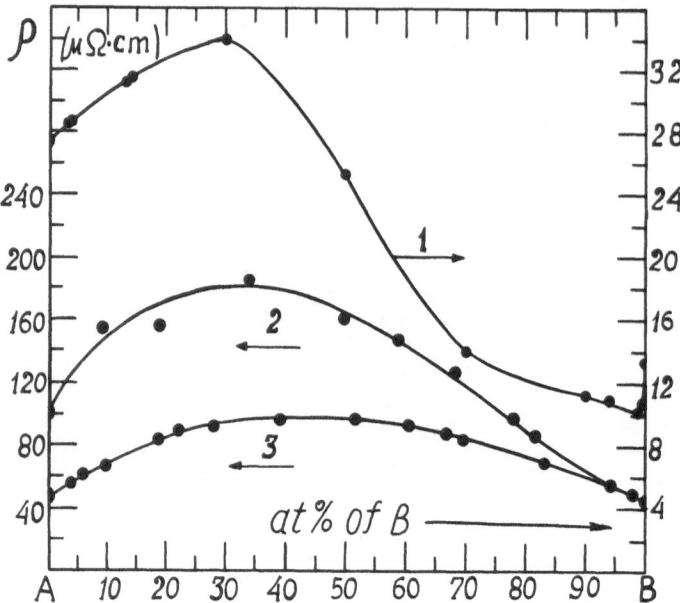

Fig. 8. Resistivity versus composition for series of CSS at 293 K:
1. V-Cr (right scale); 2. Sc-Zr (left scale); 3. Ti-Zr (left scale)
(A: V, Sc, Ti; B: Cr, Zr). Present authors.

of alloys Cr-V (Fig. 8) depends on composition in a complicated way;
the character of this dependence is different near by Cr and V. We
can see there that the curve of resistivity of Sc-Zr is similar to
Cr-V, but not to Ti-Zr. Another argument is more interesting, because it comes from comparison of the temperature dependences of
thermopower of CSS which were examined very little until now. The
temperature dependences of thermopower of the alloys Cr-Mo are shown
in Fig. 6. The curves for both pure metals have characteristic
convex form and are similar. And the temperature dependences of
alloys Cr with Mo have the same characteristic shape, too. But the
temperature dependences of pure Sc and Zr are sharply different
(Fig. 7). Would the shape of these curves change continuously with
the change of composition? Fig. 7 shows that it is not so. On addition of Sc to Zr or Zr to Sc the curve of temperature dependence of
thermopower of alloy distinctly maintains the characteristic shape
of temperature dependence of the host metal. The sharp transformation from one shape to another takes place in a rather narrow interval of concentrations: in the vicinity of 85 at.% of Zr. The
alloy with 87.6 at.% of Zr has intermediate type temperature
dependence of thermopower. Hence, the "singularity" of the curve
1 in Fig. 5 lies near by 85 at.% of Zr. Presently we do not have
enough data about temperature dependences of thermopower of CSS, to

make a general conclusion about their change with composition. But
the available data permit us to suggest that, perhaps, in all cases
when the isotherm of thermopower has a singularity, the diagram
thermopower-composition divides into two parts. There is a narrow
transition range of compositions between them. The temperature
dependences, similar to metal A, are characteristic for one part;
while for the other part the B type dependences are characteristic.
Careful analysis shows that, apparently, the only common factor for
series of CSS with thermopower isotherm without singularity is the
fact that they all are formed by isoelectronic metals. To the
second group of alloys belong those formed by nonisoelectronic metals. The analysis of resistivity isotherms of 22 series of CSS
confirms the division of all CSS into two groups according to whether
the metals forming the CSS are isoelectronic or nonisoelectronic.
The detailed investigation of temperature dependences of thermopower
of CSS is of great interest. Indeed, due to the complicated, characteristic shape of these dependences we can follow the change of
alloy's intrinsic structure more surely.   In the invited paper of
M. V. Vedernikov at this Conference a suggestion is made about a
close connection of the type of temperature dependence of thermopower with the electronic structure of the metal. Hence, for CSS
Sc-Zr the character of change of thermopower permits us to suggest
that alloys with content of Zr from 0 up to 80 at.% are close to the
electronic structure of Sc, while the alloys with 0-10 at.% of Sc
only are close to Zr. The transformation of one electronic structure
to another takes place in a narrow range of compositions in the
vicinity of 85 at.% of Zr. If it is really so, it must be interesting for the theory of solid solutions. Thus, the detailed investigation of temperature dependences of thermopower of alloys is
necessary (for a larger number of compositions and in a wider interval of temperature). We carry out such an investigation of CSS
now.

CALCULATED HIGH-TEMPERATURE THERMOPOWER OF NIOBIUM AND MOLYBDENUM*

P. B. Allen and B. Chakraborty

Department of Physics
State University of New York
Stony Brook, New York 11794 USA

ABSTRACT

The electron-diffusion thermopower is calculated for Nb and Mo at temperatures 200 K < T < 1600K. The energy dependent conductivity $\sigma(\varepsilon) = N(\varepsilon)v^2(\varepsilon)\tau(\varepsilon)$ is taken from APW and KKR band calculations. It is assumed that $N(\varepsilon)\tau(\varepsilon)$ is a smooth function of ε. Agreement with experiment for Nb is not good. This is ascribed to failure of the Boltzmann equation.

In 1936 Mott[1] showed for Pd that both the thermopower S and the high temperature electrical resistance ρ (in particular the tendency for $\rho(T)$ to fall below the Bloch-Grüneisen linear-in-T formula) could be explained using his s-d model, with a rapidly varying d density of states $N_d(\varepsilon)$ at the Fermi energy ε_F. Recently it has been observed[2] that a negative deviation from linearity (d.f.ℓ.) of $\rho(T)$ occurs in virtually all-known high temperature superconductors. It is argued in ref. 3 (and disputed in ref. 4) that the d.f.ℓ. in strong-scattering metals (such as Nb) cannot be explained by the semiclassical Boltzmann equation (SBE). The most popular explanation using the SBE is the Fermi smearing effect resulting from strong energy dependence of electronic parameters. References 3 and 4 reach opposite conclusions about the success of this explanation for $\rho(T)$ in Nb. This motivates consideration of the thermopower S(T) which depends sensitively on the Fermi smearing effect. If it is

* Work supported in part by U.S. National Science Foundation Grant No. DMR76-82946.

possible to explain the d.f.ℓ. in $\rho(T)$ using Fermi smearing arguments, then $S(T)$ should be calculable using the same assumptions.

The calculations we have done are similar to Mott's original work.

$$\sigma(T)S(T) = -(e/T) \int d\varepsilon\ \sigma(\varepsilon)(\varepsilon-\mu)(-\partial f/\partial \varepsilon) \tag{1}$$

$$\sigma(T) = \int d\varepsilon\ \sigma(\varepsilon)(-\partial f/\partial \varepsilon) \tag{2}$$

$$e^{-2}\sigma(\varepsilon) \equiv N(\varepsilon)v_x^2(\varepsilon)\ \tau(\varepsilon) \equiv \sum_k v_{kx}^2\ \tau_k\ \delta(\varepsilon-\varepsilon_k) \tag{3}$$

$$N(\varepsilon)v_x^2(\varepsilon) \equiv \sum_k v_{kx}^2\ \delta(\varepsilon_k-\varepsilon) \equiv N_{eff}/m \tag{4}$$

$$N(\varepsilon) \equiv \sum_k \delta(\varepsilon_k-\varepsilon) \tag{5}$$

In these equations, k is shorthand for wavevector and band index of the electron state of energy ε_k and velocity $v_{kx} = \partial\varepsilon_k/\partial k_x$. Also f is the Fermi function and σ is the conductivity. In principle the relaxation time τ_k should be found by solving the SBE with the full electron-phonon collision term; this guarantees that eqs. (1)-(5) are exact consequences of the SBE. This challenging task has recently been undertaken with modest accuracy by Yamashita and Asano[5] who find τ_k surprisingly independent of k at room temperature. We thus (in the spirit of Mott) make the high temperature model

$$\hbar/\tau(\varepsilon) = 2\pi\ \lambda_{tr}(\varepsilon)\ k_B T \tag{6}$$

$$\lambda_{tr}(\varepsilon) = N(\varepsilon)\langle I^2(\varepsilon)\rangle_{tr}/M\langle\omega^2\rangle \tag{7}$$

When evaluated at the Fermi energy, $\lambda_{tr}(\varepsilon)$ is closely related[6] to the mass enhancement parameter and superconducting coupling constant[7] λ; the notation of eq. (7) is from McMillan's work[7]; the subscript tr denotes "transport". So far the only approximations are to neglect anisotropy in τ_k and to drop the phonon frequency ω from the energy conserving delta function in the collision integral. These approximations probably affect $\sigma(T)$ and $S(T)$ by less than 2%.

Mott[1] made the following model:

(a) $\lambda_{tr}(\varepsilon)$ is replaced by $C_s N_s(\varepsilon) + C_d N_d(\varepsilon)$ where the s
     density of states N_s is assumed small compared to N_d.

(b) N_{eff}/m (eq. 4) is assumed to be dominated by s- electrons
     and is independent of ε.

(c) In the absence of more detailed information, a parabolic band model is used for $N_d(\varepsilon)$.

Our model differs from Mott's in the following way.

(a) In common with Mott, we neglect the ε dependence of $<I^2(\varepsilon)>_{tr}$. However, $N(\varepsilon)$ is not divided into s and d parts.

(b) N_{eff}/m is known[5,6] for Nb and Mo to contain roughly equal contributions from all Fermi surface electrons. The Fermi velocity is not extremely anisotropic, and separate "s" and "d" Fermi surface electrons cannot be identified. Consequently, N_{eff}/m is not independent of ε.

(c) We use state-of-the-art electronic band structures[8,9,6] to calculate $N(\varepsilon)$, $v_x^2(\varepsilon)$, and thus N_{eff}/m. These are plotted in fig. 1 for Mo and in ref. 3 for Nb.

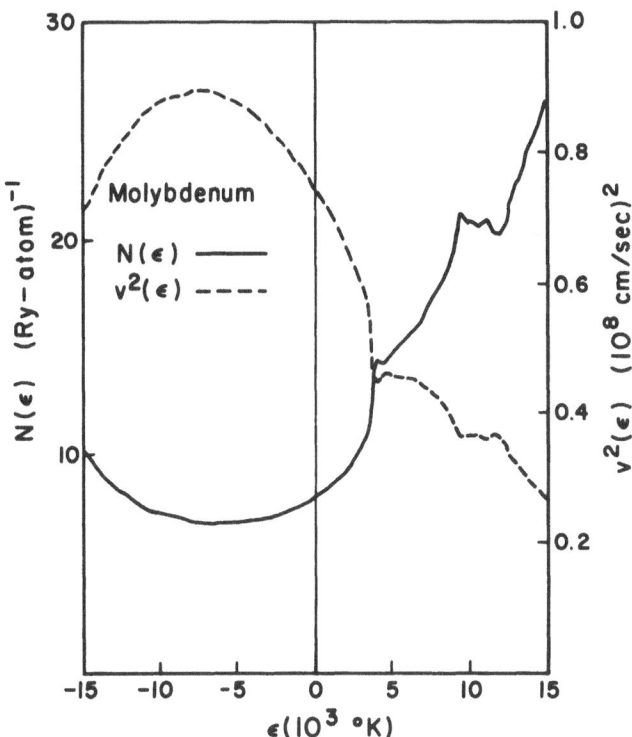

Fig. 1. Density of states $N(\varepsilon)$ and mean square Fermi velocity for Mo calculated from the APW bands of ref. 8 and plotted for energies near the Fermi energy (which is set to 0).

The thermopower S(T) has then been calculated using eq. (1) and is shown in Fig. 2. The data are from R. K. Williams[10,11]. It is important in these calculations that the chemical potential μ(T) be recalculated at each temperature. The correct sign is found for both Nb and Mo; the order of magnitude is correct for Mo but too large for Nb; in both cases the curvature of the data is not reproduced by the theory. The reasons for the discrepancy remain unknown. Two different types of corrections have been attempted, but no noticeable improvement was obtained.

First, it has been suggested in ref. 4 that a relocation of the Fermi level by either -0.013 or +0.019 Ry would allow an explanation of the d.f.ℓ. in ρ(T) of Nb using our model and eq. (2). We have verified the calculations of ref. 4, although we do not believe that

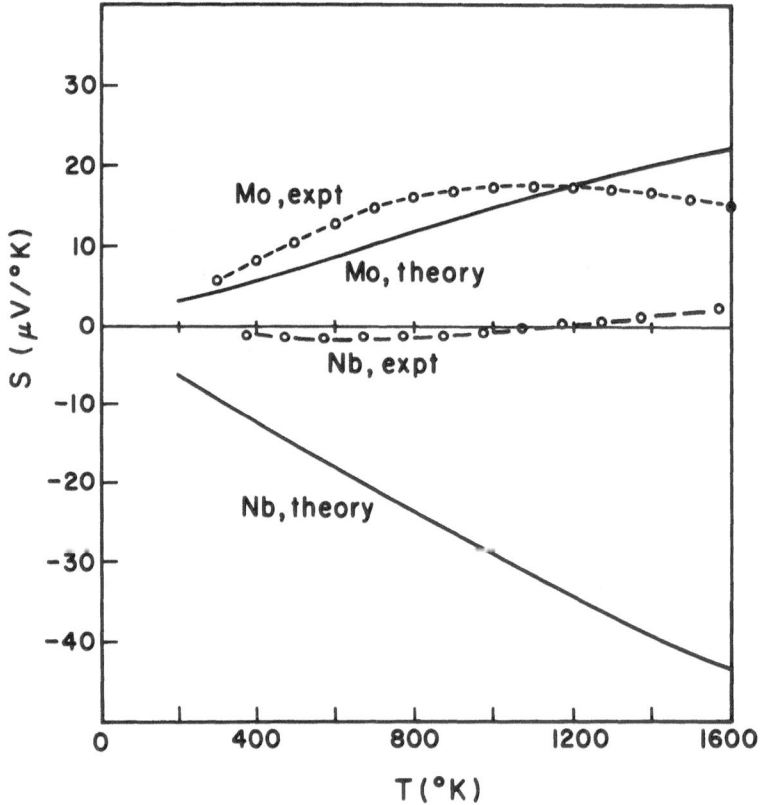

Fig. 2. Calculated thermopower as a function of temperature, and experimental data from refs. 10, 11.

such large adjustments are consistent with the known Fermi surface topology. However, moving the Fermi level has no such beneficial effect on the calculated values of S(T). No choice of ε_F between -0.013 and +0.019 caused any improvement. Similar adjustments of ε_F were tried for Mo, without creating any improvement in either S(T) or the d.f.ℓ. of ρ(T).

A second correction is the possibility of significant ε-dependence of $\langle I^2(\varepsilon)\rangle_{tr}$. In fact we believe this is the only serious objection to our method of approximating the exact solution (1) of the SBE. Butler[12] has calculated the superconducting analog $\langle I^2(\varepsilon)\rangle$ for several energies in Nb, namely near ε_F, and at 0.009 Ry below and 0.032 Ry above ε_F. Fitting to a linear behavior,

$$\langle I^2(\varepsilon)\rangle = \langle I^2(\varepsilon_F)\rangle \ (1 + \alpha \ (\varepsilon - \varepsilon_F) \ ) \qquad (8)$$

the average slope α is about 5 Ry^{-1}. Butler's values do not appear strictly linear in ε, but with only three points it does not seem warranted to complicate eq. (8) further. The calculations of S(T) and ρ(T) were repeated with values of α equal to 2.5, 5.0, and 7.5. No significant improvement was found for Mo. For Nb, the largest value of α (7.5) gave a 50% improvement in agreement between theory and experiment for S(T). However, the disagreement in the d.f.ℓ. of ρ(T) was made slightly worse.

In ref. 3 it was argued that the d.f.ℓ. of ρ(T) of Nb lies outside the SBE. The same case can now be made for S(T). The strongest argument for this, which we still think correct, is that the mean free path at 1000K is only ≃ 10 Å. Since the SBE is known to be correct only to leading order in $(k_F\ell)^{-1}$, it seems clear that corrections to the SBE ought to be significant. The arguments given here would definitely strengthen the case of ref. 3 if Mo were ignored. However, in Mo the mean free path is larger by 3 because of weaker scattering and higher Fermi velocities. The SBE should be correspondingly better. However, the calculated d.f.ℓ. and S(T) are still not agreeing well with experiment. We suspect it may require consideration of electron-electron Coulomb scattering to explain the positive d.f.ℓ. of ρ(T) in Mo using the SBE. It is difficult to know how this would affect S(T).

ACKNOWLEDGEMENTS

We thank R. K. Williams and W. H. Butler for help, and M. L. Cohen for hospitality at U. C. Berkeley where much of this work was done.

REFERENCES

1. N.F. Mott, Proc. Roy. Soc. (London) A156, 368 (1936).
2. See for example Z. Fisk and G.W. Webb in Superconductivity in d- and f-Band Metals (D.H. Douglass, ed., Plenum, New York, 1976) p. 545.
3. P.B. Allen, Phys. Rev. Lett. 37, 1638 (1976).
4. M.J. Laubitz, C.R. Leavens, and R. Taylor, Phys. Rev. Lett. 39, 225 (1977).
5. J. Tamashita and S. Asano, Prog. Theor. Phys. 51, 317 (1974).
6. B. Chakraborty, W.E. Pickett, and P.B. Allen, Phys. Rev. B 14, 3227 (1976).
7. W.L. McMillan, Phys. Rev. 167, 331 (1968).
8. L.F. Mattheiss, Phys. Rev. B 1, 373 (1970).
9. I. Petroff and C.R. Viswanathan, Phys. Rev. B 4, 799 (1971).
10. R.K. Williams, in Oak Ridge National Laboratory Metals and Ceramics Division Annual Report, June 30, 1976 (U.S. Govt. Rept. # ORNL 5182, Siegried Peterson, editor).
11. R.K. Williams, J.P. Moore, and D.L. McElroy, in Oak Ridge National Laboratory Metals and Ceramics Division Annual Report, June 30, 1968 (U.S. Govt. Rept. # ORNL 4370).
12. W.H. Butler, Phys. Rev. B 15, 5267 (1977).

DISCUSSION

J. Garland: You have done your calculations on a horrendously complex metal. I wonder if you have considered looking at a simple metal like aluminum which has a melting point of about 1000°K. You could go to two or three times the Debye temperature before the aluminum melted. It seems to me your velocities and density-of-states expressions would be much easier to work with.

P.B. Allen: We are making a style of approximation which is appropriate to a metal like niobium, but not a simpler metal. We are assuming, for one thing, isotropic scattering. Aluminum has less isotropic scattering than niobium. We are assuming that the density-of-states and mean square velocity have rather rapid energy dependences compared with the energy dependence of some of the other quantities involved. That would surely no longer be true in aluminum. A calculation for aluminum would be a lot easier than for niobium, but we could not make the same approximations and it would not serve as a valid test of what we have done.

FIRST PRINCIPLES CALCULATION OF THE HIGH-TEMPERATURE THERMOPOWER

OF SODIUM AND POTASSIUM

C. R. Leavens and Roger Taylor

Physics Division, National Research Council of Canada

Ottawa K1A 0R6

ABSTRACT

Our previous first-principles calculations of the electrical
and thermal resistivities of Na and K have now been extended to
the high temperature thermopower of these elements. Again, the
results are in very good agreement with experiment. Included in
the calculation is the effect of the energy dependence of the
electron-phonon scattering, which is directly related to the non-
locality of the bare pseudopotential. For Na this effect is
fairly small, as expected, but it is of critical importance for K.
It is also crucial to the calculation of the volume derivatives
of the thermopower, thus making it very clear that a local pseudopotential description of K is not appropriate for consistent calculations of all the transport properties.

An essential ingredient in calculating the transport properties of a crystalline metal is the transition probability for
the process in which an electron scatters by the emission or
absorption of a phonon. It is usual to approximate this probability by keeping only the term which is second order in the electron-
ion pseudopotential. However, a number of workers[1] have pointed
out that the neglected higher order terms appear to be significant
even in the alkali metals. In particular Rasolt and Taylor[2] demonstrated this clearly for the charge density problem. They suggested
that the higher order terms in the pseudopotential could be sim-
ulated by a suitable adjustment of the parameters defining the

pseudopotential. This was carried out by fitting the first order perturbation theory calculation of the charge density induced about an isolated ion in an electron gas to Dagens'[3,4] full non-linear calculation of the same quantity. This effectively folds in all multiple scattering events at a single ion site. This procedure for determining the bare electron-ion interaction used in combination with the dielectric function of Geldart and Taylor[5] leads to phonon frequencies in excellent agreement with experiment for both Na[2] and K.[4,6] With the further inclusion in the transition probability of Rasolt's form for the vertex function describing non-locality in the electron-electron interaction excellent agreement with experiment was obtained for the thermal resistivity of both these metals over the entire temperature range.[8] Very good results were also obtained for the electrical resistivity above 20K.[9] Below 20 K the situation is somewhat uncertain. Taylor, Leavens and Shukla[10] (TLS) obtained values of the electrical resistivity for potassium in this range which agreed fairly well with experiment over the range 2K$\leq$T$\leq$20K in the Bloch limit. But in the phonon drag limit, the agreement was very poor, the calculated results being a factor of 4 too low at T=2K. If, as some authors suggest,[11] the experimental data exhibit the _full_ phonon drag effect at T=2K, then the results of TLS indicate that their potassium form factor is too small by a factor of 2 at momentum transfer K=2k_F. This is unexpected in view of the striking success of the calculations of references 8 and 9 but cannot be ruled out arbitrarily. Hence a further test of the form factor used in our earlier calculations is clearly desirable and, as it turns out, the thermoelectric power provides such a test. This is because the value of the form factor at 2k_F occurs prominently in the well-known formula for the high temperature thermoelectric power.

$$S(T > \theta_D) = \frac{-\pi^2 k_B^2 T}{3|e|E_F} \xi,$$

where the thermoelectric parameter is given by

$$\xi \equiv 3 - 2q - \frac{1}{2} r.$$

q is directly proportional to the square of the Fermi surface form factor at 2k_F, i.e.

$$q \propto v^2 \quad (K = 2k_F, \ E=E_F).$$

r involves the energy derivative at the Fermi surface of the form factor for all momentum transfers from 0 to $2k_F$, i.e.

$$\left(\frac{dV(K,E)}{dE}\right)_{E=E_F} \qquad (0 \leq K \leq 2k_F)$$

For a local, energy independent pseudopotential r is identically zero. However both non-locality and explicit energy-dependence of the bare pseudopotential can give rise to a non-zero value of r. Our calculations were performed using the non-local energy-independent pseudopotentials employed in our earlier calculations[8-10].

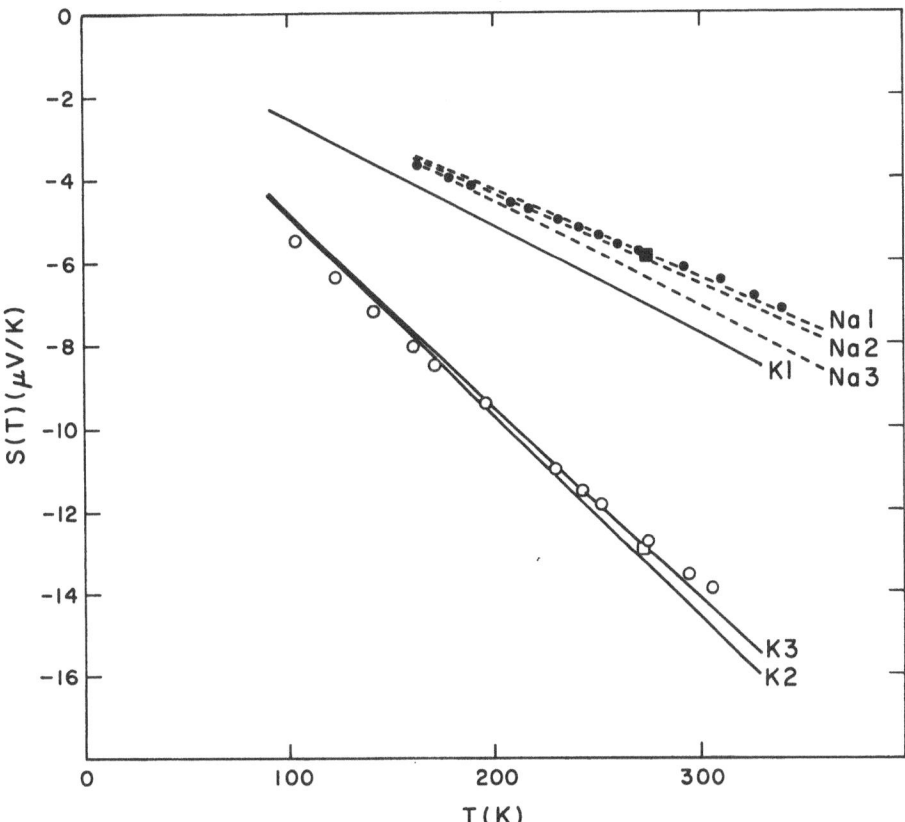

Fig. 1. A comparison of calculated and experimental high temperature thermopowers for Na and K. Experimental: ■ and □, Dugdale and Mundy[14] for Na and K respectively; ●, Cook et al.[12] for Na; o Cook and Laubitz[13] for K. Calculated: 1, at constant volume with r=0; 2, at constant volume with r≠0; 3, at constant pressure with r≠0.

Fig. 1 shows our calculated results for the high temperature thermoelectric power of Na and K. The circles indicate the experimental results of Cook et al.[12,13] and the squares those of Dugdale and Mundy[14]. The results of our constant volume calculations are labelled 1 and 2, those at constant pressure, 3. The curves labelled 1 were calculated ignoring the energy dependence of the form factor, i.e. taking r to be zero.

Let us look at Na first. Comparing 1 and 2 it is obvious that the effect of the energy dependence of our form factor is very small. However, as a glance at Table 1 will show, it should not be ignored because $2q$ and $\frac{1}{2}r$ are comparable in size and of opposite sign so that there is considerable cancellation between the two.

The indirect contribution to the thermopower arising from electron-electron scattering is not included in our calculation and probably accounts for most of the discrepancy between theory and experiment at the higher temperatures.

The calculated value of $d\ln S/d\ln V$ at 0°C of 1.5 compares favourably with Dugdale and Mundy's[14] experimental result of 2.1.

TABLE 1

A selection of constant pressure results for the thermoelectric parameter ξ in Na and K.

	$T(K)$	$2q(T)$	$\frac{1}{2}r(T)$	$\xi(T)$
Na	160	0.11	−0.07	2.96
	200	0.09	−0.07	2.98
	250	0.06	−0.07	3.01
	295	0.04	−0.07	3.03
K	100	0.82	−1.96	4.14
	200	0.82	−1.86	4.04
	300	0.84	−1.76	3.92

Now, for potassium. Here the energy dependence of the form factor leads to a factor of 2 effect on the thermopower and cannot be ignored. In a sense this is unfortunate since it lessens the sensitivity of the calculated thermopower to the value of the form factor at $2k_F$ (this enters via the q term). If r were in fact zero we would have a very sensitive test for the magnitude of $V(2k_F)$. Nevertheless, if our screened pseudopotential is in serious error then only a remarkable cancellation of errors between the large q and r terms can account for the very good agreement between theory and experiment shown in the figure.

The calculated logarithmic derivative (dlnS/dlnV) at 0°C of -0.6 compares quite well with the experimental result[14] of -0.4. (dlnS/dlnV) is difficult to calculate accurately from first principles for K because it involves competing small quantities. For example, ξ decreases by 1.5% and E_F^{-1} increases by 0.8% at 0°C when the lattice parameter is changed from 5.31 to 5.33Å. If we ignore the energy dependence of ξ then it only decreases by 0.6% and the logarithmic derivative changes sign. This underscores the importance of the energy dependence of the form factor in K and hence the non-locality of the K pseudopotential.

We see no evidence from these calculations that our form factor is in error by anything approaching a factor of 2 at $K=2k_F$. Clearly a calculation of the thermopower at low T would give further information on this question and we hope to be able to comment on this at some later date.

REFERENCES

1. For example see C.M. Bertoni, V. Bortolani, C. Calandra and F. Nizzoli, J. Phys F 4, 19 (1974) and M. Hasegewa, J. Phys. F. 5, 649 (1976).
2. M. Rasolt and R. Taylor, Phys. Rev. B. 11, 2717 (1975).
3. L. Dagens, J. Phys. C 5, 2337 (1972).
4. L. Dagens, M. Rasolt and R. Taylor, Phys. Rev. B 11, 2726 (1975).
5. D.J.W. Geldart and R. Taylor, Can. J. Phys. 48, 167 (1970).
6. G. Dolling and J. Meyer, J. Phys. F 7, 775 (1977).
7. M. Rasolt, J. Phys F 5, 2294 (1975).
8. C.R. Leavens, J. Phys. F 7, 163 (1977).
9. R.C. Shukla and R. Taylor, J. Phys. F 6, 531 (1976).

10. R. Taylor, C.R. Leavens and R.C. Shukla, Solid State Comm.
    19, 809 (1976).
11. See for example M. Kaveh and N. Wiser, Phys. Lett. 49A, 47
    (1974) and H. van Kempen, J.S. Lass, J.H.J.M. Ribot and
    P. Wyder, Phys. Rev. Lett. 37, 1574 (1976).
12. J.G. Cook, M.P. Van Der Meer, M.J. Laubitz, Can. J. Phys 50,
    1386 (1972).
13. J.G. Cook and M.J. Laubitz, Thermal Conductivity 14,
    edited by P.G. Klemens and T.K. Chu, Plenum Press, New York
    (1976), 105.
14. J.S. Dugdale and J.N. Mundy, Phil. Mag. 6, 1463 (1961).

DISCUSSION

J. Bass: I want to get clear the significance of what you are saying about the form factor. Do I gather that it is your feeling that
the evidence you now have suggests that your form factor is sufficiently good that you are convinced there is no significant evidence
for phonon drag in the resistivity of potassium at low temperatures?

C. Leavens: I am not convinced one way or the other, yet.

CALCULATION OF THERMOPOWER IN NOBLE METALS USING dHvA DATA

R. R. Bourassa and Shing-Yah Wang

University of Oklahoma

Norman, Oklahoma USA

ABSTRACT

Results for the change in area of the Noble metal Fermi surfaces with electron energy evaluated at the Fermi energy are calculated using the dHvA effective mass data on copper, silver, and gold of Lengeler et. al.[1] Local values for the Fermi velocities deduced from the effective mass data provide the full energy gradient in k-space at all points on the Fermi surface.

The precision measurements of cyclotron masses in the noble metals by de Haas-van Alphen techniques by Lengeler et. al.[1] make possible very accurate phenomenological calculations of several parameters often used in thermopower calculations. The results of such calculations are presented and compared with various theoretical and experimental values for these parameters.

The diffusion component of the thermopower is given by

$$S = \frac{-\pi^2 k^2 T}{3|e|\varepsilon_F} \left.\frac{\partial \ln\sigma(\varepsilon)}{\partial \ln\varepsilon}\right|_{\varepsilon_F} = -\frac{\pi^2 k^2 T}{3|e|\varepsilon_F}\xi \tag{1}$$

where $\sigma(\varepsilon)$ is the energy dependent electrical conductivity which is written as

$$\sigma(\varepsilon) = \frac{e^2}{12\pi^3\hbar}\int_A \tau(\vec{k},\varepsilon)v(\vec{k},\varepsilon)dA(\vec{k},\varepsilon) \tag{2}$$

137

where $\tau(\vec{k},\varepsilon)$ is the electron-phonon relaxation time, $v(\vec{k},\varepsilon)$ is the fermi velocity and $dA(\vec{k},\varepsilon)$ is an element of area on the fermi surface. It is common to introduce an average mean free path for the electron such that

$$\sigma(\varepsilon) = \frac{e^2}{12\pi^3 h} \overline{\ell(\varepsilon)} \, A(\varepsilon) \tag{3}$$

which amounts to an assumption that the anisotropy in the product $\tau(\vec{k},\varepsilon)v(\vec{k},\varepsilon)$ is not important compared to the anisotropy in the fermi surface itself. This assumption, of course, is often invalid. Using (3) in (1) we can write

$$S = -\frac{\pi^2 k^2 T}{3|e|\varepsilon_F} \left[\frac{\partial \ln \overline{\ell(\varepsilon)}}{\partial \ln \varepsilon} + \frac{\partial \ln A(\varepsilon)}{\partial \ln \varepsilon} \right]_{\varepsilon_F.} \tag{4}$$

We refer to the first term in the bracket as U and the second term as V. The second term can be written as

$$V = \frac{\varepsilon}{A} \frac{\partial A(\varepsilon)}{\partial \varepsilon}\Big|_{\varepsilon_F} \simeq \frac{\Delta A}{A_F} \Big/ \frac{\Delta \varepsilon}{\varepsilon_F} \tag{5}$$

where ΔA is the difference in area between two constant energy surfaces separated in energy by $\Delta \varepsilon$, near ε_F, and A_F is the area of the fermi surface. The results of Halse,[2] Coleridge and Watts,[3] and recently Lengeler et. al.[1] in measuring radius vectors[2] and effective masses[1,3] are used in this work and make the determination of V straightforward.

A deconvolution scheme on dHvA orbital area measurements results in a model of the fermi surface parameterized by a set of coefficients.[2,4]

Effective mass measurements using the dHvA technique are then used to determine the fermi velocities and thus the gradient to the fermi surface at each point on the surface. This allows one to determine two other sets of coefficients which parameterize neighboring surfaces at constant energy $\varepsilon_F \pm \Delta \varepsilon$.[1,2,4] The value of $\Delta \varepsilon / \varepsilon_F$ chosen by Lengeler et. al.[1] was for example 6×10^{-4}. A computer calculation using these two sets of coefficients to specify two surfaces has been made to determine the area of each surface and the difference in area giving the value of $\Delta A/A_F$. Values of V found for each of the Noble metals are listed in Table I along with other values found in the literature.

TABLE I

Source	V_{Cu}	V_{Ag}	V_{Au}
This work using coefficient of ref. 1 2 4	0.605 + 0.009 0.530 + 0.080 0.584 + 0.029	0.445 + 0.007 0.430 + 0.043	0.552 + 0.008 0.521 + 0.052
Free electron theory	1.0	1.0	1.0
Nearly free electron theory	0.06		
Hasagawa and Kasuya[5]	0.61		
Moreland and Bourassa[6] (exp)	-1.20 + 0.11	-1.91 + 0.09	-1.00 + 0.08
Huebener (exp)[7]			-1.05 + 0.19
Lin and Leonard[8] (exp)			-0.85 + 0.04
Yu and Leonard[9] (exp)	-1.43	-2.45	
Worobey et.al.[10] (exp)	-0.98 + 0.20		
Angus and Dalgliesh[11] (exp)	-3.8	-3.8	-2.5
Gouault[12] (exp)	-3.6		

The agreement between this work and Hasagawa and Kasuya[5], in a band theory calculation using the work of Burdick,[13] is excellent. The discrepancies between our result determined from low temperature (1 - 4K) dHvA measurements and the high temperature (300 K) experimental results listed in Table I are extreme. The low temperature value of V = 0.605 $\pm$ 0.009 for copper and the high temperature value of V = -1.20 $\pm$ 0.11 seem now firmly established. Possible explanations include effects of temperature dependence, residual phonon drag at high temperatures which would strongly effect the interpretation of the size effect measurements, and perhaps reexamination of the isotropy assumptions inherent in the theory used to interpret the size effect results.

Another common approximation in thermopower studies is to assume an average relaxation time $\bar{\tau}(\epsilon)$ so that Eq. (2) can be written as

$$\sigma(\epsilon) = \frac{e^2}{12\tau^3 \hbar} \bar{\tau}(\epsilon) \int v(\vec{k},\epsilon) dA(\vec{k},\epsilon) \tag{6}$$

This assumption works best at high temperatures where τ is considered isotropic. Then we have

$$S = -\frac{\pi^2 k^2 T}{3|e|\epsilon_F} \left[\frac{\partial \ln \bar{\tau}(\epsilon)}{\partial \ln \epsilon} + \frac{\partial \ln \int v(\vec{k},\epsilon) dA(\vec{k},\epsilon)}{\partial \ln \epsilon} \right]_{\epsilon_F} \tag{7}$$

Call the second term in the brackets W, then

$$W \simeq \frac{\Delta[\int v(\vec{k},\epsilon) dA(\vec{k},\epsilon)]}{[\int v(\vec{k},\epsilon) dA(\vec{k},\epsilon)]_{\epsilon_F}} \bigg/ \frac{\Delta\epsilon}{\epsilon_F} \tag{8}$$

where $\Delta[\int v(\vec{k},\epsilon) dA(\vec{k},\epsilon)]$ is then the difference between the integrals taken over the two neighboring surfaces differing by $\Delta\epsilon$. We can write the energy dependence of $v(\vec{k},\epsilon)$ as

$$v(\vec{k},\epsilon) = v_F(\vec{k},\epsilon) + \frac{\partial v_F}{\partial \epsilon} \Delta\epsilon + 0\left(\frac{\Delta\epsilon}{\epsilon}\right)^2$$

$$= v_F(\vec{k},\epsilon) + v_F \frac{\partial \ln v_F}{\partial \ln \epsilon} \frac{\Delta\epsilon}{\epsilon} + 0\left(\frac{\Delta\epsilon}{\epsilon}\right)^2$$

TABLE II

SOURCE	Cu X	Cu Y	Cu W	Au X	Au Y	Au W	Ag X	Ag Y	Ag W
This work using coefficients from reference (1)	0.750±0.006			0.610±0.005			0.647±0.005		
(2)	0.716±0.054			0.612±0.031			0.671±0.033		
(3)	0.744±0.019								
Nearly Free Electron Theory		0.31	1.06*						
Hasagawa and Kasuya (5)		0.35	1.10*						
Free Electron Theory		0.5	1.25*		0.5	1.11*		0.5	1.147*
Hasagawa and Kasuya (5) (Burdick's bands)		1.27	2.02*						
Free Electron Theory			1.5			1.5			
Williams and Davis (15)			1.81						1.5
Abelskii and Irkhin (16)			3.89			1.98			2.34
Abarenkov and Verdenikov (14)			(negative) −1.29 to −4.12						

*These values were obtained by adding the value of X from this work using coefficients from Ref. 1 to the indicated values of Y.

Thus

$$W \simeq \frac{\Delta[\int v_F(\vec{k},\varepsilon_F)dA(\vec{k},\varepsilon)]}{[\int v(\vec{k},\varepsilon)dA(\vec{k},\varepsilon)]_{\varepsilon_F}} \bigg/ \frac{\Delta\varepsilon}{\varepsilon_F} + \frac{\overline{\partial\ln v_F}}{\partial\ln\varepsilon} \qquad (9)$$

For simplicity we call the first term X and second term Y. Then

$$W = X + Y \qquad (10)$$

The first term in W can be calculated accurately from dHvA data. Table II shows our results for the first term, X, found for each of the Noble Metals using the coefficients of references 1, 2 and 4. Unfortunately, the second term Y is not well known and very difficult to calculate accurately as was discussed by Hasagawa and Kasuya.[5] Values range from 1.27 as calculated by Hasagawa and Kasuya[5] using Burdick's[13] bands to negative values used in the model of Abarenkov and Vedernikov.[14] Table II lists some of the theoretical values of Y. Theoretical values of W are also listed. Note that some values of W are obtained by adding theoretical values of Y to our determined value for X using the coefficients of Lengeler et. al.[1] Due to the ambiguity in the value of Y, no firm conclusion can be drawn about the value of W. If W lies between 1.0 and 2.0 in copper, this implies from equation (7) that

$$-3.7 < \frac{\partial\ln\tau}{\partial\ln\varepsilon}\bigg|_{\varepsilon_F} < -2.7$$

We conclude that dHvA effective mass measurements allow us to determine with high precision, surfaces of constant energy near the fermi surface. From these surfaces for copper, silver and gold we have accurately determined the parameters

$$V = \frac{\partial\ln A}{\partial\ln\varepsilon}\bigg|_{\varepsilon_F} \text{ and } X = \frac{\partial\ln\int v_F(\vec{k},\varepsilon_F)dA}{\partial\ln\varepsilon}\bigg|_{\varepsilon_F}.$$

The parameter V has also been determined in a band theory calculation for copper,[5] and the values agree exactly. The parameter X, although now known precisely, is not useful until a more accurate value of

$$\frac{\overline{\partial\ln v}}{\partial\ln\varepsilon}\bigg|_{\varepsilon_F}$$ is obtained.

REFERENCES

1. B. Lengeler, W.R. Wampler, R.R. Bourassa, K. Mika, K. Wingerath and W. Uelhoff, Phys. Rev. B15, 5493 (1977).
2. M.R. Halse, Phil. Transact. Roy. Soc. London, 265, 507 (1969).
3. P.T. Coleridge and B.R. Watts, Can. J. Phys. 49, 2379 (1971).
4. P.T. Coleridge, G.B. Scott, and I.M. Templeton, Can. J. Phys. 50, 1999 (1972).
5. A. Hasegawa and T. Kasuya, J. Phys. Soc. Japan 25, 141 (1968).
6. R.F. Moreland and R.R. Bourassa, Phys. Rev. B12, 3991 (1975).
7. R.P. Huebener, Phys. Rev. A136, 1740, (1964).
8. S.F. Lin and W.F. Leonard, J. Appl. Phys. 42, 3634 (1971).
9. W.F. Leonard and H.Y. Yu, J. Appl. Phys. 44, 5320 (1973) and H.Y. Yu and W.F. Leonard, J. Appl. Phys. 44, 5324 (1973).
10. W. Worobey, P. Lindenfeld, and B. Serin, Phys. Lett. 16, 15 (1965); in Basic Problems in Thin Films, edited by A. Niedermayer and H. Mayer (Vandenhoech and Ruprecht, Gottingen, 1966) p. 601.
11. R.K. Angus and I.D. Dalgliesch, Phys. Lett. A31, 280 (1970).
12. J. Gouault, J. Phys. (Paris) 28, 931 (1967).
13. G.A. Burdick, Phys. Rev. 129, 138 (1963).
14. I.V. Abarenkov and M.V. Vedernikov, Soviet Phys. Solid State 8, 186 (1966).
15. R.W. Williams and H.L. Davis, Phys. Lett., A28, 412 (1964).
16. S.S. Abel'skii, Y.P. Irkhin, Soviet-Phys. Solid State, 11 2231 (1970).

DISCUSSION

I.M. Templeton: I may have missed something in this, but it seems to me that nowhere in your calculation is there any indication that you are measuring the variation of scattering with energy. But the point is that in getting thermoelectric powers you surely need the derivative of scattering with energy. Now from de Haas-van Alphen measurements you can get the shape of the Fermi surface from frequency measurements, you can get the velocity distribution from effective mass measurements. You can get details of the change of the shape of the Fermi surface from de Haas-van Alphen phase shift measurements. And you can get details of scattering anisotropy from Dingle temperature measurements. But there is no way in which you can determine scattering gradients with energy, which is surely absolutely essential.

R.R. Bourassa: Let me say that Dr. Templeton is quite correct. Perhaps the confusion is in the title of the talk; where it says "Calculation of Thermopower" it should be "Calculation of Thermopower Parameters". The main idea is simply that often these parameters like dlogA/dlogE are used in thermopower calculations and we can

determine them very accurately. There is no attempt here to calculate the actual thermopower. The energy dependence of scattering,
of course, cannot be determined from de Haas-van Alphen measurements.

R.I. Boughton: You mentioned that the discrepancy with the thermoelectric size effect data was perhaps due to some erroneous element
in the analysis of that data. Could you just mention what you think
it is.

S. Ya Wang: For example the assumption of a mean free path, that
is one possibility. Another assumption is, for example, the phase
of the thermoelectric size effect, and the size of the sample should
not change the Fermi surface; that is a basic assumption of the
thermoelectric size effect.

ON CALCULATION OF THERMOPOWER OF GROUP I METALS

M.V. Vedernikov, I.V. Abarenkov, I.L. Korobova and

J. Marton

A.F. Ioffe Physico-Technical Institute

Leningrad, 194021, USSR

ABSTRACT

The anomalous positive thermopower is observed in metals of group I for which the Fermi surfaces are relatively distorted. This anomaly is partially explained here by the immediate influence of deviations of the energy spectrum in the vicinity of Fermi surface from the simple dependence $E \sim K^2$.

Let us consider Table I, where transport properties of the first group metals are compared with the geometry of their Fermi surfaces (FS).

Here n_{eff} is calculated from the experimental values of R within the free electron approximation (FEA). "Distortion of FS" is the ratio $(k_{max}-k_{min})/k_o$, where k_{max} and k_{min} are maximum and minimum radii of the experimental FS while k_o is the radius of the free-electron Fermi sphere. The "negative" ("positive") thermopower means the thermopower is negative (positive) and the temperature derivative is negative (positive). "Contact" means that the FS touches the Brillouin Zone.

From this table one can see that in every case R is quite close to its FEA value. However, this model is entirely consistent with only Na, K, and Rb cases (both S and R are of the same, negative, sign). For these three metals the FS distortion is the smallest. When FS distortion approaches approximately 4% the thermopower becomes, and remains, positive. It permits one to

TABLE I

Metal	Thermopower S	Hall Constant R	Effective number of electrons per atom n_{eff}	Distortion of FS (per cent)
FEA	negative	negative	1.0	0
Na	negative	negative	0.9	0.15
K	negative	negative	0.9	0.25
Rb	negative	negative	1.0	1.5
Cs	intermediate	negative	1.1	4.5
Li	positive	negative	1.3	5.0-6.0
Cu	positive	negative	0.8	contact
Ag	positive	negative	0.8	contact
Au	positive	negative	0.7	contact

assume that the change of the thermopower sign is due, in the first place, to the deviation of the metal electronic structure from FEA. Following the paper of Ziman[1] we will assume the isotropy of the relaxation time τ. Then:

$$S = \frac{\pi^2 k^2 T}{3e} \frac{\partial \ln \int \tau v dS_F}{\partial E} \Big|_{E=E_F} = \frac{\pi^2 k^2 T}{3e} \frac{1}{E_F} \left(\frac{\partial \ln \tau}{\partial \ln E} + \frac{\partial \ln \int v dS_F}{\partial \ln E} \right) \tag{1}$$

$$\frac{\partial \ln \int \tau v dS_F}{\partial \ln E} = \xi \qquad\qquad \frac{\partial \ln \int v dS_F}{\partial \ln E} = \xi_1$$

We will also employ the simplified expression for R obtained by Ziman within the isotropic τ approximation. The dimensionless term ξ_1 depends only on the energy spectrum of the metal. In the FEA case ξ_1 is positive. To estimate the immediate influence of the FS distortion on the thermopower, it is expedient to calculate ξ_1. For this purpose we shall use the Slater-Koster expression (SKE)[2].

In the case of noble metals the SKE with only two terms can provide a good approximation to the FS. In Fig. I a sample of FS obtained in our paper[3] is shown. With the use of 4 - 6 terms in SKE it is possible to obtain very good approximations to the FS of noble metals[4]. Our main assumption is: the SKE with the

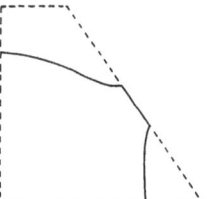

Fig. 1: One of the central cross-sections of the Fermi surface
for noble metals (one quarter). The Brillouin zone is shown by
the dashed line.

cofficients chosen to fit the experimental FS is considered as the
energy spectrum E(k) in the vicinity of FS. With the use of such
chosen E(k) it is not difficult to calculate ξ_1 and R values. In
[3] it is shown that the two terms E(k) gives $\xi_1 < 0$, $R < 0$, and
$n_{eff} \approx 1$. We have calculated ξ_1 and R with the use of the more
complicated 4 and 6 terms E(k) and obtained ξ_1 and R values close
to that obtained with two terms E(k).

It was thought that it is impossible to obtain almost spherical
FS of alkali metals with the use of SKE with a small number of
terms [5]. Nevertheless we made an attempt to apply the method
discussed to the alkali metals as well. We used not more than three
terms in SKE for the b.c.c. lattice in the following form:

$$E(\vec{k}) = C_1(1-\cos\tfrac{1}{2}k_x a\ \cos\tfrac{1}{2}k_y a\ \cos\tfrac{1}{2}k_z a) + C_2(3-\cos k_x a-$$

$$\cos k_y a-\cos k_z a) + C_3(3-\cos k_x a\ \cos k_y a-\cos k_y a\ \cos k_z a \qquad (2)$$

$$-\ \cos k_z a\ \cos k_x a)$$

It was found that with the use of only the two first terms of (2)
it is possible to obtain a FS for which deviations from a sphere
are comparatively small. In the case $C_1=1$, $C_2=0.14$, $C_3=0$ the FS
distortion is approximately 12%. With the use of three terms of
(2) a FS quite close to the experimental FS of Cs and Li can be
obtained. The smallest FS distortion obtained was about 3%. This
FS ($C_1=1$, $C_2=0.165$, $C_3=0.055$) is shown in Fig. 2. The calculation
of ξ_1 and R values for this E(k) gives $\xi_1 < 0$, $R < 0$, and $n_{eff} \approx 1$. It
seems that 3% is about the minimum FS distortion which can be
obtained with the use of three-terms expression (2).

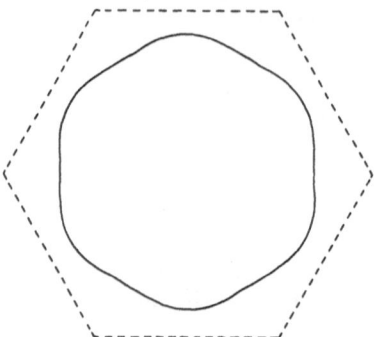

Fig. 2: One of the central cross-sections of the Fermi surface
for alkali metals. The Brillouin zone is shown by the dashed line.

       The validity of our choice of E(k) needs special investigation,
which we are going to carry out in the future. We believe however,
the present calculations show that in the metals of group I the
essential part of the thermopower anomaly is due to the immediate
influcence of the energy spectrum details while the Hall constant
is influenced weakly. The separate calculations were made which
show that the integrand in the expression for ξ_1 is negative in
every point of FS. Therefore the thermopower anamaly is due to the
general behavior of E(k) and not the distortion of the FS in some
special places. Our calculations show also that the closeness of
FS to the sphere is not necessarily evidence of the applicability
of the free electron approximation. A spectrum of totally different
nature can result in a surface of constant energy of almost spherical shape. On the other, it does not prove the validity of the
tight binding approximation in the case of alkali metals. The
coefficients in (2) were found by fitting the experimental shape
of the FS and, therefore, our expression for energy does not connect
directly with LCAO method.

REFERENCES

1. J.M. Ziman, Adv. Phys. <u>10</u>, 1 (1961).
2. J.C. Slater and G.F. Koster, Phys. Rev. <u>94</u>, 1498 (1954).
3. I.V. Abarenkov and M.V. Vedernikov, Fizika Tverd. Tela <u>8</u>, 236
   (1966). English trans.: Soviet Phys. Solid St. <u>8</u>, 186 (1966).
4. D.J. Roaf, Phil. Transact. Roy. Soc. (London) <u>255</u>, 135 (1962).
5. J.M. Ziman, Electrons and Phonons (Clarendon Press, Oxford,
   1962), p.110.

THERMOELECTRIC EFFECTS IN SUPERCONDUCTORS

G. F. Zharkov

Lebedev Institute

Moscow, U.S.S.R.

It is a pleasure and an honour for me to speak before this highly esteemed audience. The subject of my talk is devoted to thermoelectric effects in superconductors. The question of whether thermoelectric phenomena can be observed in metals in the super-conducting state was raised more than 50 years ago, but first attempts to observe such effects as Seebeck electromotive force, or Peltier heat, or Thomson heat, failed to reveal any trace of the existence of these effects in superconductors. In any case, it was definitely shown that these effects, if they exists are weaker by many orders of magnitude, than the analogous effects in normal metals. This was especially clearly demonstrated by the experiment of Steiner and Grassman in 1935,[1] who tried to observe a Seebeck electromotive force in superconductors. These authors suspended a superconducting bimetallic ring on a torsion wire placed in a weak external magnetic field and heated the ring. If a Seebeck thermo-electromotive force would arise, as it arises in a case of normal metals, then this force would induce acceleration of superconducting electrons and the current in the ring would increase with time, leading to a torque on the wire, which it might be possible to observe.

However, long observation failed to reveal any torque in the system and this made it possible to assert with great accuracy that no accelerating force exists in a superconducting ring. The result of this and of other experiments lead to the conclusion formulated, for instance, in Shoenberg's book "Superconductivity",[2] that "there is no dispute about the disappearance of all thermoelectric effects in superconducting state". This conclusion remained generally accepted up to recent times and was reproduced in a number of text-books.[3-5] Moreover, theoretical considerations of this question,

based on a thermodynamical approach,[3-6] as well as microscopic
arguments,[7] also supported the conclusion about the disappearance
of practically all thermoelectric effects in superconductors.

Nevertheless, thermoelectric effects in superconductors by no
means vanish and in principle they can be observed. This was
indicated more than 30 years ago by Ginzburg[8] on the basis of the
two-fluid model of superconductivity. In accordance with this
model the total current density in a superconductor is a sum of
two quantities $\underline{j}=\underline{j}_s+\underline{j}_n$, where $\underline{j}_s$ and $\underline{j}_n$ are the densities of superconducting and normal currents. When a temperature gradient is
present, the normal current in general is non-zero, but it can be
compensated by the superconducting current (so that the total current vanishes, $\underline{j}=0$). The total compensation occurs only in the
most simple cases (e.g. in an isotropic and homogeneous superconductor). However it was shown by Ginzburg[8] that in the case of
inhomogeneous superconductors (e.g. in a bimetallic plate, Fig. 1)
in the vicinity of a joint the resulting total non-zero current I
and the corresponding magnetic field H_T should appear. The existence of this effect follows from considering the expression for
the total current in a superconductor $\underline{j}=\underline{j}_s+\underline{j}_n$,

$$\underline{j}_s = - \frac{e^*}{m^*c} n_s^* \underline{A} \qquad , \qquad \underline{j}_n = \sigma_n \alpha_n \nabla T. \qquad (1)$$

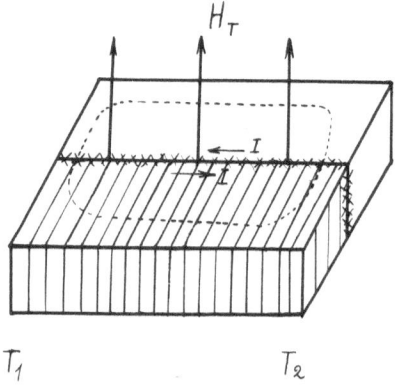

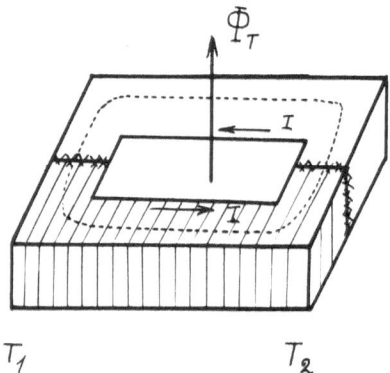

Fig. 1. A thermoelectric circuit consisting of a superconducting bimetallic plate.

Fig. 2. A bimetallic super-conducting ring.

The superconducting current is proportional to the vector potential $\underline{A}$; e*=2e and m*=2m are doubled electron charge and mass, in accordance with the concept of bound Cooper pairs of electrons in superconductors, n_s^* is the number of pairs $n_s^* = \frac{1}{2}n_s$, $n_s = |\psi|^2 n$ is the density of "superconducting" electrons, n being the total electron density, ψ is the superconducting order parameter, or macroscopic wave function of the superconductor. In the case of a singly-connected superconductor or in the absence of vortices in the system ψ can be taken to be a real function. The normal current j_n is proportional to temperature gradient, σ_n and α_n being the conductivity and normal differential thermopower coefficients.

If we consider the closed circuit, denoted by dotted line in Fig. 1, which lies deep in the bulk of the superconducting plate, then the total current on this circuit vanishes because of the Meissner effect

$$\underline{j} = -\frac{e^*}{m^*c} n_s^* \underline{A} + \sigma_n \alpha_n \nabla T = 0 \quad . \tag{2}$$

Integrating Eq.(2) over this circuit gives the result

$$\oint \underline{A} d\underline{\ell} = \int \underline{H} d\underline{S} = \Phi_T = \frac{m^*c}{e^*2} \oint \frac{\sigma_n \alpha_n \nabla T}{n_s^*} d\underline{\ell} \neq 0 \quad . \tag{3}$$

Thus, according to Eq.(3) inside the circuit there exists a magnetic flux Φ_T, which is different from zero, provided the material constant $\sigma_n \alpha_n / n_s^*$ depends on co-ordinates, as in the case of a bimetallic plate. Ginzburg[8] derived the result (3) by a slightly different procedure and he showed that the magnetic field H_T is concentrated in vicinity of a joint (see Fig. 1) and reaches the **magnitude of** $H_T \sim 10^{-4}$ gauss. The flux Φ_T and the field H_T are accompanied by the superconducting current I circulating around the joint.

The main attention in Ginzburg's paper[8] was devoted to the other effect, also predicted by him. He showed that inside a homogeneous but anisotropic superconductor the magnetic field should appear, if the temperature gradient does not coincide with crystalline axes of symmetry. The attempt to observe such a field was undertaken a few years ago by Selzer and Fairbank[9] but I cannot here go into details to describe this effect and shall concentrate on a case of an isotropic, but inhomogeneous superconductor, which attracted a lot of attention recently.[10-14]

As is clear from Fig. 1 it is not an easy task to observe a

small magnetic field $H_T\sim10^{-4}$ gauss in vicinity of a bimetallic
plate. It was definitely a very difficult joint undertaking 30
years ago, when modern quantum interference magnetometers were not
available. So the question rested in peace until recently. The
interest in thermoelectric phenomena in superconductors was reviewed
only in 1973 when Leningrad physicists Galperin, Gurevich and Kozub
and independently a little later Garland and Van Harlingen[10] pointed
out that if one takes a bimetallic superconducting ring (Fig. 2)
then in the presence of a temperature gradient, the magnetic flux
will appear inside the ring. Indeed, if we consider the closed
circuit shown by dotted line in Fig. 2 and integrate Eq. (2)
(j=0) around this circuit, we again arrive at the result (2). The
estimate of thermoelectric flux Φ_T in Eq. (3) gives the value
$\Phi_T\sim10^{-9}$ gauss.cm^2, or $\Phi_T\sim10^{-2}\Phi_o$, where $\Phi_o = 2.10^{-7}$ gauss.cm^2 is
the quantum of flux. Later the result (2) was experimentally con-
firmed by Savaritsky[13] and investigated in more detail in a number
of papers[14a,b]. In experiments it is convenient to use a somewhat
different, but topologically equivalent thermoelectric circuit, as
in Fig. 3, and observe directly not the flux in the ring, but the
field of a circulating current I, which can be made to pass through
the subsidiary solenoid L (Fig. 3). Some questions remain to be
clarified concerning the experimentally observed magnitude of the
effect,[15a,15b,15c] but it seems reasonable to believe that this
effect has been observed already.

It should be stressed that thermoelectric effect in a superconducting bimetallic plate[8] and thermoelectric effect in a
superconducting bimetallic ring[10-12] are physically completely
equivalent.[12] Indeed, if the hole inside a ring is made to vanish,

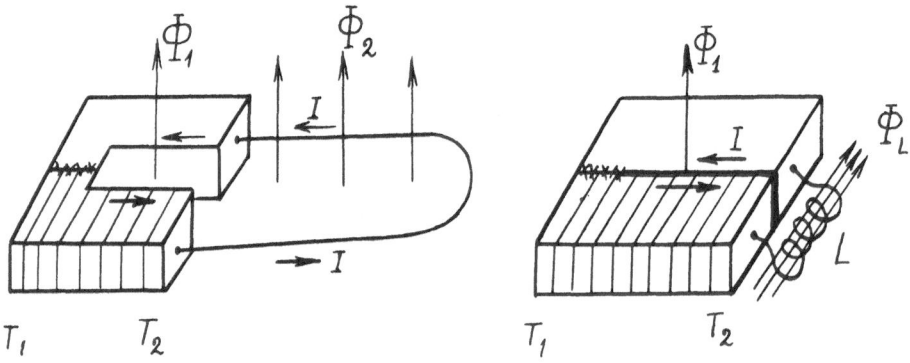

Fig. 3. Schematic arrangement for observing a thermoelectric
circulating current.

we are passing continuously to the case of a bimetallic plate, the
field localizing in the vicinity of a joint in a region characterized
by the London penetration depth $\lambda_L \sim 10^{-5}$ cm, and the magnetic field
reaching the magnitude Φ_T/λ_L 1 cm$\sim 10^{-4}$ gauss as was estimated by
Ginzburg.[8] But the use of a toroidal geometry, as proposed by
Galperin, Gurevich, Kosub[10] and Garland and Van Harlingen[11] has
led to success in observing the effect.

At first glance the appearance of a thermoelectric current in
a nonuniformly heated bimetallic superconducting ring looks very
much like the usual thermoelectric current produced by Seebeck
electromotive force in a normal metal. So one may raise a question
whether there exists a contradiction between the statements about
the total disappearance of all thermoelectric effects in superconductors, which I mentioned above, and the appearance of a magnetic
flux Φ_T and of a corresponding circulating current I in a bimetallic
superconducting ring.

Before answering this question, the physical nature of the
resulting thermoelectric current should be revealed, i.e., we should
find out what the forces are that drive the superconducting component into motion. The origin of such a current is easy to understand
if Eq. (1) for $\underline{j}_s$ is rewritten in a form

$$\underline{j}_s(t) = \frac{e^*}{m^*} n^*_s \int_0^t \underline{E}dt' \ , \ \underline{E} = - \frac{1}{c} \frac{\partial \underline{A}}{\partial t} \ . \tag{4}$$

It can be argued that in the case of sufficiently slow processes
the role of the forces due to the inequilibrium in a superconducting subsystem can be neglected, so we put $\nabla(\mu_s + e\phi) = 0$, where
$\mu_s + e\phi$ is the pair electrochemical potential; see (16) for more
details. From Eq. (4) the following can be seen: If at some
initial moment t=0 the current and the vector potential were absent
($\underline{j}_s = \underline{j}_n = \underline{A} = 0$ when $T_1 = T_2$), then at a later time the superconducting
current may appear only due to the induction force $\underline{E} = -c^{-1}\partial \underline{A}/\partial t$,
acting in the system, i.e., due to nonstationarity. The only non-
stationarity in the situation considered is caused by applying the
temperature gradient, i.e., by changing the temperature on one of
the sides of the specimen, say, $T_2(t)$. We can argue then, that
when the specimen is heated and the temperature $T_2(t)$ varies with
time, the inequilibrium appears in a normal subsystem. The force
proportional to the temperature gradient acts on a normal subsystem
and sets it into motion, while a weak alternating electric field
$\underline{E}(t)$ is induced in the ring. This field acts on the superconducting component and produces its acceleration. The resulting
superconducting current at the moment t is determined by Eq. (4),
which coincides with London's expression (1). The presence of the

magnetic flux (3) in the ring then follows from Eq. (2) as before.

The electric field E(t) itself is negligibly small (the esti-mate gives its value $\varepsilon(t) = \oint Ed\ell \sim c^{-1} \partial \Phi / \partial t \sim 10^{-24}$ V when $\Phi \sim \Phi_T \sim 10^{-9}$ gauss.cm^2 and $dT/dt \sim 10^{-2}$ K/sec.); however, it is just this weak field that acts during the whole heating cycle (integrally in time!) and that accelerates the superconducting component. An integral character of the action of these forces corresponds to the fact that the superconducting component moves without collisions with the lattice and therefore the influence of the forces is summed in time. After reaching its final value (4) or (1) the superconduct-ing component moves further on by inertia without losses. Thus in the final steady state there is no force acting on the superconduct-ing component.

The qualitative arguments presented above permit us to express the opinion that the total circulating superconducting current which results in an unevenly heated bimetallic superconducting ring is of inductive origin and moves by inertia. In this respect there is an obvious difference from an analogous effect in a normal circuit (Seebeck effect), where thermoelectric current is, so to speak, of diffusive nature, and is directly connected with dissipative proc-esses and requires the presence of an electromotive force to maintain the current.

Thus, answering the question raised above, we can say that there is no contradiction between the opinions expressed in differ-ent papers concerning the existence of thermoelectric effects in superconductors, because in reality different effects were consid-ered in these papers. Thermodynamic arguments presented in a number of papers[2],[4-6] show, indeed, that the diffusion thermoelectromotive force and other thermal effects in superconductors vanish, but those arguments do not forbid the existence in a closed ring of a non-dissipative inertial current, produced by inductive electric forces. The negative result of the experiment of Steiner and Grassman[1] becomes clear from the point of view presented above. Indeed, the electric current increasing in time could not be observed by them because the electromotive force in stationary conditions was absent, $\nabla(\mu_s + e\phi) = 0$. However, if the sensitivity of their experiment were higher, then the constant current and magnetic flux connected with the electrons already accelerated and moving by inertia could be registered. Such a current was observed much later by Savaritsky[13] and Falco.[14]

Additional arguments in favor of the inductive origin of a thermoelectric current in a bimetallic superconducting ring can be obtained if we analyze the following situation. Consider a thermo-electric circuit as shown in Fig. 4, and introduce inside it a superconducting core, which will force out the field into a

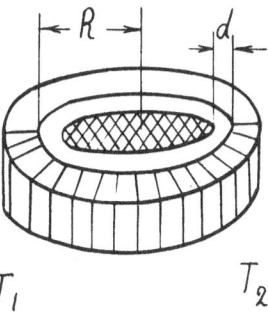

Fig. 4. A toroidal superconducting thermoelectric circuit with a
superconducting core.

narrow slit between the main thermocircuit and the core. Because
the total flux $\Phi=\Phi_T$ is held constant, the field in the slit will
increase by R/2d times (proportionally to the ratio of the total
surface of the hole to the surface of a remaining slit, R being
the inner radius of a ring, d is the width of a slit). In the case
of a very narrow slit, the magnetic field and the screening current
circulating the circuit can be increased by several orders of magnitude (for instance, if we take R~1 cm and d~10^{-3}cm the ratio
R/2d~10^3). It is obvious that the possibility of increasing the
thermoelectric current by introducing the superconducting core
inside the circuit **supports** the inductive nature of this current.

Another possibility of increasing a thermoelectric current
is to use a thermoelectric battery,[17] as shown in Fig. 5, the battery consisting of a closed superconducting circuit with a number
of consecutive links, the ends of which are held at different temperatures. It is easy to see that the resulting current in the
circuit with m links is m times larger than in a single one. For
a sufficiently big battery the current in the circuit and the flux
can be substantially increased.

I shall now touch on another question. As is clear from
Eq. (3) the magnetic flux inside a closed circuit is proportional
to a temperature difference; or more exactly, if we evaluate the
integral in (3), using the temperature dependence $\eta_b(T)=\eta_s(0)/(T_c-T)$,
the result will be

$$\Phi_T = B\ell n\frac{T_c-T_1}{T_c-T_2} \quad , \quad B\sim 10^{-2}\Phi_o\sim 10^{-9} \text{ gauss.cm}^2 \quad , \tag{5}$$

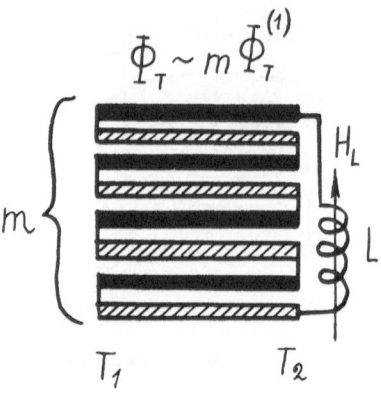

Fig. 5. Schematic of a superconducting thermoelectric battery.

where B is some constant and we consider the case when the critical
temperatures of superconductors I and II strongly differ,
$T_{cII} \ll T_{cI} = T_c$. For a small temperature difference $\Delta = T_2 - T_1 \ll 1$ the
flux is linearly proportional to $\Delta = T_2 - T_1$ and vanishes if $\Delta = 0$. But
when $T_2 \to T_c$ the flux Φ_T diverges logarithmically. The linear dependence was observed by Savaritskii,[13] as well as a sharp rise of Φ_T
near T_c, the logarithmic dependence was traced in detail by Falco.[14]
As was pointed out by Pegrum and Guénault, some peculiarities of
the dependence $\Phi_T(T)$ found by Falco can be explained by taking into
account the possibility that the frozen-in flux can be present
inside the closed superconducting circuit. In this case the total
flux will be

$$\Phi = \Phi_T + n\Phi_o \ , \quad \Phi_o = hc/2e \ , \quad n = 0,1,2, \qquad (6)$$

(Φ_o is the flux quantum), so when the temperature difference is
reduced to zero, we are left with the constant value $\Phi = n\Phi_o$ for the
flux inside the ring, as was probably observed by Falco.[14]

It is appropriate to mention here that the flux is a smooth
function of the temperature difference and, consequently, is not
quantized, in contrast to a frozen-in flux $n\Phi_o$, which, in the case
of a bulk superconductor, is quantized. The absence of quantization
of the thermoelectric flux Φ_T does not contradict the well-known
London quantization theorem

$$\oint (\underline{V}_s + \frac{e^*}{m^*c} \underline{A}d\underline{\ell}) = \frac{\hbar}{m^*} \oint \nabla\theta = 2\pi n \frac{\hbar}{m^*} \ , \qquad (7)$$

because, according to this theorem, it is not the magnetic flux that is quantized, but the so-called fluxoid, or in other words, the number of vortices present in a superconductor is quantized (θ in (7) is the superconducting phase). When the superconductor is in an off-equilibrium state (the normal current $\underline{j}_n$ is present, due to temperature difference), Eq. (7) takes the form

$$\oint \frac{e^*}{m^*c} A\underline{d\ell} = 2\pi n \frac{\hbar}{m^*} - \oint \frac{\underline{j}-\underline{j}_n}{e^*n_s} \underline{d\ell} \quad , \tag{8}$$

so if $\underline{j}=0$ (Meissner effect) we have

$$\Phi = \frac{\hbar c}{e^*} \oint \nabla\theta\underline{d\ell} + \oint \frac{\underline{j}_n}{e^*n^*_s} \underline{d\ell} = n\Phi_o + \Phi_T \quad . \tag{9}$$

So the number of vortices in the system is always quantized, but the off-equilibrium contribution to the flux Φ_T is not quantized. Only when the superconductor is in equilibrium ($\underline{j}_n=0$) is the total flux in a bulk superconductor quantized.

Sometimes the resulting thermoelectric flux Φ_T is connected with the order parameter phase difference, which is assumed to appear between any two points of a nonuniformly heated supercon-ductor.[10] In this terminology the effect under discussion acquires quantum-mechanical meaning and the "phase" itself emerges as a real physical quantity as the cause of the current appearance. From Eq. (9) it is evident, however, that the frozen-in flux $n\Phi_o$ can be associated with a phase, but the thermoelectric flux Φ_T is not connected with a phase and has a classical interpretation, being the result of the action of the inductive forces in the sys-tem. (Note, by the way, that Planck's constant $\hbar$ does not enter Eqs. (3) or (9) for Φ_T.) This methodical remark does not affect, however, the essence of the results obtained by using the notion of a "phase difference".

It is interesting to note that if the thermoelectric circuit includes two weak links, as shown in Fig. 6 (so-called quantum interferometer), then this device can be used for measuring the thermoelectric flux. Indeed, the maximum stationary current, pass-ing through the device, depends on the external flux in the ring $\Phi=\Phi_e$ according to the law

$$I_{max} = 2I_o \left| \cos(\pi\Phi/\Phi_o) \right| \quad . \tag{10}$$

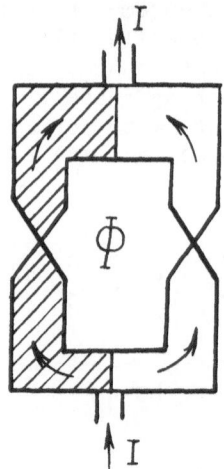

Fig. 6.   Thermoelectric circuit with two weak links.

When the temperature gradient is present, the total flux is $\Phi = \Phi_e + \Phi_T$ and the interference curve (10) would shift proportionally to the temperature difference $\Delta T = T_2 - T_1$ (see Fig. 7). As the positions of the minima in Fig. 7 can be found with great accuracy, such a device can be used, in principle, as a precision thermometer.[1]

I shall spend the remainder of my time discussing some other nonequilibrium thermoelectric phenomena. Suppose the superconducting subsystem is in a quasi-equilibrium state, i.e., the pair electrochemical potential is constant: $\mu_s + e\phi$ = const. However, the normal electrons' electrochemical potential is not necessarily constant: $\mu_n + e\phi \neq$ const. There will then be an additional

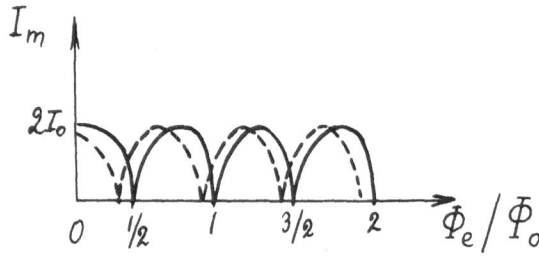

Fig. 7.   Shift of the interferometer current with a change in temperature difference.

contribution to the normal current

$$\underline{j}_n = \sigma_n(\alpha_n \nabla T + \frac{\nabla \delta \mu}{e}) \ , \ \delta\mu = \mu_s - \mu_n = \text{adiv}\underline{j}_n \ , \tag{11}$$

where $\delta\mu$ is the off-equilibrium chemical potential difference appearing in the regions where the mutual transformation of normal and superconducting currents takes place.[18] The constant a is proportional to the branch imbalance relaxation time between the populations of the electron-like and hole-like branches of quasi-particle exitation spectra in the superconductor.[19] From (11) the equation follows

$$\frac{d^2 \delta\mu}{dx^2} - \frac{1}{\ell_b^2} \delta\mu = \beta \frac{d^2 T}{dx^2} \ , \ \beta = -e^*\alpha_n \ , \tag{12}$$

which determines the space distribution of the chemical potential difference $\delta\mu(x)$. Here ℓ_b is a characteristic length connected with branch relaxation time. According to the theoretical estimates[20] the length ℓ_b in pure superconductors with energy gap can reach the value $\ell_b \sim 1$ mm, i.e., it can be very large. As can be seen from Eq. (12) the contribution $\delta\mu(x)$ arises at points where $d^2T/dx^2 \neq 0$; that is, at points where the heat flows in and out of the specimen (div $\nabla T \neq 0$); $\delta\mu$ falls exponentially over the characteristic length ℓ_b when we go away from these points.[20,21] A number of effects are associated with the off-equilibrium contribution $\delta\mu$. For instance, an electric field and a noncompensated electric charge arise in a nonuniformly heated superconductor;[20] these effects are analogous to effects arising at the border between normal and superconducting metals, where the current is flowing and the reciprocal transformation of normal and superconducting currents occurs.[22]

The other effect is the appearance of a small contribution to a thermoelectric flux in the ring. Indeed, as can be seen from Eqs. (11), (9) the total flux is now

$$\Phi = \Phi_T + \oint \frac{\sigma_n \nabla \delta\mu}{en_s} d\underline{\ell} = \Phi_T + \delta\Phi \ , \tag{13}$$

where Φ_T is the usual thermoelectric flux, produced by a temperature gradient, and $\delta\Phi$ is an additional flux due to the off-equilibrium contribution $\delta\mu$. This additional flux is estimated[20,21] as small

($\delta\Phi/\Phi_T < 10\%$), but in principle it can be distinguished from the flux Φ_T, if careful measurements are made.

I have no time to consider many other thermoelectric effects in superconductors, such as an additional thermoflow, small Peltier, Thomson and Joule heat contributions appearing when $\delta\mu \neq 0$, as well as very interesting thermoelectric and thermomagnetic effects connected with the movement of vortices in superconductors. However, it is evident that thermoelectric phenomena in the super- conducting state by no means vanish, but are rather many-sided. It is true that the corresponding effects are, in general, very small, as compared with those in the normal state. But they belong to a very interesting and important group of phenomena where the non- equilibrium processes play a significant role. The study of such nonequilibrium processes is now attracting continuously increasing attention. In view of this, there is no doubt that theoretical and experimental investigation of thermoelectric phenomena in super- conductors will intensify in the future.

REFERENCES

1. K. Steiner and P. Grassman, Phys. Zeit, 36, 527 (1935).
2. D. Shoenberg, Superconductivity, Cambridge University Press (1965), Chap. 3, 7.
3. A.C. Rose-Innes and E.H. Rhoderick, Introduction to Superconductivity, Pergamon Press (1969), Chap. V & 5.
4. W.F. Vinen in Superconductivity, ed. by Parks, Vol. 2, Chap. 20.
5. S.I. Putterman, Superfluid Hydrodynamics, North Holland/American Elsevier (1974).
6. J.M. Luttinger, Phys. Rev., A136, 1481 (1964).
7. M. Stephen, Phys. Rev., 139, 197 (1965).
8. V.L. Ginzburg, Zh. Eksp. Theor. Fiz. 14, 177 (1944); Journ. Phys., USSR 8, 148 (1944).
9. P.M. Selzer and W.M. Fairbank, Phys. Lett., 48A, 279 (1974).
10. Yu.M. Galperin, V.L. Gurevich and V.I. Kozub, Pisma Zh.E.T.P. 17, 687 (1973); Zh.E.T.P. 66, 1387 (1974).
11. J.G. Garland and D.J.Van Harlingen, Phys. Lett. 47A, 423 (1974).
12. V.L. Ginzburg, G.F. Zharkov and A.A. Sobjanin, Pisma Zh.E.T.P. 20, 223 (1974).
13. N.V. Zavaritsky, Pisma Zh.E.T.P., 19, 205 (1974).
14a. C.M. Pegrum, A.M. Guénault, G.R. Pickett, Proceedings of 14th Int. Conf. Low Temp. Phys., Otaniemi, Finland (1975), Vol. 2, p. 513.
14b. C.M. Falco, Solid State Comm. 19, 623 (1976).
15a. C.M. Pegrum, A.M. Guénault, Physics Lett., 59A, 393 (1976).

15b. D.J. Van Harlingen and J.C. Garland, Proceedings of Int. Conf. on Thermoelectricity, East Lansing, USA (1977), p. 173 (in print).

15c. C.M. Pegrum and A.M. Guénault, Proceedings of Int. Conf. on Thermoelectricity, East Lansing, USA (1977), p. 179 (in print).

16. V.L. Ginzburg and G.F. Zharkov, Thermoelectric Effects in Superconductors, Preprint N 115, Physical Lebedev Inst. Moscow (1977).

17. G.F. Zharkov and A.A. Sobjanin, Pisma Zh.E.T.P. 20, 163 (1974).

18. T.J. Rieger, D.J. Scalapina and J.E. Mercerna, Phys. Rev. Lett. 27, 1781 (1971).

19. M. Tinkham and J. Clarke, Phys. Rev. Lett., 28, 1366 (1972). M. Tinkham. Phys. Rev., B6, 1747 (1972).

20. S.N. Artemenko and A.F. Vokov, Zh.E.T.P. 70, 1051 (1976).

21. A.M. Gulian and G.F. Zharkov, Short Comm. in Physics, (in press). Proceedings of the Conference on Thermoelectric Phenomena in Metals, East Lansing, USA, 1977 (to be published).

22. M.L. Yu and J.E. Mercerau, Phys. Rev. B12, 4909 (1975).

THERMOELECTRIC EFFECT IN A NONEQUILIBRIUM SUPERCONDUCTOR*

Charles M. Falco

Argonne National Laboratory

Argonne, Illinois 60439

As Ginzburg[1] first noted, there exists in a superconductor the possibility of a simultaneous flow of a normal current of density $\vec{j}_n = L_T(-\vec{\nabla}T)$ and a supercurrent $\vec{j}_s = -\vec{j}_n$. Recently[2] calculations based on the two fluid model have predicted that this flow of normal current in a superconductor will give rise to a non-quantized contribution to the magnetic flux in a loop made up of two different superconductors. Experimental data[3-5] indicate the existence of such a magnetic flux with a value of as much as five orders of magnitude[5] larger than predicted by theory. However, more recent work[6] has suggested that this magnetic flux may be due to the temperature dependence of the penetration depth. This discrepancy, coupled with the opportunity to study quasiparticle transport and relaxation processes in a nonequilibrium super-conductor, prompted us to make measurements of the pair and quasi-particle electrochemical potentials in a superconductor held in a temperature gradient. It has been predicted[7] that in nonequilibrium situations in which the electron and hole branches of the quasi-particle excitation spectrum are unequally populated, the quasi-particles in a superconductor may be described by a different electrochemical potential than that which describes the pairs. This paper reports initial results showing experimental evidence for a pair-quasiparticle electrochemical potential difference in a super-conductor in a temperature gradient.

---

* Work performed under the auspices of the U.S. Energy Research and Development Administration.

A normal metal tunnel junction and superconducting metal
probe were placed in a nonequilibrium region of a thin film superconductor. It has been shown[12] that this allows a direct measurement of the quasiparticle and pair potentials to be made. A strip
of 99.999% purity Sn in the form of a "T" (Fig. 1) of width $\sim$ 2 mm,
length $\sim$ 2 cm and thickness $\sim$ 2000–4000 Å was evaporated onto a
1 mm thick sapphire substrate and oxidized. A small rectangle of
Cu ($\sim$ 2.5 mm x 0.6 mm x 1.5 μm) was then evaporated in the center
of the strip to form the normal probe. Finally, a 3000 Å Pb film
was deposited over the Cu to reduce the normal probe resistance.
This was necessary in order to reduce the Johnson noise in the
sample and thus enable very low voltage measurements to be made.
The high thermal conductivity of the sapphire ensured that the
temperature gradient in the vicinity of the junctions was constant
to within 0.1%.

Additional leads attached to the sample enabled the resistance
of the normal probe to be measured by applying a current between
X and Z (Fig. 1) and measuring the voltage developed between X and
Y. Typical junctions had a resistance of 1–4 x 10^{-5} Ω immediately
below the T_c of Sn which, as shown in Fig. 2, increased approximately at the rate expected for a Normal-Superconductor tunnel junction.
The voltage between X and Y was measured with a SQUID voltmeter in
a feedback configuration. The resolution of this system was
approximately 2 x 10^{-13} volts rms with a 1 sec averaging time. It
was also possible to inject a current through the normal probe to
verify that Josephson tunneling was not taking place through the

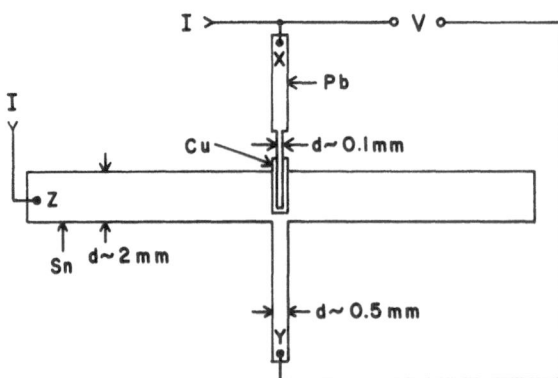

Fig. 1. Sample configuration. The Sn film is deposited on a
sapphire substrate and oxidized; followed by the Cu electrode and
Pb strip. 0.08 mm diameter Nb wires connected to a SQUID voltmeter allow the potential difference between points X and Y to be
measured.

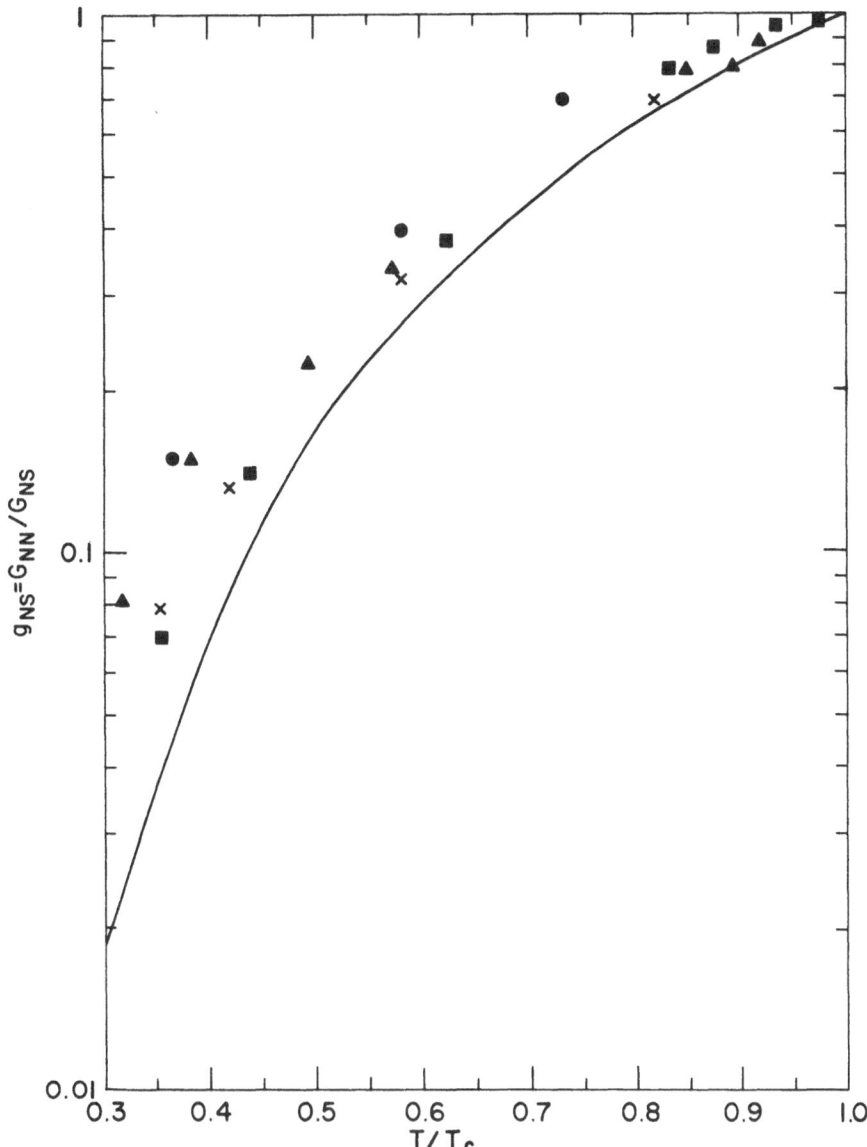

Fig. 2. Normalized low voltage conductance g_{NS} for four samples as a function of temperature compared with theory (solid line).

Cu. Such tunneling would couple the pair electrochemical potentials in the Sn and Pb making the measurements described here impossible.

All measurements were conducted in a screened room with mu-
metal and superconducting shields reducing the field in the sample
vacuum chamber to less than 10^{-6} T.   The temperatures of both ends
of the sample were measured to within an uncertainty of less than
1 mK using calibrated Ge resistors.   Any leaks in the vacuum
chamber (which would reduce the temperature gradient along the
sample) could be readily detected as a reduction of the thermal
time constant of the system (approximately 15 minutes).

The voltage developed per unit temperature difference for one
sample is shown in Fig. 3.   All samples which showed no evidence
of a superconducting short through the normal probe exhibited the

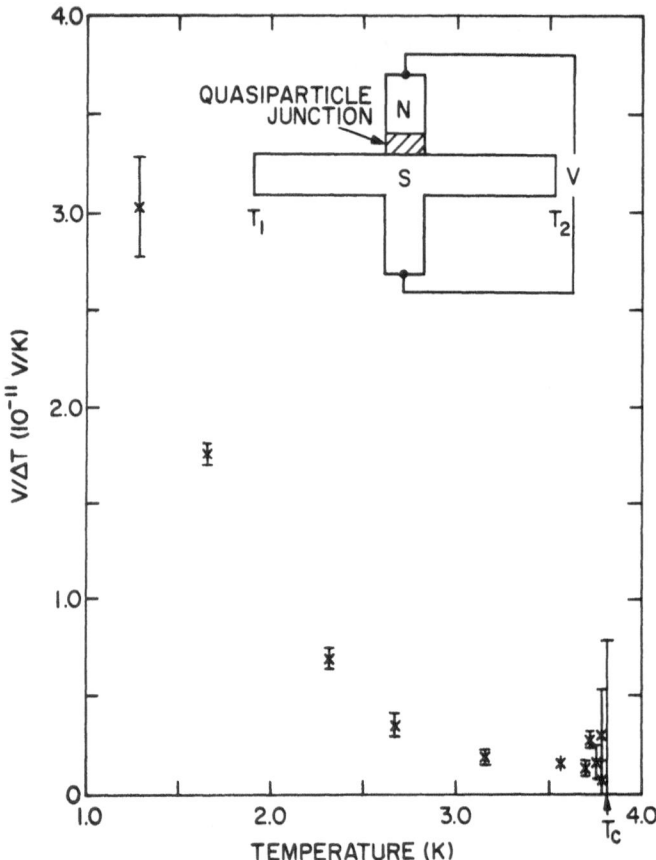

Fig. 3.   Potential difference between X and Y per unit temperature
difference across the sample versus temperature for a 2510 Å Sn
film.

same behavior shown in this figure, except that the absolute
magnitudes varied by as much as a factor of 8. This is not too
surprising as we are measuring a quantity proportional to a thermal
transport coefficient. This coefficient is extremely sensitive to
impurities and can be expected to vary widely for thin film specimens. The data shown in Fig. 3 have the lowest magnitude of those
measured. This thermally induced potential is linear in ∇T
over at least a factor of ten change in the temperature gradient.
Reversing the sign of T does not reverse the sign of the observed
voltage and leaves the magnitude unchanged to within 15-30% as in
as shown in Fig. 4. This failure to reproduce the magnitude of
V for opposite signs of ∇T is perhaps due to a small normal
thermoelectric voltage contribution from the Cu. Such a conventional thermoelectric potential would change sign with ∇T and could
cause the observed asymmetry.

This potential is reminiscent of that observed by Clarke[8] in
his branch mixing experiment. In that experiment an imbalance in
the population of the electron and hole branches of the excitation
spectrum created by tunnel injection led to a measurable potential
difference between normal and superconducting probes. Similar
effects have also been observed near superconductor/normal boundaries carrying a transport current[9] and within Josephson weak links
excited by rf radiation.[10] In all of these experiments a flow of

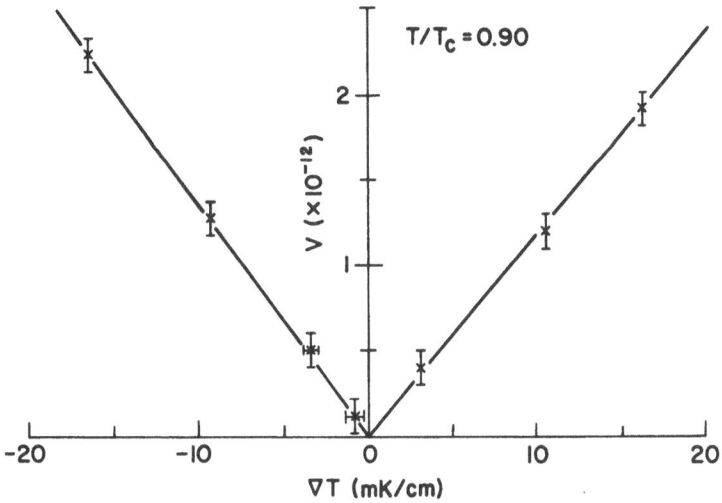

Fig. 4. Measured potential difference as a function of ∇T for a
2890 Å Sn film. $T/T_c = 0.90$.

normal excitations is accompanied by a difference between the quasi-particle and pair electrochemical potentials. Tinkham has shown[7] that such quasiparticle-pair electrochemical potential differences can only arise from an electron-hole branch imbalance. These results, together with the data presented here, is evidence that such a branch imbalance induced by a thermally generated quasiparticle current is the cause of the potential measured in the present experiment.

To first order, a branch imbalance should not arise in a superconductor in a thermal gradient since the increase in electron-like excitations and decrease in hole-like excitations on one side of the Fermi surface should be balanced by the opposite effect on the other side of the Fermi surface. However, there is not exact electron-hole symmetry since the two branches of the excitation spectrum have opposite curvature. This lack of symmetry, coupled with an energy dependent relaxation time,[11] will give rise to a branch imbalance.

If we assume that such a branch imbalance exists in our experiment, it is possible to extract information on the thermal transport coefficient for the quasiparticles from the data of Fig. 3. The quasiparticle current produced by the temperature gradients involved in this experiment should lead to a small asymmetry near the bottom of the electron and hole branches of the excitation spectrum. Another way a similar asymmetry could be produced is by injecting quasiparticles into a superconductor at very low voltages via a tunnel junction. Although to obtain detailed information will require a theory directly applicable to this experiment, the data of Fig. 3 can be interpreted using the theory of Tinkham[7] for tunneling generation of an electron-hole imbalance in the appropriate limit of small injection voltages. In this limit the voltage is given by

$$V \le \frac{If(\Delta)}{2e^2 N(0)\Omega g_{NS}^2} \tau_Q \qquad (eV_{inj} \ll \Delta(T)) \qquad (1)$$

where I is the injected quasiparticle current, $f(\Delta)$ is a Fermi factor, $N(0)$ is the one spin density of states at the Fermi surface, Ω is the volume sampled by the voltage probes, g_{NS} the normalized low-voltage conductance of the NS junction (Fig. 2) and τ_Q is the branch imbalance relaxation time.[8] The equality sign in Eq. (1) is valid near T_c. Since there exist both theory[11] and experiments[12] for τ_Q in Sn, and since all of the other quantities in Eq. (1) are known or can be measured, it is possible to use Eq. (1) to extract a value for the transport coefficient L_T by assuming $I = L_T(-\nabla T)$. The results of such an analysis for three different samples are shown in Fig. 4.

Due to the small temperature gradients necessary and the concomitant small voltages developed, it is not possible at present to obtain information very near T_c; at lower temperatures we find the data to fall off smoothly, apparently to zero. In absolute magnitude, $L_T = 2 - 16$ A/cm-K near T_c. Noting that in the normal state $L_T = \sigma S$ where σ is the electrical conductivity and S is the thermoelectric power, this transport coefficient corresponds to normal state thermoelectric powers between 3 and 24 μV/K. This is about what would be expected from measurements on the normal state thermoelectric power in Sn just above T_c.[13]

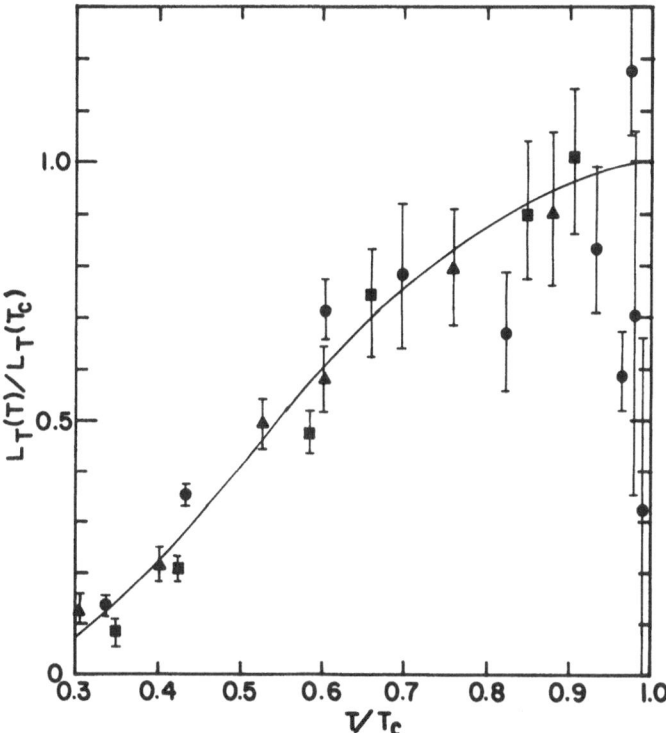

Fig. 5. Plot of the transport coefficient L_T normalized to its value at T_c versus temperature for three samples. The solid line is from a theory by Gal'perin et al[14] in which $L_T(T_c)$ is taken as the value in the normal state just above T_c.

A theory for L_T has been developed by Gal'perin et al.[14] This theory, which does not explicitly take into account the character of the electron and hole branches of the excitation spectrum, describes quasiparticle relaxation using the same mean free path appropriate to the normal state. The result of this theory with a one parameter fit to the data at T_c is shown as a solid line in Fig. 5. As can be seen, the fit with experiment is good. However, without a reliable theory for interpreting the experimentally measured voltages in terms of a transport coefficient, one cannot comment further on this agreement.

In summary, we have observed a difference between the quasiparticle and pair electrochemical potentials in thin film superconductors held in a temperature gradient. This potential diverges at low temperature and, within the resolution of our data, approaches a constant at T_c. It is possible to use our data to extract a value for the thermal transport coefficient L_T appropriate for the normal excitations in the superconductor. Unlike the experiments on bi-metallic superconducting loops,[3-5] this analysis leads to values for L_T near T_c which are in rough agreement with those in the normal state. We also find L_T to fall off to zero at low temperatures. However, in the absence of a satisfactory theory to interpret these observations, this agreement with normal state values must be regarded as very approximate. In particular, a satisfactory theory for thermoelectric effects in nonequilibrium superconductors must take into account the directional dependence of the distortion of the Fermi surface by a thermal gradient as well as the various relaxation mechanisms appropriate in the nonequilibrium state.

ACKNOWLEDGMENTS

I would like to thank Drs. K. E. Gray, R. A. Sacks, and I. K. Schuller for numerous discussions and T. R. Werner for a critical reading of the manuscript.

REFERENCES

1. V.L. Ginzburg, J. Phys. USSR 8, 148 (1944).
2. J.C. Garland and D.J. Van Harlingen, Phys. Letters 47A, 423 (1974).
3. C.M. Pegrum, A.M. Guénault, and G.R. Pickett, Low Temperature Physics LT-14, edited by M. Krusius and M. Vuorio (North-Holland, Amsterdam, 1975), Vol. II, p. 513.

4. C.M. Falco, Solid State Commun. 19, 623 (1976).

5. D.J. Van Harlingen and J.C. Garland, to be published.

6. C.M. Pegrum and A.M. Guénault, Phys. Lett 59A, 393 (1976).

7. M. Tinkham, Phys. Rev. B 6, 1747 (1972).

8. J. Clarke, Phys. Rev. Letters 28, 1363 (1972).

9. M.L. Yu and J.E. Mercereau, Phys. Rev. B 12, 4909 (1975).

10. M.L. Yu and J.E. Mercereau, Phys. Rev. Letters 37, 1148 (1976).

11. S.B. Kaplan, C.C. Chi, D.N. Langenberg, J.J. Chang, S. Jafarey, and D.J. Scalapino, Phys. Rev. B 14, 4854 (1976) and Phys. Rev. B 15, 3567 (E) (1977).

12. J.Clarke and J.L. Paterson, J. Low Temp. Phys. 15, 491 (1974).

13. G.T. Pullan, Proc. Roy. Soc. London 217A, 280 (1953).

14. Yu. M. Gal'perin, V.L. Gurevich, and V.I. Kozub, Zh. Eksp. Teor. Fiz. 66, 1387 (1974) [Sov. Phys. - JETP 39, 680 (1974)].

THE USE OF TOROIDAL GEOMETRIES IN INVESTIGATIONS OF THERMOELECTRICITY IN SUPERCONDUCTORS

D. J. Van Harlingen and J. C. Garland

The Ohio State University

Columbus Ohio   43210   USA

ABSTRACT

An analysis is given of a toroidal sample configuration used to study thermoelectricity in superconducting metals. The toroidal geometry is shown to be highly effective in discriminating against extraneous magnetic flux while having a high sensitivity to the desired signal flux. Details of the flux transformation equations are given, and it is shown that flux amplification can take place if circuit parameters are properly chosen.

The thermoelectric flux effect in superconductors proposed recently by Garland and Van Harlingen[1] and independently by Galperin et al.[2] provides in principle a mechanism for studying quasiparticle relaxation and scattering processes in superconductors. Quasiparticles in a superconductor are driven by a temperature gradient much as the conduction electrons are in a normal metal. In the superconductor, however, charge conservation in an isolated sample is maintained not by the establishment of an electric field but by a counterflow of supercurrent. Since a supercurrent generates a phase gradient in a superconductor, the thermoelectric response of the quasiparticles may be made to manifest itself in a connected geometry as a result of the long-range phase coherence of the superconducting state. In particular, an unquantized magnetic flux Φ_T is generated in a bimetallic superconducting loop when the junctions are at different temperatures; the flux is produced by surface currents which flow on the inner walls of the ring (the net current in the bulk of each superconductor is zero). Theoretically, Φ_T is expected to diverge at the transition temperature of the superconducting component

with the lower T_c and to be within the flux resolution of Josephson
effect devices near T_c. Experimentally, Φ_T may either be measured
directly using a magnetometer or, if the total loop inductance is
known, by inserting a galvanometer in the loop and measuring the
surface currents responsible for Φ_T.

Previous experiments to attempt to observe the thermoelectric
flux effect in superconductors by Zavaritskii[3], Pegrum et al.[4], and
Falco[5] utilized the galvanometer technique in the configuration
shown in Fig. 1.

A wire of the superconductor to be studied forms one component
of the loop while the superconducting signal coil of a SQUID galvanometer constitutes the second component. The current I_T which
generates the thermoelectric flux when one junction is heated relative to the other is detected by the SQUID.

The loop galvanometer method is unfortunately subject to a
number of spurious signal sources. External magnetic fields couple
strongly into the loop circuit and produce circulating currents
which are also detected by the SQUID galvanometer. Such fields may
arise from electric currents in heaters and thermometers, circulating
surface currents in superconducting flux shields, and thermoelectric

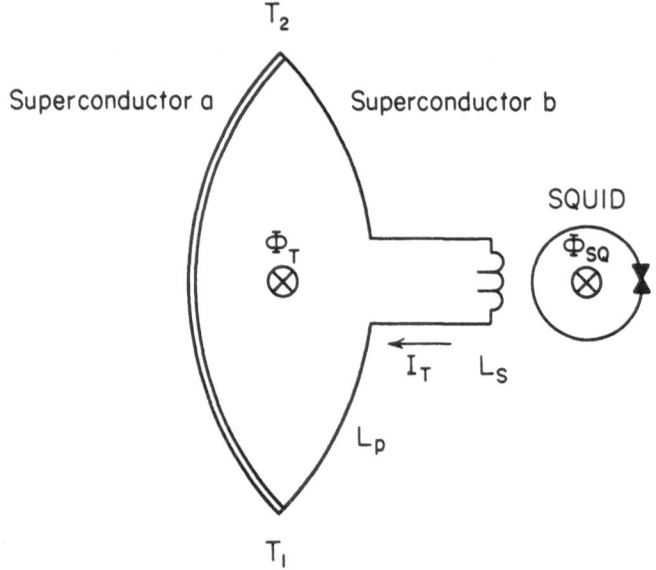

Fig. 1. In the loop galvanometer technique, the surface currents
I_T that generate the thermoelectric flux Φ_T are measured by a SQUID
galvanometer. The superconducting signal coil comprises one side
of the bimetallic loop.

currents in normal metal heater supports in the vicinity of the sample. A more serious problem is caused by penetration depth changes in the superconductors comprising the loop. As was pointed out by Pegrum and Guénault[6], a change in the penetration depth of the sample changes the total self-inductance of the superconducting loop. If the bimetallic ring contains any trapped magnetic flux, as will almost surely be the case in practice, the inductance change alters the magnitude of the circulating current in the circuit which is detected by the SQUID. Since the superconducting penetration depth diverges at the transition temperature, the signal from this mechanism is also expected to diverge and thus is difficult to distinguish from the true thermoelectric effect.

For our experiments, we have developed a technique for measuring the thermoelectric flux effect in superconductors which we believe provides both greater sensitivity to the thermoelectric flux and increased rejection of spurious signals. We have used a toroidal geometry that is a topological extension of the simple bimetallic loop. The toroid may be envisioned as a solid of revolution formed by rotating the double loop in Fig. 2(a) around the center superconductor; the solid generated in this way is shown in Fig. 2(b) -- one superconductor forms the center post of the toroid while the other makes up the outer surface. The thermoelectric flux Φ_T circulates within the toroidal cavity, generated by surface currents flowing on the inner walls of the cavity when one face of the toroid is heated with respect to the other.

The flux in the cavity is detected by a toroidal transformer coil wound around the sample through a hole in the center superconductor, and is coupled into a SQUID magnetometer by means of a superconducting flux transformer. This configuration is illustrated in Fig. 3. The thermoelectric flux couples strongly into the flux transformer, linking each turn of the primary winding. On the other hand, flux from extraneous magnetic field sources is greatly attenuated for two reasons. First of all, the toroidal geometry of the primary winding causes a first order cancellation of uniform external fields. Secondly, the toroidal sample forms a diamagnetic core for the primary winding which screens the transformer from external field signals --- this may be viewed as a reduction in the effective cross-sectional area into which flux can be coupled.

The toroidal magnetometer method is also less sensitive to temperature dependent changes of the superconducting penetration depth than is the loop galvanometer technique. Since the total magnetic flux in the toroidal cavity is measured, the method is completely impervious to penetration depth changes on the inner walls of the toroidal cavity. These changes may redistribute the flux in the cavity spatially and may alter the surface currents, but they cannot change the magnitude of the flux. Spurious signals may,

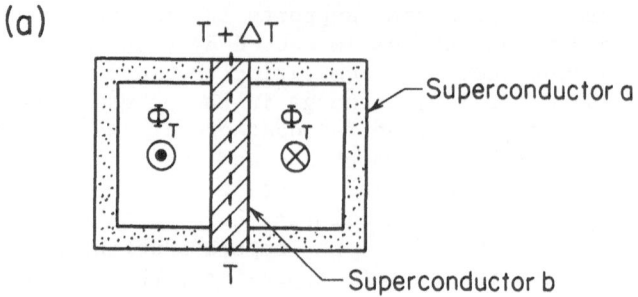

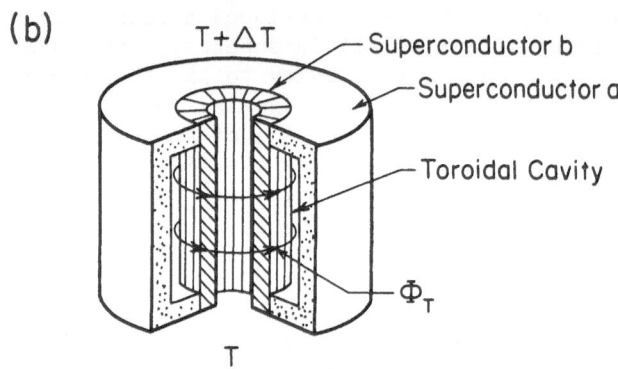

Fig. 2. (a) In the bimetallic double loop, the same flux (but oppositely directed) is generated in each side in the presence of a
temperature gradient. Rotation about the center superconductor
forms the toroidal sample shown in (b). The thermoelectric flux
circulates within the toroidal cavity when one face is heated relative to the other.

however, be produced by penetration depth changes on the outer surfaces of the sample which create a second-order change in the
screened or effective self-inductance of the flux transformer primary coil. As a practical matter, we have not found this source
of penetration depth signals to be particularly troublesome. In our
experimental configuration, the sample may be heated uniformly, i.e.
with no temperature gradient. This feature provides a means for
testing for spurious effects that depend on the absolute temperature of the sample rather than on a temperature gradient.

An additional advantage of the toroidal magnetometer method is
an increased flux transfer efficiency. This efficiency we may
characterize by a flux transfer factor f defined by

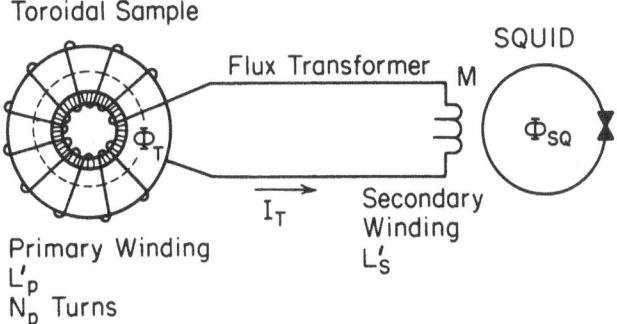

Fig. 3. In the toroidal magnetometer method, the thermoelectric
flux circulating in the toroidal sample cavity is coupled into the
SQUID via a superconducting flux transformer. M is the mutual
inductance of the secondary to the SQUID while L_p' and L_s' are the
screened self-inductances of the primary and secondary windings
respectively.

$$f = \delta\Phi_{SQ}/\delta\Phi_T \qquad\qquad (1)$$

so that f is the fraction of thermoelectric flux that is ultimately
coupled into the SQUID. The total flux in the flux transformer is
constant, composed of flux generated by circulating supercurrents
I as well as by flux picked-up from external sources (such as the
thermoelectric flux) so that

$$\Phi = \text{constant} = \Phi_{ext} + (L_p' + L_s')I, \qquad\qquad (2)$$

where L_p' and L_s' are the screened self-inductances of the primary
and secondary coils of the flux transformer respectively (cf. Fig.
3). The inductances are reduced from their unscreened values by
the diamagnetic supercurrents that flow on the surfaces of the
toroidal sample and SQUID body in response to the fields generated
by current circulating in the flux transformer. The external flux
produced in the transformer by the thermoelectric effect is $N_p\Phi_T$,
where N_p is the number of turns in the primary winding. If M is
the mutual inductance of the SQUID signal coil to the SQUID detector, then the flux transfer factor is given by

$$f = \frac{N_p M}{L_p' + L_s'} \qquad\qquad (3)$$

Since $L_p' \sim N_p^2$, f is optimized for $L_p' = L_s'$ in which case f = $(N_p/2)(M/L_s')$. The quantities M and L_s' are intrinsic properties of the SQUID sensor used, so that in practice f may only be enhanced by increasing N_p subject to the constraint $L_p' = L_s'$. Although L_s' is very small, typically 1μH, N_p can be made large ($\sim$ 50) by winding the turns tightly on the toroidal sample in order to keep the screened self-inductance per turn small.

By comparison, the flux transfer factor for the loop galvanometer technique is given by

$$f_{loop} = \frac{M}{L_p' + L_s'} \tag{4}$$

which has an optimum value of M/L_s' for $L_p' \ll L_s'$. Thus in principle the magnetometer method provides a flux transfer gain of $N_p/2$ relative to the galvanometer method; in practice, the enhancement is usually even larger because of the difficulty in obtaining $L_p' \ll L_s'$ in the simple loop geometry.

In the toroidal geometry configuration, it is important to place the superconductor with the lower transition temperature on the outside for several reasons: (1) so that magnetic flux which penetrates the sample in the normal state will be expelled to the outside during cool-down, (2) because penetration depth changes in the hole of the center superconductor have a greater effect on the screened inductance of the primary coil, and (3) to facilitate determination of the transition temperature of the sample. This latter may be accurately obtained to $\pm$ .2 mK by observing the sharp decrease in sensitivity to sample current when the outer surface of the toroid becomes superconducting. We should also note that the material used to form the toroidal cavity is not important since the total flux in the cavity is determined completely by the constraint on the phase of the superconducting wavefunction.

REFERENCES

1. J.C. Garland and D.J. Van Harlingen, Phys. Lett. 47A, 423 (1974).
2. Yu. M. Gal'perin, V.L. Gurevich, and V.I. Kozub, Zh. Eksp. Teor. Fiz. 66, 1387 (1974). [Sov. Phys. - JETP, 39, 680 (1974)].
3. N.V. Zavaritskii, ZhETF Pis. Red. 19, 205 (1974). [JETP Lett. 19, 126 (1974)].
4. C.M. Pegrum, A.M. Guénault, and G.R. Pickett, Low Temperature Physics - LT 14, edited by M. Krusius and M. Vuorio (North-Holland, Amsterdam, 1975), Vol. II, p. 513.
5. C.M. Falco, Solid State Commun. 19, 623 (1976).
6. C.M. Pegrum and A.M. Guénault, Physics Lett. A59, 393 (1976).

THERMOELECTRIC FLUX GENERATION IN SUPERCONDUCTING LOOPS

A.M. Guénault, C.M. Pegrum* and K.A. Webster

Physics Department
University of Lancaster
Lancaster, LA1 4YB, United Kingdom

ABSTRACT

Several pitfalls are discussed concerning experiments designed to detect thermoelectric generation of flux in a bimetallic superconducting loop. In particular, thermally-induced changes in the superconducting penetration depth can produce significant effects.

In a recent letter (Pegrum and Guénault[1]) we have discussed in some detail how thermally-induced changes in the superconducting penetration depth can produce magnetic flux changes in a SQUID which is coupled in to part of a superconducting loop. Certainly we have demonstrated experimentally that, in an attempt to measure the non-quantised magnetic flux generated thermoelectrically in a superconducting bimetallic loop, it is important to be able to change:
(1) the magnitude of the residual flux trapped in the loop, and
(2) the sign of the temperature gradient, before one can be at all confident that one is observing a true thermoelectric effect. We believe that these unwanted effects were present at least to some extent in our early measurements (Pegrum, Guénault and Pickett[2]) on Nb-Ta loops and also in our later work[1] on Pb-Sn loops, leading us only to be able to set an upper limit upon the thermoelectric effect.

---

* Present address: Department of Applied Physics, University of Strathclyde, Glasgow. G4 0NG, UK.

In our recent work, we have concentrated on measurements on
superconducting loops made entirely of Sn, so that the true thermoelectric effect should be absent. We have used an arrangement in
which a well-defined portion of the loop can be heated and in which
the trapped flux in the loop can be readily varied. The measurements, really outside the scope of this conference, confirm our
earlier calculations and observations concerning the importance of
the penetration depth effects; and in fact gives a reliable and
fairly straightforward method for determining the temperature dependence of the penetration-depth of a superconductor.

REFERENCES

1. C.M. Pegrum and A.M. Guénault, Phys. Lett. <u>59A</u>, 393 (1976).
2. C.M. Pegrum, A.M. Guénault and G.R. Pickett, Proceedings of
    LT14, Vol. 2, p 513 (1975).

DISCUSSION

C.M. Falco: I have a couple of questions. How do you do your
magnetic shielding? Do you shield the whole loop, the whole area?
What is the residual field, that is really the question.

A.M. Guénault: The field is first reduced over the whole experiment
with a mu-metal shield, to the value of 10^{-7} tesla, which is one
milligauss in U.S. units; then the experiment itself is inside a
very long lead can. The individual bits of it, so they don't talk
to each other, are in lead foil, and, in fact, in some of our
experiments there were three superconducting cans inside the experiment.

C.M. Falco: Do you actually measure the residual field?

A.M. Guénault: It's rather difficult to do inside the superconducting cans.

C.M. Falco: The point is that if you cool the lead can rapidly to
low temperature, you set up circulating currents, and you can
actually generate much larger fluxes and trap them.

A.M. Guénault: Certainly you can, and that's why a mu-metal shield
is so important here. You really want to compensate the field to
zero before you start fooling around with superconducting cans.

C.M. Falco: But even if it is zero to begin with, when you cool
the lead can rapidly it will generate its own thermoelectric currents
which then get trapped as it goes superconducting. And you can
thereby trap hundreds of milligauss.

A.M. Guénault: But I think you can estimate this, and from our
results on injected flux you can tell how much it is. The figure
of about one milligauss is about right. The superconducting shields
serve to keep that constant. They also serve to keep the noise
level down. I agree with you; I don't think they get rid of the
flux much.

C.M. Falco: Do you have an estimate of the inductance of the loop?

A.M. Guénault: 70 nanohenries, speaking from memory.

C.M. Falco: Is that calculated, or from the injected flux?

A.M. Guénault: Both; the one I have given you is the calculated
value.

J. Garland: I just want to make one brief comment to any unwary
persons in the audience who may be thinking about getting into this
area. As you can probably gather from the questions, it is very
easy to get data in this kind of experiment, but it is very difficult to know where the data are coming from. I would just like to
stress the importance of being extremely careful. One really cannot
be too careful in taking steps to minimize the sources of extraneous
flux signals. One must consider the shield partitions, the way the
wire leads are anchored, the normal metals in the vicinity of the
sample. These experiments look very simple when you draw them in
schematic fashion and show them on a slide, but as Guénault, Falco
and I can all attest, they are really extremely difficult.

LAMINAR INTERMEDIATE STATE IN SUPERCONDUCTING In AND Sn

R. H. Dee*, A. M. Guénault, and E. A. Walker

Physics Department
University of Lancaster
Lancaster, LA1 4YB, United Kingdom

ABSTRACT

We are making measurements of the transport properties in the laminar intermediate state of superconducting indium and tin using a SQUID picovoltmeter arrangement which incorporates a simple low-pass filter. Unexpectedly large thermoelectric effects are observed which are thought to arise from the N-S interface region.

We are currently measuring the transport properties (thermal and electrical resistance, and thermopower) in the intermediate state of type I superconductors. The geometry employed is that of a cylinder in a transverse magnetic field of magnitude $H_c/2 < H < H_c$, so that the intermediate state consists of a stack of normal (n) and superconducting (s) laminae of thicknesses a_n and a_s. Our latest measurements are made on oriented single crystal cylinders, with an electro-polished surface condition, 1-2 mm in diameter and about 60 mm long. The sample is mounted vertically between an electro-magnet on a rotating base, enabling a rather perfect intermediate state to be set up by slowly rotating the transverse field.[1] Temperature differences of about 30 mK are measured by carbon resistors; and a SQUID system using a low-pass filter and a superconducting switch is used to measure the small voltages generated. This system is described in detail elsewhere.[2] Simultaneous measurements are made of electrical resistance, thermoelectric power and thermal resistance.

---

*Present address: Physics Department, University of British Columbia, Vancouver, British Columbia.

There are two distinct motives for doing these measurements:

(i) At temperatures well below the transition temperature T_c
     $(T/T_c \lesssim 0.1)$ a large maximum in the thermal resistance is
     observed[3]. This anomalously high resistance is explained on
     the basis of a reduction in the effective phonon mean free
     path, λ_g, in the superconducting regions compared with the
     bulk value and is most important when $\lambda_{gs} > a_s > \lambda_{gn}$. The
     thermoelectric power in this region is of considerable
     interest[4] since the electronic component, S_e, should disappear
     rapidly below H_c as most of the temperature gradient will
     appear across the s laminae (a result of $W_s \gg W_n$). However,
     the phonon drag component of the thermoelectric power, S_g,
     should remain at its normal state value, since the phonon
     diffusion current generated in the S regions will predominantly
     scatter against electrons in the n regions. This should enable
     a direct measurement of S_g and hence differentiate between the
     two contributions to the normal state thermoelectric power.

(ii) Nearer T_c $(0.3 < T/T_c < 1)$ a large maximum in the thermal re-
     sistance is still observed[1], which cannot be explained as in
     (i). In this case $\lambda_{gs} \ll a_s$ due to scattering from electron
     quasiparticles in the superconducting regions. However, the
     high resistance has been explained by Andreev[5] in terms of the
     reflection of quasi-particles at the n-s interfaces which
     impedes the flow of heat but is consistent with the passage
     of an electric current between the two phases. At tempera-
     tures very near T_c this maximum in W is reduced as the Andreev
     reflection coefficient falls below unity due to a reduction
     in the energy gap in the superconductor. In this case the
     quasi-particles (under the influence of an electric field)
     enter the superconductor unimpeded and suffer scattering there
     leading to an anomalous increase in the electrical resistance.
     This excess resistance has been studied extensively by a
     number of authors[6].

The thermoelectric power in this temperature region has been
recently discussed by Dzhikaev[7] and by Artmenko and Volkov[8]. These
authors suggest that the thermoelectric power of a normal region
is unaffected by its superconducting neighbours, but that an addi-

tional thermoelectric field exists which decays exponentially with distance into the superconducting region. Dzhikaev[7] has calculated thermoelectric parameters for the intermediate state when the Andreev reflection coefficient is high (i.e. when T is not too near T_c). His result depends on the angular derivatives of the transition probability for passage of an excitation across an interface. It therefore depends on the geometrical arrangement of the interfaces. Artmenko and Volkov's analysis[8] is only valid at temperatures very near T_c since they assume a uniform temperature gradient along the sample. In both theories an additional thermo-electric power, S_A, is predicted, and in the geometry under con-sideration here the theories imply $S_A < S_n$.

So far, our results have been confined to region (ii) although we are extending the study to lower temperatures also. We have measured indium samples with RR up to about 16,000 and tin with RR in excess of 20,000. In these pure samples the condition $\lambda_e \gtrsim a_n$ is satisfied, and large additional thermal resistance is observed, similar to that seen by other authors[1].

However, even at temperatures of about 0.4 or 0.5 T_c, a very large additional thermoelectric power S_A is observed in these pure samples, equal to about $2S_n$. In contrast, the theory of Dzhikaev[7] mentioned above predicts $S_A \lesssim 0.05\ S_n$ at this temperature (since the Andreev reflection coefficient is so high). That S_A does arise from the interface region is evident from comparison with our results for a less pure sample of indium (RR 2600, for which $\lambda_e < a_n$), in which case the maxima in W and S are suppressed. The unexpectedly large magnitude of S_A is as yet unexplained, but it could indicate that there is an invalid assumption in the theory[7], namely that the conduction properties of the quasi-particles in the interface region are the same as those in the normal state. This would be quite an attractive explanation, particularly since the anomalously high results of the thermo-electric flux experiments (discussed earlier in this conference) seem to be pointing towards a similar questioning of the conduc-tion properties of quasi-particles in a superconductor.

Further investigations are continuing at Lancaster.

REFERENCES

1. A.J. Walton, Proc. Roy. Soc., A289, 377, (1965).
2. R.H. Dee, A.M. Guénault and G.R. Pickett, J. Phys. E9, 807,
    (1976).
3. e.g.K. Mendelssohn and J.L. Olsen, Phys. Rev., 80, 859, (1950).
4. A.M. Guénault, Phil. Mag., 9, 331, (1964).
5. A.F. Andreev, J.E.T.P., 19, 1228, (1964).
6. e.g. A.B. Pippard, J.G. Shepherd and D.A. Tindall, Proc. Roy.
    Soc., A324, 17, (1970).
7. Y.K. Dzhikaev, J.E.T.P., 41, 144, (1975).
8. S.N. Artmenko and A.F. Volkov, J.E.T.P., 43, 548, (1976).

DISCUSSION

K. Böning: I have two questions. The first is an experimental one.
You have been placing carbon resistors in a magnetic field. Is there
any magnetic field dependence of the calibration of these carbon
resistors? The second question concerns the interpretation of your
measurements for the low temperature case. You have been assuming
that the phonons penetrate from the normal region through the superconducting material into the normal again. Is there any possibility
that the phonons are scattered at the interface and can one experimentally check the assumption that they are not scattered?

A.M. Guénault: The first question about thermometers: The fields
we are using are really quite small, and these are ordinary Allen-
Bradley resistors. There is data on that, but I think the magnetoresistance of the thermometers is very small. We can check this;
but we are measuring differentially with two thermometers, so that
is really not a significant effect. However, the other question,
yes, that's a thing you would like to know. I think that everybody believes the phonons do go through the boundary region and are
scattered just within the normal region. It could be that we could
tell the difference between actual interfacial scattering and a
scattering within the normal region. But, I believe that they go
through and are scattered within the normal region, in which case
we should see the phonon drag. In the event of interfacial scattering I don't know what would happen.

B.R. Coles: You spoke of the phonon drag term being generated in
the superconducting lamellae and detected in the normal ones. I'm
not quite clear whether you mean that, in fact, the actual drag
effects between phonons and quasi-particle excitations are taking
place in the superconducting lamellae or whether the phonons generated in the superconducting lamellae are in fact producing the drag
effect in the normal lamellae. If it is the second, I don't think
it is true to say that the phonon drag is generated in the superconducting lamellae.

A.M. Guénault: I stand corrected with my English, Brian. What I meant was that the phonon diffusion current is generated within the superconducting region. It then goes into the normal region and there drags and thus produces the thermopower. Sorry, I didn't make that clear.

C. Uher: I would like to ask, Tony, would you expect the same effect if you had a sandwich of a superconductor and a normal metal. That means, the enchanced dragging in a normal metal, if it were part of a sandwich of a normal metal and superconductor.

A.M. Guénault: Well, it is difficult to set up, isn't it, because you have mismatches of all sorts there. You are only going to see this effect when you are sufficiently below T_c so that the superconducting regions are transparent to phonons. So, it is possible, but I think you introduce a lot more difficulties doing that sort of experiment. You introduce some with ours because you can't control the number of interfaces that you've got.

D. Greig: Referring back to the second part of the first question - Cr. Böning's questions. Presumably the size of the intermediate regions varies with magnetic field. Are there any conditions in the experiment when the dominant phonon wave length is either smaller than these regions or bigger than these regions, and does this condition vary throughout the range of the experiment?

A.M. Guénault: Well, the point I was just making really is that it is true that the scale of the intermediate state changes with field. But you have no control over that, which is a bit unfortunate.

D. Greig: Yes, I was just thinking that the answer to the question of whether phonons are reflected or not at the boundaries does depend to some extent, I think, on whether the phonons are well characterized within a region or whether the dominant wavelength covers regions of normal and superconducting lamellae.

A.M. Guénault: Yes, that is an interesting point. Still, I don't know that you have got that problem, have you, because you do have lattice continuity as Brian Coles was saying. I think, from that point of view, you'd expect the phonons to be like the normal bulk ones.

ON THE THERMOELECTRIC EFFECT IN A SUPERCONDUCTING RING

A. M. Gulian and G. F. Zharkov

P. N. Lebedev Physical Institute

Moscow

Thermoelectric effects in superconductors attracted much attention recently (see review paper[1]). In particular, when the temperature gradient in a superconducting bimetallic ring is present, the magnetic flux and thermoelectric current arise, which have inductive origin.[1] In Ref. 2 it was pointed out that an additional contribution to the magnetic flux arises if the mutual transformation of the superconducting and normal currents is taken into account. This additional effect is considered in more details in the present paper.

We shall use the concept[3] of inequilibrium chemical potential difference $\delta\mu$, which arise in a circuit when $\mathrm{div}\mathbf{j_s} \neq 0$ (see also ref. 1):

$$-\frac{\sigma_n}{e\ell_b^2}\,\delta\mu = \mathrm{div}\underline{\mathbf{j}}_s \;, \quad \underline{\mathbf{j}}_s = -A/c\Lambda. \tag{1}$$

Here σ_n is a normal excitation conductivity for a superconductor, ℓ_b is a characteristic length, connected with the relaxation time between the branches of quasiparticle spectrum for electron-like and hole-like excitations,[4,5] $\underline{A}$ is the vector potential, $\Lambda = m/e^2 n_s$, n_s is the number of superconducting electrons, which is a function of the temperature.

In the presence of a temperature gradient the normal current exists in a bimetallic ring (Fig. 1a)

$$\underline{\mathbf{j}}_n = \sigma_n(\alpha_n \nabla T + \nabla\delta\mu/e) \;, \tag{2}$$

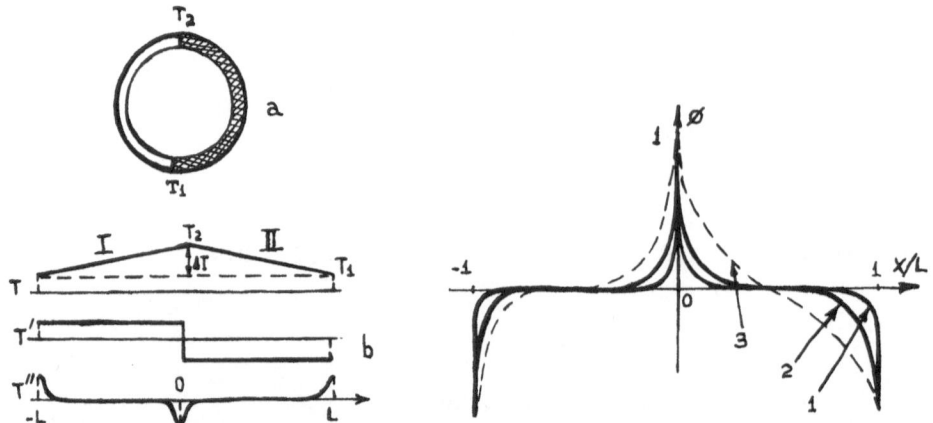

Fig. 1. a) A bimetallic ring. b) The temperature distribution in
a ring (the ring is unfolded in a segment $-L \leqslant x \leqslant L$).

Fig. 2. The solutions (6)-(8) of Eqs. (4), (5) (normalized to the
value $(\alpha_n e\Delta T/L) \ell_1\ell_2(\ell_1 + \ell_2)^{-1}$; $\ell_1 = 0.1$, $\ell_2 = 0.2$). 1) $L = 10$,
2) $L = 3$), 3) $L = 1$.

where α_n is the differential thermopower for a normal current. In
stationary conditions

$$\text{div}(\underline{j}_s + \underline{j}_n) = 0 \tag{3}$$

and from Eqs. (1-3) the equation for the off-equilibrium chemical
potential difference $\delta\mu \equiv \emptyset(x)$ follows (we take σ_n = const):

$$\emptyset'' - \frac{1}{\ell_b^2(x)}\emptyset = \beta T''(x) , \tag{4}$$

where $\ell_b(x) = \ell_1\theta(-x) + \ell_2\theta(x)$, $\theta(x) = 0$ when $x < 0$, $\theta(x) = 1$ when
$x > 0$, $\beta = -\alpha_n e$, $\ell_{1,2}$ are the characteristic lengths for the superconductors 1 and 2. By using the kinetic equation[6] it was shown
in Ref. 2 that for a pure superconductor with an energy gap the
length ℓ_b near T_c can be very large, reaching $\ell_b \sim 1$ mm (compare[3,5]).
Accepting the temperature distribution in the ring as shown in
Fig. 1b (the ring is unfolded into a segment $-L \leq x \leq L$) we have

$$T''(x) + \frac{\Delta T}{L} [\delta(x + L) + \delta(x-L) - 2\delta(x)] . \qquad (5)$$

The analytic solution of Eqs. (4), (5), which satisfies the continuity condition for a potential $\phi'(x)$ at points $x = 0$ and $x = \pm L$, is

$$\phi(x) = \alpha_n e \frac{\Delta T}{L} \left\{ A(\ell_1, \ell_2)\theta(-x) + A(\ell_2, \ell_1)\theta(x) \right\} \, \text{sh} \, \frac{L-2|x|}{2\ell_b(x)} , \qquad (6)$$

$$A(\ell_1, \ell_2) = 8\text{sh} \frac{L}{2\ell_2} [\ell_1 \text{sh} \frac{L}{2\ell_2} \, \text{ch} \frac{L}{2\ell_1} + \ell_2 \, \text{sh} \frac{L}{2\ell_1} \, \text{ch} \frac{L}{2\ell_2}] \, x$$

$$[(\frac{\ell_1}{\ell_2} + \frac{\ell_2}{\ell_1}) \, \text{sh} \frac{L}{\ell_1} \, \text{sh} \frac{L}{\ell_2} + 2 \, (\text{ch} \frac{L}{\ell_1} \, \text{ch} \frac{L}{\ell_2} - 1)]^{-1} . \qquad (7)$$

If the lengths are equal $\ell_1 = \ell_2 = \ell_b$, the solution simplifies:

$$\phi(x) = \frac{\alpha_n e \Delta T}{L} \, \ell_b \left[\text{ch} \frac{L-|x|}{\ell_b} - \text{ch} \frac{x}{\ell_b} \right] \Big/ \text{sh} \frac{L}{\ell_b} \qquad (8)$$

The curves in Fig. 2 represent the solutions (6), (8) for several values of ℓ_1 and ℓ_2. It is seen, that off-equilibrium contribution $\delta\mu = \phi(x)$ arises in the points, where the heat is flowing in or out of the specimen; $\phi(x)$ falls exponentially over the characteristic length ℓ_b of the corresponding material when we go away from these points.

As the total current density in the bulk of a superconductor vanish (Meissner effect), then integrating equation $\underline{j} = \underline{j}_s + \underline{j}_n = 0$ around this circuit and using (1), (2), we get

$$\Phi = \oint \underline{A} d\underline{\ell} = \Phi_T + \delta\Phi$$

$$\Phi_T = \oint c\Lambda\sigma_n \alpha_n \nabla T d\underline{\ell}, \quad \delta\Phi = \int c\Lambda\sigma_n \nabla\delta\mu d\underline{\ell} \qquad (9)$$

Here Φ_T is the usual thermoelectric flux in a superconducting ring due to a temperature gradient,[1,7,8] and $\delta\Phi$ is an additional flux due to the off-equilibrium contribution $\delta\mu$.

The integrands of Eq. (9), which are proportional to currents $\underline{j}_T = \sigma_n\alpha_n\nabla T$ and $\delta j = \sigma_n\nabla\delta\mu$, are shown in Fig. 3 (normalized to the quantity $c\Lambda_2 j_{T2}$). The difference of the shaded areas is equal to the usual thermoelectric flux Φ_T, the difference of the unshaded areas under the solid curves is equal to the additional flux $\delta\Phi$

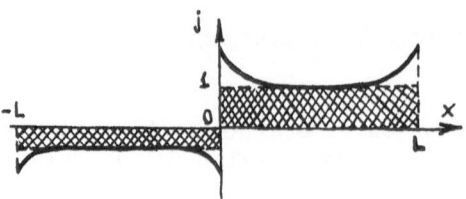

Fig. 3. The distribution of the normal currents in a ring with
L = 1, ℓ_b = 0.1 (normalized to $c\Lambda_2 j_{T2}$). The broken lines are proportional to the thermoelectric currents j_T without the off-
equilibrium contribution, the solid lines are propertional to the
total currents $j_T + \delta j$.

which is due to the additional current δj. As can be seen from
Fig. 3 and Eqs. (6-8) the additional flux is smaller than the main

flux Φ_T, roughly speaking, by the factor $\ell_b/L(\frac{2}{j\ell})$. If $\ell_2 \sim 0.1$ cm

L $\sim$ 1 cm, we have $\delta\Phi/\Phi_T \sim 5 - 10\%$. However, the fluxes Φ_T and $\delta\Phi$
depend differently on the parameters of the superconductors and in
principle there is a possibility to separate both components from
the experimentally observed values. In particular, if the ring
materials are such, that $\sigma_n \alpha_n \Lambda$ = const. but $\sigma_n\Lambda \neq$ const. then
Φ_T = 0 (because the integrand of Φ_T in (9) is now a full differential), while $\delta\Phi \neq 0$; in this case all thermoelectric magnetic flux
in the ring is produced by the off-equilibrium chemical potential
difference $\delta\mu$.

If the dependence σ_n on temperature in Eq. (2) is taken into
account, then new small contributions to a thermoelectric flux arise.
Indeed, as $\sigma_n = \sigma_n(T) = e^2\tau m^{-1}n_n(T)$, while $T = T(x)$, then in the
left side of Eq. (4) a new additional term appears and this equation
takes the form

$$\frac{d^2\phi}{dx^2} + \frac{1}{n_n}\frac{dn_n}{dx}\frac{d\phi}{dx} - \frac{1}{\ell_b^2}\phi = \frac{d^2T}{dx^2} \tag{10}$$

For estimating the magnitude of $1/\ell(x) = n_n^{-1}|dn_n/dx|$ we use the
known empiric relation[9] $n_s = n[1 - (T/T_c)^4]$ and the condition
$n = n_s + n_n$, then $n_n = n(T/T_c).^4$ In this case

$$\frac{1}{\ell} = \frac{4}{T}\left|\frac{dT}{dx}\right| = \frac{4}{T}\frac{\Delta T}{L} \quad .$$

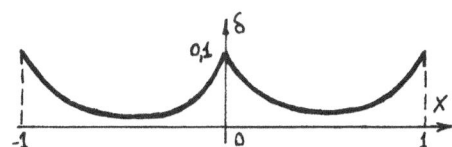

Fig. 4. The correction $\delta = (\Phi - \Phi^*)1\Phi$, which is due to the temperature dependence $n_n(T)$. Here Φ is the solution of Eq. (4), Φ^* is the solution of Eq. (10). L = 1, $\ell_1 = \ell_2 = 0.1$, $\Delta T/T = 0.9$.

Thus, when $\Delta T \ll T$, the inequalities hold: $\ell \gg L \gg \ell_b$ and the term $\sim \ell^{-1}d\phi/dx$ in (10) leads only to small corrections. These corrections increase, however, when $\Delta T \sim T$. The analytic solution of Eq. (10) is possible, but it is rather cumbersome and shall not be reproduced here. Fig. 4 shows the difference of the solutions of Eqs. (4) and (10) when L = 1, $\ell_1 = \ell_2 = 0.1$ cm, $\Delta T/T = 0.9$. It is clear that these corrections are small.

REFERENCES

1. V.L. Ginzburg and G.F. Zharkov, Preprint FIAN, N 115, Moscow (1977).
2. S.N. Artemenko and A.F. Vokov, Zh.E.T.P., 70, 1051 (1976).
3. T.J. Rieger, D.J. Scalapino, J.E. Mercereau, Phys. Rev. Lett. 27, 1787 (1971).
4. M. Tinkham and J. Clarke, Phys. Rev. Lett. 28, 1366 (1972).
5. M. Tinkham, Phys. Rev. B6, 1747 (1972).
6. A.G. Aronov, V.L. Gurevich, Fis. Tverd. Tela, 16, 2656 (1974).
7. Yu. M. Galperin, V.L. Gurevich and V.N. Kozub, Pisma Zh.E.T.P. 10, 687 (1973); Ah.E.T.P. 66, 1387 (1974).
8. J.C. Garland and D.J. Van Harlingen, Phys. Lett. 47A, 423 (1974)
9. P.G. DeGennes, "Superconductivity of Metals and Alloys" (Benjamen, New York, 1966).

THE THERMOELECTRIC POWER OF LIQUID METALS

J. E. Enderby

H. H. Wills Physics Laboratory

Tyndall Avenue, Bristol BS8 1TL

ABSTRACT

In this review two types of systems where measurements of
thermoelectric power, S, have proved to be particularly rewarding,
will be considered.

(a) Pure Liquid Metals and Alloys

Within the nearly free electron theory two contributions to S
can be identified. The first of these arises from the product of
the structure factor and the scattering (pseudopotential) evaluated
at the fermi energy. The second is significant when there is marked
energy dependence of the pseudopotential. Carefully planned experiments on both pure and alloy systems are able to yield direct information on the relative contributions of these effects. It is
essential to invoke energy dependent pseudopotentials for the
divalent metals. Surprisingly, S for the noble metals appear to be
well understood in terms of energy independent pseudopotentials.

(b) Liquid Semiconductors

Although a range of alloys which show semiconducting behaviour
over part of the composition range are known, the 'metal + metal'
systems like liquid Mg-Bi and liquid Li-Bi are of particular interest
to the theorist. Recent results for the thermoelectric power, for
liquid lithium alloys of the type Li_xM_{1-x} where $1>x>0$ and M represents a metal whose valence, Z, is in the range 2 to 5 will be
described. The results for these experiments, together with other

evidence of a chemical, thermodynamic and structural nature are
discussed within a unified framework in which the notion of impurity
ionicity is central.

1.   The Thermoelectric Power of Pure Liquid Metals

1.1 Simple Liquid Metals. The theoretical and experimental
situation for simple liquid metals (i.e. those in which d- or f-like
resonance behaviour is well removed from the Fermi energy E_F) has
not materially changed since the review article by Cusack (1) was
written a decade ago. Tentative generalisations made at that time
have, with the passage of time, become generally accepted. The
nearly free electron picture of liquid metals as evidenced by Hall
coefficient, electrical resistivity and thermoelectric power is now
well established and for an up-to-date review of the present situation the reader is referred to the book by Faber (2). The Ziman (3)
formula for the electrical resistivity, involving as it does the
pseudopotential and the liquid structure factor, has received considerable experimental support in recent years. It remains generally true that a detailed comparison between theory and experiment
is often not possible because the input parameters (i.e. the structure factor and the pseudopotential) are insufficiently well defined.

1.2. Liquid Alkaline Earth Metals. The group IIa elements
(Mg, Ca, Sr and Ba) are important because they represent, in the
condensed states, metals which are precursors of the transition
series. The effect of the empty d-states above E_F will therefore
begin to affect the electrical properties and to this end a study
of the resistivity and thermoelectric power of liquid alkaline earth
metals has now been completed (Van Zytveld, Enderby and Collings
(4)). The resistivity measurements were interpreted in terms of
resonance scattering from the vacant d-states. The thermoelectric
power results indicated the increasing role of the energy dependence
of the pseudopotential as one proceeds from Mg to Ba. Calculations
by Ratti and Evans (5) in which the effect of the d-bands is incor-
porated into an energy dependent pseudopotential confirm this trend.

1.3. Liquid Transition Metals. A review of thermoelectricity
in these interesting systems will be given at the Conference on the
Physics of Transition Metals, Toronto (6).

2.   The Electrical Properties of Liquid Alloys

In 1964 Faber and Ziman (7) showed that if scattering of
electrons in a binary alloy is treated in Born approximation, the
resistivity and thermoelectric power can again be evaluated in terms
of pseudopotentials and structure factors.

Let us define a generalised form factor $F(q)$ given by:

$$F(q) = |w_1|^2 [c(1 - c) + c^2 a_{11}] +$$
$$+ |w_2|^2 [c(1 - c) + (1 - c)^2 a_{22}]$$
$$+ 2w_1 w_2 c(1 - c) (a_{12} - 1).$$

Here the pseudopotentials of the two components are written as w_1 and w_2, while c is the atomic fraction of component 1; a_{11}, a_{22} and a_{12} represent the three partial structure factors the definition and determination of which has been discussed in ref (2). It then follows that the resistivity ρ of a binary liquid alloy can be expressed as

$$\rho = \frac{3\pi^2}{e^2 \hbar v_F^2 \Omega} \ < F(q) > \qquad \qquad \cdots\cdots \qquad (1)$$

where

$$< F > \ = \ \int_0^{2k_F} F(q) q^3 dq$$

v_F and k_F are respectively the Fermi velocity and Fermi wave number and Ω is the atomic volume.

The thermoelectric power, S, can readily be derived from (1) using the Mott formula (ref 2).

$$S = - \frac{\pi^2 k_B^2 T}{3 |e| E_F} \quad x$$

where

$$x = - E_F \left[\frac{d\ell n \ \rho}{d \ E} \right]_{E_F}$$

We now consider a hypothetical liquid binary alloy whose structure accords with the substantial model of Faber and Ziman(7) and for which the Fermi wave number, k_F, and atomic volume, Ω, are independent of composition. Under these circumstances, as was demonstrated by Howe and Enderby (8) the Ziman theory of electron transport in liquid metals and alloys yields a simple relationship between the S for the alloy and that of the pure components provided the energy dependence of the pseudopotential can be neglected.

This relationship is independent of the detailed form of w(q) or of the partial structure factors and provides a useful indirect check on the validity or otherwise of the basic theory; it avoids the difficulties associated with a direct comparison between the calculated and the experimental electron transport parameters, referred to above.

There are five equivalence alloys which approximate closely to the system postulated above. The first of these to be studied was the monovalent system, liquid Ag-Au(8) and the measure of agreement secured between theory and experimenta was very satisfactory. Subsequently four other systems (Mg-Cd, In-Tl, Pb-Sn, and Bi-Sb) were studied(9) so covering the whole range of metallic valencies.

For Ag-Au, Pb-Sn, and Bi-Sb, there were no significant disagreements between theory and experiment. By contrast, the composition dependence of S for Mg-Cd (and to a lesser extent In-Tl) was found to be very different from that expected on theoretical grounds. Small departures from the substitutional model and the condition of constant k_F could be allowed for with the aid of Percus-Yevick structure factors and the Animalu-Heine model potentials did. However, such corrections did not alter the theoretical curve by more than a few percent. In particular, the discrepancies for liquid Mg-Cd were well outside the experimental errors and the uncertainties introduced by departures from ideal behaviors.

Enderby et al(9) concluded that the energy dependence of w must be included for the divalent metals. Indeed, the quantitave calculations which followed this empirical approach (see, for example Evans et al(10)) are able to generate quite respectable theoretical values for the thermoelectric power.

A thoroughgoing comparison of theory and experiment is not in general possible because information about the partial structure factors $a_{\alpha\beta}$ is limited. However, for one case, (liquid Cu-Sn alloy) $a_{\alpha\beta}$ has been determined directly by combining the technique of neutron diffraction with isotopic substitution. The predicted thermoelectric power for this system - again employing energy independent pseudopotentials - is in excellent agreement with experiment(11).

3. Liquid Semiconductors

There exists a second group of liquid alloys whose electronic properties are different from those of metallic (i.e. Faber-Ziman) systems and are rather similar to those of semiconductors. Such liquids are usually referred to as 'liquid semiconductors'. Two main types of liquid semiconductors have been identified, namely,

those based on the chalcogenides (for example liquid Cu_2Te) and
those in which metals of vastly different electronegativities are
alloyed together. These latter systems are of considerable interest
because they enable the experimenter to follow continually the
transition from metallic behaviour to semiconducting behaviour.
Liquid Mg-Bi and Mg-Sb represent two of the alloys which are known
to fall into this group. Experimental results for ρ and S have
been reported for liquid Mg-Bi by Ilschner and Wagner(12) and by
Enderby and Collings(13). At the composition Mg_3Bi_2, the maximum
value in ρ is associated with a sign change in S and a negative
$d\rho/dT$. It is clear from this evidence and from the behaviour of
the excess free energy that a major change in the bonding characteristics takes place as we proceed from pure liquid Bi or Mg to the
liquid alloy Mg_3Bi_2.

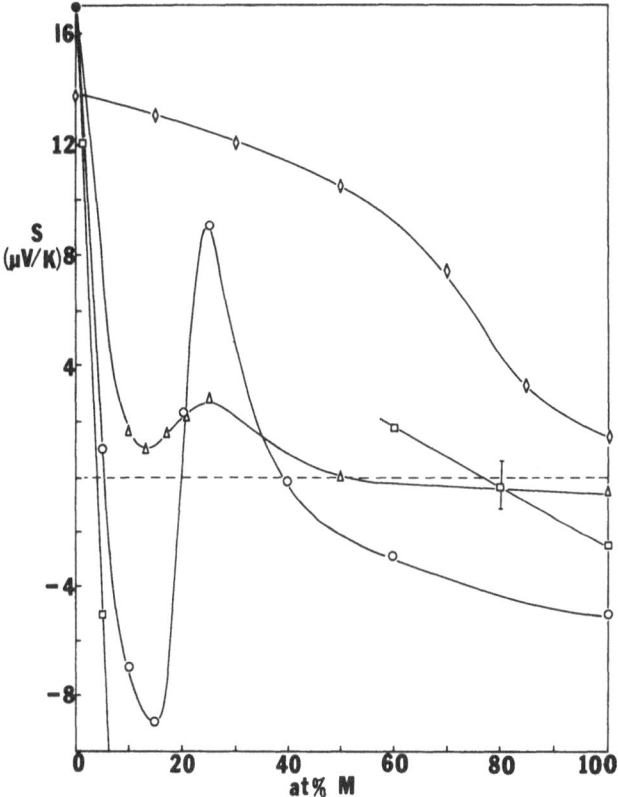

Fig. 1. Absolute thermoelectric power of liquid Li alloys as a
function of concentration.
   ◊ Li-Mg (Van Zytveld[16])
   △ Li-Tl at 800°C
   O Li-Pb at 800°C
   ☐ Li-Bi at 600°C

Some recent experiments(14) on alkali metal based liquid alloys have been completed in our laboratory. The aim was to show how the alloys became progressively less free electron like as the valence difference between the two components was increased. In Fig. 1 we show how S changes with composition for Li_xM_{1-x} alloys where M is Mg, Tl, Pb and Bi. The 'n-p transition' which occurs at Li_4Pb corresponds to a maximum in ρ.

Our basic understanding of these liquids is still in a very primitive stage, the development of a comprehensive theoretical treatment in which the various electronic and thermodynamic properties of liquid semiconductors can be discussed, has been hampered by the lack of information about the nature of the atomic order which exists in alloys of this type. Nguyen and Enderby(14) have argued that electron transfer occurs even at the dilute end provided the electronegativity difference between the solvent and the solute is sufficiently great. Thus liquid Li_4Pb can be thought of as essentially ionic. This certainly could lead to a change in sign in the derivative of the density of states at E_F and hence, in principle, a possible explanation of the n-p transition(15). How in detail, however, one should think about the transition from metallic conductivity (which does not involve, in conventional theory, the density of states) to a regime in which the density of states parameter becomes significant, remains an unsolved and outstanding problem.

REFERENCES

1. N. Cusack, Rep. Prog. Phys. <u>26</u> 361, (1963).
2. T.E. Faber, Introduction to the Theory of Liquid Metals (Cambridge University Press) 1972.
3. J.M. Ziman, Phil. Mag. <u>6</u> 1013 (1961).
4. J.B. Van Zytveld, J.E. Enderby and E.W. Collings, J. Phys. F 2 73 (1972).
5. V.K. Ratti and R.J. Evans, Phys. F 3 L238 (1973).
6. J.E. Enderby, Experimental Studies of Liquid Transition Metals (presented at Physics of Transition Metals, Toronto 1977).
7. T.E. Faber and J.M. Ziman, Phil. Mag. <u>11</u> 153 (1964).
8. R.A. Howe and J.E. Enderby, Phil. Mag. <u>16</u>, 467 (1967).
9. J.E. Enderby, J.B. Van Zytveld, R.A. Howe and A.J. Mian, Phys. Letts. <u>28A</u> 144 (1968).
10. R. Evans, D.A. Greenwood, P. Lloyd and J.M. Ziman, Phys. Letts <u>30A</u> 313 (1969).
11. J.E. Enderby and R.A. Howe, Phil. Mag. <u>18</u> 923 (1968).
12. B.R. Ilschner and C.N. Wagner, Acta. Met. <u>6</u> 712 (1965).
13. J.E. Enderby and E.W. Collings, J. Non-Cryst. Solids <u>4</u> 161 (1970).
14. V.T. Nguygen and J.E. Enderby, Phil. Mag. <u>35</u> 1013 (1977).
15. J.E. Enderby, Band Structure Spectroscopy of Metals and Alloys Academic Press, London, 1973).
16. J.B. Van Zytveld, J. Phys. <u>F5</u> 506 (1975).

DISCUSSION

W.B. Pearson: If you look, say with neutron diffraction, would the geometrical factor of say Li_4Pb show cluster arrangements in the liquid? Because certainly with Mg_2Si in, say liquid aluminum you do have physical clusters of the molecules.

J.E. Enderby: Ruppersberg has looked at liquid lithium-lead by neutron diffraction and we have also looked at systems that we know are molecular like titanium tetrachloride. The idea is that such measurements give a benchmark; there is no question that liquid Li_4Pb is ionic rather than molecular. So I think the picture that we have is that the electrons are localized on the lead. Lithium ions and lead ions then put themselves on alternate shells. Now, of course, a note of caution; that's a ground state model and to get the conductivity, as you well know, you have to know the excitations, and that is a problem that has been worked on. But I think the answer to your specific question is that the neutron diffraction studies on liquid lithium-lead are consistent with the ionic picture of that particular alloy.

K. Böning: I would like to go back to the beginning of your talk. You were mentioning this very basic model, and I'm wondering if this model is really so bad. It involves three assumptions. One was the one-electron approximation which is agreed upon, I think, in most of solid state physics. Another one was a spherical Fermi surface, but are there any indications that the Fermi surface is not spherical in this system which has, at least macroscopically, spherical symmetry? And, finally, the relaxation time approximation. The relaxation time could be defined exactly; yours was only energy dependent, that is to say, isotropic. I cannot imagine why it should not be isotropic in a system which is isotropic intrinsically. It could, of course, be energy dependent as you were saying. So my question is, what are the improvements of this basic model you had in mind?

J.E. Enderby: I had not intended to make a judgement about this. I was really stating where I intended to base my talk. But, of course, had I gone on to the transition metal problem, some of those basic assumptions with which one treats the data are very important. I mean, for example, is it right to assume that it is just the s-p-electrons that carry the current? Do you have to worry about the fact that at high temperatures the distribution might not be completely degenerate? And then, of course, if you have clustering or when you cannot decouple properly the electron motion from the ion motion, then that does call into question the whole business about whether the Fermi surface is properly spherical or if it is fuzzy. I am sorry if I gave the impression that I was making a judgement. I was really stating where I was going to begin my talk.

B.R. Coles: One point that arises from that is that yesterday we
heard that there are serious doubts about the applicability of the
Boltzmann equation approach to some metals even in the solid state
at high temperature as far as the resistivity is concerned. If
that were seriously to be the case I would have thought the thermoelectric power would be much more sensitive and that we would cease
to get your points for theory and experiment lying on the same side.
So it does suggest that perhaps there is some rescue operation that
can be mounted for the Boltzmann equation even in the solid state.
The other point that I think is worth emphasizing is that your
invocation of a k-dependence of the pseudopotential is rather a
phenomenological one, and that the actual effects you were seeking
to explain by a finite value of r are rather different in, say,
strontium where the approach of these states to the Fermi surface
is doing something, and in an alloy system like cadmium-magnesium
where the electronegativity difference is playing a bigger role.
I don't think saying, "ah, yes, we can explain this by a k-dependence
of the pseudopotential" is giving any more than a phenomenological
answer.

J.E. Enderby: I don't think electronegativity difference matters
much in liquid cadmium-magnesium; it is a very well behaved solid
solution. Electronegativity difference is surely not sufficient
to localize electrons in the sense in which it is happening in
liquid lithium-lead. The concentration dependence of the conductivity is very smooth, the structure factors are consistent with
random packing of spheres and so on. I can't quite agree with that
comment. I think that in liquid magnesium-cadmium there is fairly
clear evidence that you do have to put in a k-dependence. There
is gross disagreement between the simple theory and the experiments.
If electronegativity factors were coming in I think they would show
up in the thermodynamics or the structure, of the resistivity, and
that is not the case.

THERMOELECTRICITY IN LIQUID METALS: A REVIEW OF EXPERIMENTAL

METHODS

J. B. Van Zytveld

Physics Department
Calvin College
Grand Rapids, Michigan USA

ABSTRACT

This paper reviews various methods that have been used to date to measure the thermopowers of liquid metals, commenting on the relative advantages and disadvantages of each. Those experimental factors which are rather unique to liquid metal systems are highlighted. The methods presented span the range of temperatures from room temperature to about 1500°C.

INTRODUCTION

Measurements of thermoelectric power in liquid metals involve a number of experimental factors not generally encountered in making similar measurements on solid metals. In the present paper, we highlight especially these factors and the methods that have been used to address them. In particular, we look at: (1) containment and sample reactivity; (2) counter-electrode materials and thermopower standards; (3) contact configurations; (4) large- and small-ΔT methods; and (5) thermopower measurements in a conducting sheath.

CONTAINMENT AND SAMPLE REACTIVITY

Liquid metals tend, in general, to be fairly reactive, making the problems on containment and the maintenance of sample purity important ones. The most straightforward approach to the measurement of thermopower, S, involves holding the liquid in a nonconducting, inert container. (The problem of the conducting container will

be addressed in a separate section.) The following are a representative selection of the ceramic materials that have been used successfully to contain liquid metals: pyrex (Hg, Ga, Na, K, Rb, Cs);
quartz (Al, In, Tl, Sn, Pb, Bi, Zn, Cd, Cu, Ag, Au); high-density
alumina (Mg, Ca, Sr, Ba, most transition metals); beryllia (Li).
In the above listing, a given container material may generally be
replaced by a subsequent one (ie: quartz for pyrex, beryllia for
alumina, etc.) if a more inert material is desired, or if measurements extend beyond the useful temperature range of the material
indicated. Other ceramic materials have been used to contain liquid
metals especially at high temperatures, but these usually suffer
from one or another disadvantage (eg: Zirconia and Silicon Carbide
become conductors at high temperatures; and magnesia has a rather
high vapor pressure at elevated temperatures).

COUNTER-ELECTRODE MATERIALS AND THERMOPOWER STANDARDS

Thermopower standards at high temperature are ultimately based
on measurements of the Thomson Coefficient via the Thomson Effect,
as they are over much of the low-temperature range as well. The
thermopower of. Cu has been used above room temperature as a primary
standard, based upon the Thomson Effect measurements of Nyström[1]
and Lander[2], and on extrapolations of S(Cu), using Pb as a standard
by Gold et al[3]. These suggest that

$$S(Cu) = 0.05 + (5.45 \times 10^{-3}) \ T(°K) \ \mu V/deg$$

from room temperature to about 1000°C. Here the uncertainty may
approach ±0.3 μV/deg near the high-temperature end of the range, but
is perhaps ±0.1 μV/deg at room temperature. For the range above
1000°C, Cusack and Kendall[4] provide estimates of S(W) and S(Mo) to
2100°C from literature data; they suggest that an overall uncertainty
of about 5% is appropriate for these metals, the uncertainty being
largest near the high-temperature limit.

Recently a number of measurements of S have been reported
against thermal chromel, the authors using one leg of the chromel-
alumel thermocouple at both the hot and cold junctions as a standard
material. This secondary standard has been calibrated in our laboratory against pure Cu, giving

$$S(chromel) = 21.43 + .02406 \ T - 6.739 \times 10^{-5} \ T^2 + 3.166 \times 10^{-8} \ T^3$$

μV/deg, where T is in °C, and the temperature range is from ±0.2
μV/deg at 0°C to ±0.4 μV/deg at 900°C, including the uncertainty
in S (Cu). The above calibration is in agreement with calibration
of Cook and Laubitz[5] to well within the stated uncertainty.

CONTACT CONFIGURATIONS

In general, two types of contact configurations have been
utilized: contact probes for immersion into a pool of liquid sample;
and leads which can be attached to the exterior of the ceramic sample container making contact to the liquid sample via holes drilled
through the wall of the container. The latter method is illustrated
in Fig. 1. Here thin (~.003") W or Mo foils are drawn over holes
in the quartz or alumina container, protecting the thermocouple and
counterelectrode wires which are attached to the foil immediately
over the holes. Good seals are generally possible for quartz tubes;
the foils bound to alumina tubes often require additional sealing
material (eg. powdered graphite) under the foil.

Several types of immersion probes are illustrated in Fig. 2.
Fig. 2(a) shows schematically a probe designed for use at low temperatures with fairly reactive liquids. Its advantages are that
of a positive seal of the W foil to the pyrex sheath, and the use
of a Cu primary standard counter-electrode. This configuration
has been used extensively by N. E. Cusack and his group[6]. Fig. 2(b)
illustrates a probe assembly used in our laboratory for a wide
variety of liquids (Li, Mg, Ca, Sr, Ba) in the temperature range
from 0°C to about 900°C. The sheath is of thin-wall stabilized
stainless steel (either 321 SS or 347 SS); the chromel thermocouple

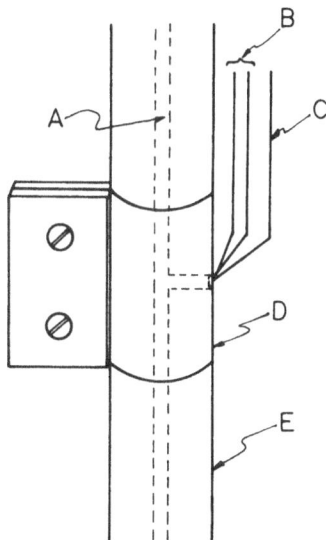

Fig. 1.   Thermopower external-contact-configuration.
   A - liquid sample column; B- thermocouple; C - copper counter-
   electrode; D - W or Mo foil; E - ceramic (quartz, alumina) tube.

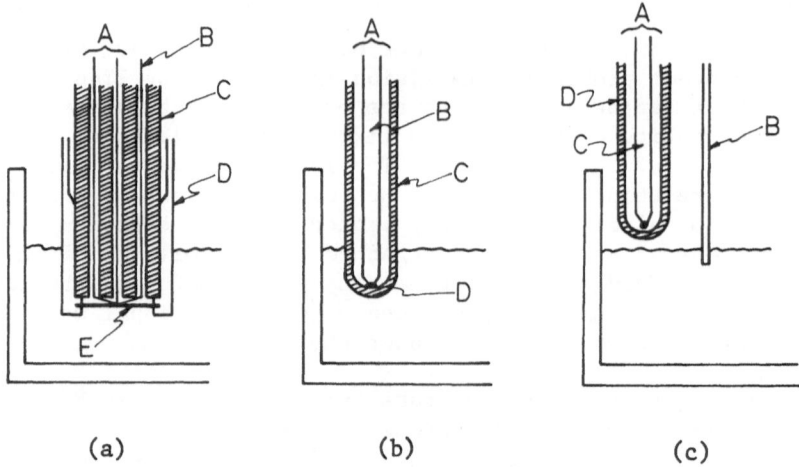

Fig. 2. Thermopower immersion probes
    (a) Low-temperature configuration. A - thermocouple; B -
copper counterelectrode; C - multibore ceramic tube; D - pyrex
sheath; E - W foil. (Illustration after Cusack et al, reference 6).
    (b) Intermediate-temperature configuration. A - chromel-
Alumel thermocouple; B - MgO insulation; C - 321 SS or 347 SS tube,
1/16" O.D.; D - welded tube-thermocouple junction.
    (c) High-temperature configuration (esp. for transition
metals). A - Pt-PtRh thermocouple; B - W wire; C - Alumina twin-
bore insulators; D - Alumina tube.

lead is also used as a counterelectrode.

    Since most liquid transition metals readily attack any solid
metal placed into contact with them, special precautions must be
taken in making measurements of S on these liquids. Fig. 2(c)
illustrates an immersion probe which has been used successfully in
this application. The thermocouple is protected from the sample by
a ceramic (alumina) sheath; the W counterelectrode is dipped into
the sample for only a very short time in order to minimize the introduction of W impurities into the sample by its attack on the electrode. It is possible by this means to keep the accumulated impurity
level to less than about 0.4 at. %W over the course of a given
experiment. This dipping method, pioneered by J. E. Enderby and his
group[7], is the only one to permit the successful measurement of S
for liquid transition metals to date.

LARGE- AND SMALL-ΔT METHODS

The standard method for measurement of thermopower involves the use of rather large temperature differences, ΔT. In this method, one holds one junction at constant temperature, T_1, and varies the temperature, T_2, of the other junction. One then records the thermoelectric voltage, E, as a function of T_2 and extracts the relative thermopower as the local slope, $\frac{dE}{dT_2}$. This method has been used in measurements on liquid metals, but because it is essentially the same as a standard method for solids, we will not discuss it further here.

Several authors have proposed methods for measuring S for liquid metals employing small temperature differences. One reason that these methods have been developed is that it is often inconvenient to obtain a liquid sample in the extended configuration required to sustain large temperature differences. Bradley[8] applied both W and Ta counterelectrodes to each end of his liquid sample and measured the average temperature of the sample independently. By measuring the thermoelectric voltage for a given small ΔT using each set of counterelectrodes successively, he obtained the liquid thermopower, S_L, as

$$S_L = S_W - \Delta E_W \frac{S_W - S_{Ta}}{\Delta E_W - \Delta E_{Ta}} .$$

Here E_i is the thermoelectric voltage measured utilizing the "i" counterelectrode material. Implicit in this method (as it is in all small ΔT methods proposed) is the assumption that S_i and S_L are constant over ΔT (usually about 5°C). An advantage of the above method is the fact that $S_W - S_{Ta}$ is only weakly dependent upon temperature. (This advantage is also shared by the next method discussed, as S(Chromel) - S(Alumel) is also nearly temperature-independent.) Bradley's method does have a disadvantage, however, in requiring the accurate knowledge of two secondary thermopower standards (W and Ta), effectively using these materials as thermocouples for measuring ΔT, as well as using W as a standard against which to measure S_L.

Van Zytveld, et al[9] describe a small ΔT method in which the chromel leg of each temperature-measuring thermocouple is used as a counterelectrode. Small constant (over ΔT) parasitic thermoelectric voltages and thermocouple errors are eliminated by making two measurements at different temperature differences, ΔT_i, and corresponding thermoelectric voltages ΔV_i, at each average temperature. The average thermopower S_{rel}, relative to S(Chromel), free from constant (over ΔT) parasitic voltages, is given by

$$S_{rel} = \frac{\Delta V_1 - \Delta V_2}{\Delta T_1 - \Delta T_2} \quad .$$

In this method, the ΔT_i must be measured to approximately $\pm 1\%$ to obtain S_{rel} to about about $\pm 2\%$. This can be done easily, since the temperature-derivative of the thermocouple voltage, $E(T)$, namely $\frac{dE}{dT}$, is well-known and nearly constant; we then have

$$\Delta T = \Delta E(T) / \frac{dE(T)}{dT} \quad .$$

Errors in thermocouple calibration that are constant over ΔT are then removed by taking the difference ($\Delta T_1 - \Delta T_2$) in the expression for S_{rel}. This technique has proven useful in a variety of applications.

Davies[10] has also used a variation of the method just described. In his work, he has adjusted his sample temperatures to force $\Delta T_2 \equiv 0$, obtaining

$$S_{rel} = \frac{\Delta V_1 - \Delta V_2}{\Delta T_1} \quad .$$

Note that this does not simultaneously force $\Delta V_2 = 0$, since the ΔT_i here are the <u>measured</u> temperature differences, not the actual, error-free ones. While there is some small computational advantage to forcing $\Delta T_2 = 0$, it is certainly more difficult to realize this condition experimentally. For this reason, the former method is probably to be preferred.

THERMOPOWER MEASUREMENTS IN A CONDUCTING SHEATH

Because of the difficulties associated with attaching leads to a liquid column contained in a non-conducting tube, and in some cases, because of the difficulties of obtaining ceramic tubes immune to attack by certain liquids, attempts have been made to measure the thermopowers of liquid metals in conducting tubes. The containment metals that have been used most successfully include Ta, Mo, W, and stainless steel--especially types 321 and 347. The theory of such a parallel conductor arrangement has been worked out[11], and does not present any unusual problems. The arrangement is schematically illustrated in Fig. 3. Here "L" indicates the liquid sample, "c" identifies the conducting container, and "o" is the counter-electrode standard. If S_i and σ_i are the thermopower and the <u>conductance</u> (<u>not</u> conductivity) of material "i", respectively, then

$$\frac{dV}{dT} = \frac{\sigma_L S_L + \sigma_c S_c}{\sigma_L + \sigma_c} - S_o \quad .$$

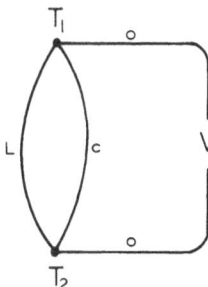

Fig. 3. Schematic illustration of a parallel thermopower measurement. (For explanation, see text.)

Since σ_c and S_c can be measured for the empty container, $\frac{dV}{dT}$ and σ_L
can be measured with the container filled, and S_0 is assumed known,
S_L can be extracted immediately.

Some very real experimental difficulties do exist, however.
Ioannides et al[12] identify two of these, namely the necessity of
establishing a uniform thermoelectrically-induced current flow in
the L-c loop, and the fact that the temperature distribution in the
sheath may be different when it is filled than when it is empty,
perhaps invalidating the measurement of $\sigma_c(T)$ (made on the empty
tube). The former difficulty requires the selection of a tube of
uniform wall thickness and diameter, and the mechanical polishing
and electropolishing of the interior surface of the tube. The latter difficulty may be overcome by careful regulation of the temperature distribution along the conducting container. Even so, while
these techniques can be utilized, the procedure remains a difficult
one, and probably one that should be employed only in cases in
which other methods are clearly inadequate.

ACKNOWLEDGMENTS

I am happy to acknowledge partial support for this study which
was provided by Research Corporation via a Cottrell College Science
Grant, and the National Science Foundation via Grant number
SMI76-82672.

REFERENCES

1. J. Nyström, Ark. Mat. Astr. Fys. 34A, No. 27 (1947).
2. J.J. Lander, Phys. Rev. 74, 479 (1948).
3. A.V. Gold, D.K.C. MacDonald, W.B. Pearson, and I. Templeton,
    Phil. Mag. 5, 765 (1960).

4. N. Cusack and P. Kendall, Proc. Phys. Soc. $\underline{72}$, 898 (1958).
5. J.G. Cook and M.J. Laubitz, Can. J. Phys. $\underline{54}$, 928 (1976).
6. N.E. Cusack, P.W. Kendall, and A.S. Marwaha, Phil. Mag. $\underline{7}$, 1745 (1962).
7. B.C. Dupree, J.B. Van Zytveld, and J.E. Enderby, J. Phys. F: Metal Phys. $\underline{5}$, L200 (1975).
8. C.C. Bradley, Phil. Mag. $\underline{7}$, 1337 (1962).
9. J.B. Van Zytveld, J.E. Enderby, and E.W. Collings, J. Phys. F: Metal Phys. $\underline{3}$, 1819 (1973).
10. H.A. Davies, Phys. Chem. Liquids $\underline{1}$, 191 (1969).
11. A.D. Wilson and H.B. Ulsh, Rev. Sci. Inst. $\underline{39}$, 346 (1968).
12. P. Ioannides, V.T. Nguyen, and J.E. Enderby, J. Phys. E: Sci. Inst. $\underline{8}$, 315 (1975).

"IT AIN'T NECESSARILY SO"*

P. L. Taylor

Department of Physics
Case Western Reserve University
Cleveland, Ohio 44106 USA

ABSTRACT

The theory of thermoelectricity is to be found in a looking-glass land where all our best-loved folk theorems are cruelly violated. Everyone knows that pseudopotentials are better than Hartree potentials in calculating scattering amplitudes, that higher-order electron-phonon interaction terms are negligibly small, and that the electronic relaxation time for phonon scattering may be approximated by a function that is symmetric about the Fermi energy. And yet ....

Thermoelectricity is a subject whose siren song has lured many an unwary traveller into errors for which history has been brutally unforgiving. Even as canny a physicist as Lord Kelvin has suffered; his beautiful intuitive grasp of the relationship between the Seebeck and Peltier coefficients[1] is seldom recalled without the attachment of some pejorative coda regretting that he did not also have the understanding of irreversible thermodynamics that Onsager[2] was to provide seventy-five years later.

The very founder of the subject, Thomas Seebeck himself, suffered from having noticed his effect within months of Oersted's discovery of the magnetic effects of an electric current. Seeing the deflection of a compass needle brought near a thermocouple, he assumed the temperature gradient to give rise directly to a magnetic field, and wasted years trying to explain the earth's

---

* Work supported by the N.S.F., and performed in part at the Aspen Center for Physics.

magnetic field in terms of the temperature difference between the
equator and the poles.[3]

The mistakes that have been made in the more recent past, and
which we will presumably continue to make into the distant future,
are made understandable (if not completely forgivable) by the
delicate nature of thermoelectric effects. The commemorative
napkin ring shown in Fig. 1, composed half of gold and half of
platinum and of a type given away by generous conference organisers
to all participants, might inadvertently be laid down in contact
with both a well chilled bottle of Chateau d'Yquem and a dish of
créme brulée. The resulting net electrical current might only be
of the order of 50 mA, but flowing around the ring would be
oppositely directed currents of close to 10 A carried respectively
by electrons with energies less than and greater than the chemical
potential. The almost complete cancellation of these opposing
currents allows ample opportunity for small omissions to lead to
major inaccuracies.

When we calculate simple conductivities we rely on a number
of techniques and facts that experience has shown to be valid.
Daily use of these facts then may tempt us to forget that they
have been derived only within a restricted framework; we elevate
them to the status of "folk theorems" of universal validity, and
recklessly apply them to the theory of thermoelectricity. A few
examples which spring to mind are:

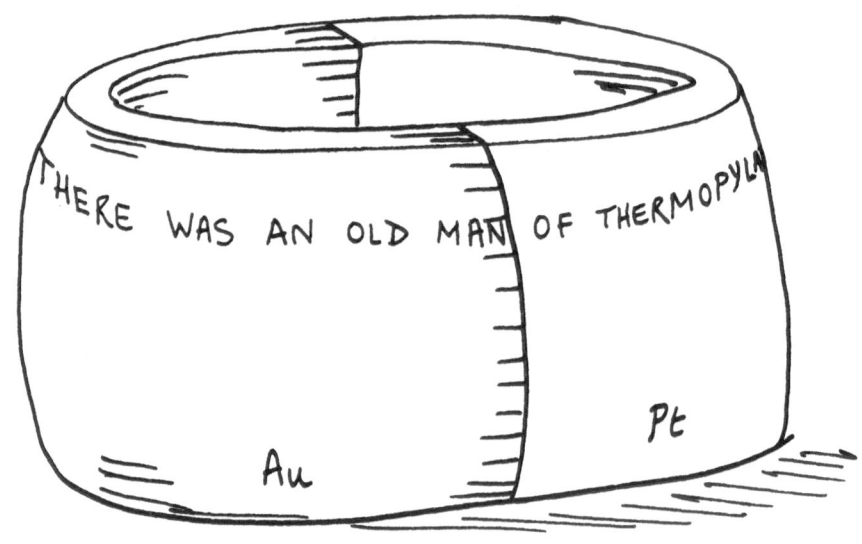

Fig. 1. Commemorative napkin ring.

1. "The heat current operator is trivially derived."
2. "Renormalization of scattering to include electron-phonon interactions of higher order only changes results by a factor of the order of the ratio of electronic to ionic masses."
3. "Pseudopotentials are a useful and reliable tool in calculating transport properties."
4. "Intrinsic two-phonon scattering processes are negligible at low temperatures."

I hope your own favorite was among those.

The heat current operator is now receiving some more of the attention that it deserves. Vidal[4], Make[5], Hu[6] and others have raised and answered some subtle questions, and Lyo in the following talk will doubtless discuss this topic. Whether the oscillatory terms in the phonon heat current[7] will play a major role in phonon-drag effects remains to be seen.

The use of pseudopotentials in calculations of thermoelectricity is an act of daring that does not seem to be passing out of fashion, even though many professional band-structure calculators consider the pseudopotential an idea whose time has come and gone with the vacuum tube computer. Not all workers, it is sad to relate, pay strict attention to the warnings voiced many years ago by Austin et al[8], who pointed out that the scattering amplitude in a dilute alloy cannot be interpreted in the Born approximation as simply the matrix element of the impurity potential between pseudo-wavefunctions, because normalization corrections occur. As a trivial illustration I have calculated the scattering amplitude for a delta-function potential in one dimension, which has a single bound "core" state, and then taken the energy derivative. One can see in Fig. 2 that the Born approximation applied to the pseudopotential lies closer to the exact result than does the Born approximation applied to the unmodified potential. On the other hand, the energy derivatives shown in Fig. 3 indicate that the pseudopotential result may be considerably worse, and even have the wrong sign. Caveat emptor!

Violations of the second and fourth quoted folk-theorems have been reported[9-11] in papers on virtual recoil in alloys and on multiple phonon scattering in pure metals. These phenomena are sometimes known as phony phonon drag or as NTDYRBTS (an abbreviation for Nielson-Taylor-do-you-really-believe-that-stuff) effects.

It is amusing that in the field of photon emission by nuclei it was the universal assumption that the emitting atom would recoil, and it took an imaginative leap by Mössbauer[12] to consider

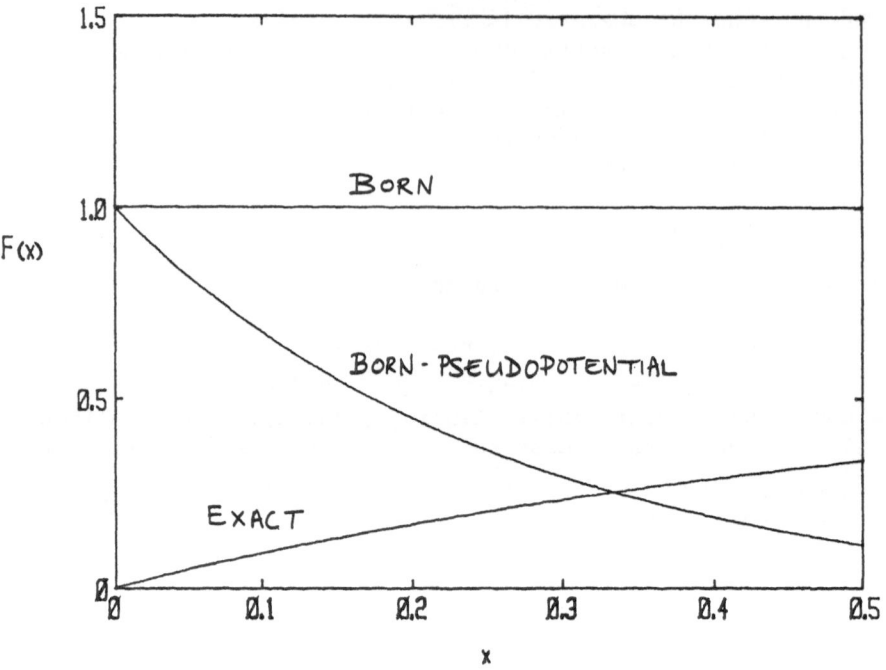

Fig. 2. Scattering probability for a one-dimensional δ-well
potential as a function of energy.

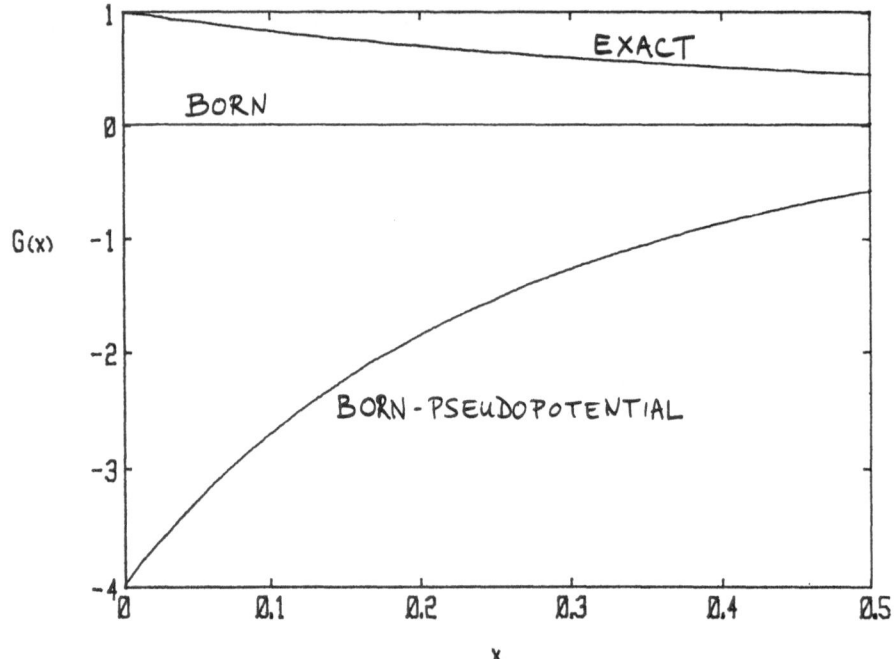

Fig. 3. Energy derivative of the scattering probability shown
in Fig. 2.

the possibility of recoil-less emission. In the theory of electron scattering by impurities in metals, on the other hand, it was the universal assumption that scattering would be elastic, and Koshino's[13] suggestion that recoil might occur was a novel one. Because the recoil of the impurity takes some energy from the scattered electron, and the Exclusion Principle limits the number of lower-energy states for the electron to drop into, the probability of such events occurring is dependent on the energy of the electron relative to the Fermi energy. The transport equations are accordingly made more complex by the existence of this effect[14].

A futher degree of subtlety is introduced when one considers virtual recoil, in which the emitted phonon describing the recoil of the impurity is reabsorbed, restoring to the electron its original energy and making the scattering process an elastic one. Because the Exclusion Principle still applies, a ghostly image of the Fermi surface remains in the scattering probability, and severe modification of the Seebeck coefficient S is caused. The contribution of this effect to the thermoelectric parameter ξ (proportional to S/T) varies quadratically with temperature at low temperatures. However, there is a more esoteric effect in which the virtual phonon emitted in the recoil of the impurity is reabsorbed in a crystal-momentum-conserving electron-phonon interaction elsewhere in the lattice[11]. The low temperature behavior of ξ due to this process is more complicated, and a maximum appears at around 0.15 T/Θ, with Θ the Debye temperature. The series expansion is of the form

$$\xi \propto 1 - \frac{8\pi^2}{5} \left(\frac{T}{\Theta}\right)^2 \ln \left(\frac{T}{\Theta}\right) - 22.78 \left(\frac{T}{\Theta}\right)^2 + \ldots,$$

but such series should always be taken with a large pinch of salt because of the essential singularities floating around in the theory; that is, there are also terms varying as $\exp - (\Theta/T)$, which have no power series expansion but which may not be negligible at any but the lowest of temperatures.

The unusual dependence of ξ on T when virtual recoil is involved (that number 22.78 contains derivatives of Riemann zeta functions as well as Euler's constant!) leads to consideration of another folk theorem - the one that states[15]

5. "The Seebeck coefficient is proportional to the logarithmic derivative of the conductivity with respect to energy."

If one looks at the derivation of this result in some excellent out-of-print text[16] or other, then one finds that it involves a
Taylor expansion of the mean free path Λ in energy about the chemical
potential ζ, and a neglect of terms in $(\varepsilon - \zeta)^3$. Such a neglect
is hardly ever justified because Λ is only rarely obliging enough
to comply with this requirement. Even for such a simple case as
a pure free-electron metal with electron-phonon scattering in
lowest order there is a rapid variation in Λ near the Fermi surface. This was explicitly demonstrated by Rhodes[17] in 1950, and
has very recently been computed in more realistic models by
Leavens[18] and coworkers. This variation arises simply from the
nature of the kernel of the Boltzmann equation, and is not a
true many-body effect; it does not occur in the lowest-order calculation of impurity scattering.

The phenomena of virtual recoil and intrinsic two-phonon
scattering, on the other hand yield an energy dependence of Λ in
the vicinity of ζ that is both rapid and ubiquitous. Even for
impurity scattering alone one finds a variation of Λ within the
thermal thickness of the Fermi surface that is too rapid to be
ignored. Much of the mathematical complexity of our 1974 paper[11]
on phony-phonon drag effects resulted from this inconvenient fact.

I would like to spend the rest of this talk in a discussion
of the calculation of the energy dependence of the mean free path
from the Boltzmann equation. However, before doing so it should
be stated that the use of the Boltzmann equation is itself a sin.
The standard derivation of this equation from more fundamental
quantum-mechanical transport theories[19] involves the assumption
that the imaginary part of the electron self-energy is much less
than ζ. For dilute alloys at low temperatures this appears at
first sight to be justified, in that both elastic scattering by
impurities and inelastic scattering by phonons will be infrequent.
The thermoelectric coefficients, however, are not proportional to
the amount of scattering in the way that the resistivities are,
but are functions of the energy dependence of the scattering;
quite different criteria are involved in the justification of the
Boltzmann-equation approach, and some of these criteria have yet
to be clearly stated.

That having been said, let me now turn to the question of
calculating the energy dependence of the mean free path. In order
to simplify the discussion I shall deal only with the classic
electron-phonon interaction, and shall not include virtual recoil
and multiphonon effects in this illustrative example. The
linearized Boltzmann equation for independent electrons interacting

with phonons states that

$$\vec{v}_k f^o_k (1-f^o_k) = \sum_{k'} P(\vec{k},\vec{k}') \, (\vec{\Lambda}_k - \vec{\Lambda}_{k'}) \tag{1}$$

with $\vec{v}_k$ the electron velocity, f_k the Fermi–Dirac function of the reduced energy $\hat{\varepsilon}_k = \varepsilon_k - \zeta$, $\vec{\Lambda}_k$ the vector mean free path, and $P(\vec{k},\vec{k}') = P(\vec{k}',\vec{k})$ the equilibrium scattering rate between states. For free electrons and isotropic Debye phonons the problem reduces to a one-dimensional integral equation of the form

$$B = \int [\Lambda_\eta - \Lambda_{\eta+\rho}(1 + \frac{\rho}{\mu+\eta})^{-1/2}(1 + \frac{\rho-b\rho^2}{2(\mu+\eta)})] \, (1 + \frac{\eta}{\mu})^{-1} G(\eta,\rho) d\rho. \tag{2}$$

Here Λ is written as a function of the dimensionless variable $\eta = \beta\hat{\varepsilon}$, with $\mu = \beta\zeta$. The range of integration of the variable ρ which is proportional to the phonon energy is $\pm\Theta/T$, and

$$G(\eta,\rho) = \frac{\rho^2}{(1+\tanh\frac{\eta}{2}\tanh\frac{\rho}{2}) \, |\sinh\rho|} \quad . \tag{3}$$

The constant B is proportional to T^{-3}, while b is proportional to T and $\beta^{-1} = k_B T$. This expression is equivalent to that given by Wilson[21], for example.

While formal (but tediously complicated) solutions to this problem can be produced in terms of ratios of determinants, things get out of hand when one adds (as one must) virtual recoil or multiphonon interactions. One can, however, obtain some relief by expanding Λ in a double power series in η and μ^{-1}. Because μ, the ratio of Fermi energy to thermal energy, is invariably large, only the lowest two orders in μ^{-1} will usually be important[22]. Much higher powers of η, on the other hand, must be retained, as it is after all the η-dependence of Λ that we are seeking. It appears desirable to retain terms up to and including η^6 in this context.

To lowest order in μ^{-1} the mean free path is an even function of η, as Eq. (2) reduces to the form

$$B = \int [\Lambda_\eta - \Lambda_{\eta+\rho} (1 - K\rho^2)] \, G(\eta,\rho) \, d\rho \qquad (4)$$

with $K \propto (T/\Theta)^2$. This is a deceptively simple-looking equation which has some hidden traps. At very low temperatures, for example, it can be shown that Λ tends to a constant, Λ_o, and so we might be tempted to substitute this in Eq. (4), put $\eta = 0$, and solve to find

$$\Lambda_o^{-1} \propto 93 \, \zeta_R(5) \, (\tfrac{T}{\Theta})^5 \qquad (5)$$

with ζ_R the Riemann zeta function. Alternatively we could multiply both sides of Eq. (4) by $-(\partial f^o/\partial \eta)$, the derivative of the Fermi function, integrate over η, and solve, but then we would find

$$\Lambda_o^{-1} \propto 240 \, \zeta_R(5) \, (\tfrac{T}{\Theta})^5 , \qquad (6)$$

which is more than two and a half times as big as the previous result! That is, we made a correct assumption about the constancy of Λ at low temperatures, but came unstuck in not being delicate enough in handling the way the vanishingly small variations from constancy of Λ are amplified in the Boltzmann integral. There is a strong message here to the effect that one should be very cautious in undertaking purely numerical solutions of the Boltzmann equation at low temperatures. Because $\Lambda \propto T^{-5}$ but $B \propto T^{-3}$ one sees in Eq. (4) that large terms must cancel to leave something smaller by a factor of order $(T/\Theta)^2$.

Perhaps some of you are wondering which of the two approximate solutions is the better one. The answer is that Eq. (6) corresponds to the lowest-order variational solution, and is consequently superior. The more general solution starts with the expansion

$$\Lambda_\eta = \Lambda_o + \Lambda_1 \eta + \Lambda_2 \eta^2 + \cdots .$$

Multiplication by $-\eta^n(\partial f^o/\partial \eta)$ and integration yields a series of equations for various n of the form

$$B = K(A_o\Lambda_o + A_2\Lambda_2 + A_4\Lambda_4 + \cdots) \quad .$$

$$B = KD_o\Lambda_o + (C_2 + KD_2)\,\Lambda_2 + \cdots \tag{7}$$

$$B = \ldots .$$

Here the A's, C's and D's are sums of Debye integrals, which tend to constants at low temperatures but have a complicated form at intermediate temperatures.

The thermoelectric effects arise principally from the components of Λ that are odd in η. To evaluate these we must return to Eq. (2) and expand the integrand in powers of μ^{-1} retaining at least the first-order terms. We then find a set of equations of which the first is

$$(F_1 - KG_1)\Lambda_1 + (F_3 - KG_3)\Lambda_3 + \cdots$$
$$= \mu^{-1}[KI_o\Lambda_o + (H_2 + KI_2)\Lambda_2 + \cdots] \quad . \tag{8}$$

The odd terms are clearly smaller than the even terms by a factor of μ, which equals $\zeta/k_B T$, and thus are easily overlooked in a numerical computation. Rhodes[17] found Λ to be even to within one per cent, and then proceeded with his calculation under the assumption that it was exactly even.

The set of variational equations was solved in this model by Sondheimer[23]. He found a considerable temperature variation of ξ; at high temperatures it was close to factor of three larger than at low temperatures. Despite the fact that anisotropy and many-body effects were ignored it remains one of the most detailed of the calculations of naive models yet to be performed.

The theoretical situation in thermoelectric effects now seems to be that we have a large number of specialists, each busy with his or her own pet effect, and rather dim prospects for any all-encompassing calculations that will include all the phenomena now known to be important. We have calculations[23] that solve the Boltzmann equation with precision for independent free electrons and Debye phonons, we have calculations[11] of many-body effects in the same simple model but in which the Boltzmann equation is solved less rigorously, and we have calculations[24] in which the band structure is treated carefully but which use only the Mott relation for independent electrons. To bring these elements together in a precise calculation appears a formidable enough

task even in the absence of phonon drag or of effects that lie
outside the framework of the Boltzmann equation. I won't even
mention the possibility of applying strong magnetic fields to the
sample.

So where can we expect progress in the near future? The most
pressing need is for some understanding of the effects of anisotropy on the many-body contributions, and for a closer examination
of the validity of the Boltzmann equation in the presence of virtual recoil and intrinsic two-phonon interactions. We should also
expect to see before too long some more detailed discussions of
the energy dependence of the vector mean free path when actual
(as well as virtual) recoil of impurities is considered. These
last will not be simple calculations, because the many-body effects
spoil some of the convenient cancellations that occur in the
variational treatment of transport, and leave some very messy
integrations to be performed.

By the time the next international conference on thermoelectricity in metals is convened there should be a large amount
of detailed work completed and - who knows? - perhaps some new
physical insights achieved.

REFERENCES

1. Kelvin, Lord (W. Thomson) Proc. Roy. Soc. Edin. 3 255 (1854).
2. L. Onsager, Phys. Rev. 37 405, 38 2265 (1931).
3. E. Frankel, Dictionary of Scientific Biography (Scribner's,
    New York, 1975) and S. Angrist, Direct Energy Conversion
    3rd Ed. (Allyn and Bacon, Boston, 1976).
4. F. Vidal, Phys. Rev. B 8, 1982 (1975).
5. K. Maki, Phys. Rev. Lett. 21 1755 (1968).
6. C.-R. Hu, Phys. Rev. B 13 4780 and 14 4834 (1976).
7. P.L. Taylor, A Quantum Approach to the Solid State (Prentice-
    Hall, Englewood Cliffs N.J. 1970) p. 94.
8. B.J. Austin, V. Heine, and L.J. Sham, Phys. Rev. 127 276.
9. P.E. Nielsen and P.L. Taylor, Phys. Rev. Lett. 21 893 (1968).
10. P.E. Nielsen and P.L. Taylor, Phys. Rev. Lett. 25 371 (1970).
11. P.E. Nielsen and P.L. Taylor, Phys. Rev. B10 4061 (1974).
12. R. Mössbauer, Z. Physik 151 124 (1958).
13. S. Koshino, Progr. Theoret. Phys. (Kyoto) 24 484 (1960).
14. P.L. Taylor, Phys. Rev. 135 A1333 (1964).
15. This is sometimes called the Mott relation, and is found in
    the 1936 book that he wrote with H. Jones. If I remember
    correctly, Nevill Mott tells the story of how he and R. V.
    Jones traveled in Germany near the end of the war to study
    the war-time progress of German science. Their approach was

frequently heralded with the cry "Hier kommt der Mott und Jones", a phrase one is sometimes tempted to reiterate on seeing yet another current paper making inappropriate application of this result.

16. Ref. 7, p. 278.
17. P. Rhodes, Proc. Roy. Soc. A 202 466 (1950).
18. C.R. Leavens, J. Phys. F 7 163 (1977), X, XXX (1977).
19. There are, it is rumored, more pleasurable ways of sinning than solving Boltzmann equations, and so no workers should be tempted into these calculations on this account alone.
20. Ref. 7, section 8.5.
21. A.H. Wilson, The Theory of Metals 2nd Ed. (Cambridge Univ. Press 1953) p. 263.
22. Please do not make this statement into another folk theorem: each case must be tested on its merits!
23. E.H. Sondheimer, Proc. Roy. Soc. A203 75 (1950).
24. J. Yamashita and S. Asana, Prog. Theoret. Phys. 50 1110 (1974).

DISCUSSION

M. Harrison: Phil, can you tell anything about the binding of the impurity from these extra effects as opposed to just its mass? Are they fine enough to tell you anything about the nature of the chemical binding of the impurity in the crystal?

P. L. Taylor: I guess the information is buried in there somewhere, but the calculation is very hard to do without making a number of assumptions; Bill Hartmann has hit some of these difficulties, too. The first assumption that is convenient is that the impurities have the same mass as the solvent. The next assumption is that the spring constants are the same all the way along, and even then you have already a fairly rough problem to calculate in all detail. Then to undo these approximations would be complicated and still leaves you with just one number. As John Enderby was saying, you cannot get a lot of information out of one number; it's nice to have a function of something. Well, you do have a function of temperature, but for reasons similar to the ones he gave, it is not as informative as it could be. So I would hate to do all that processing and rely on what you get out of one number.

H.J. van Daal: This is a question of a simple experimentalist who doesn't want to be sitting on one of your fences. You said in one of your last remarks that no theoretical work has been done on the thermopower due to paramagnon scattering. Is that true?

P.L. Taylor: No, I don't think I said that. I said that there may be some unanswered questions about the heat current operator in the presence of magnetic interactions.

Just this year several papers on this topic have appeared.
When you think of the heat current operator, you say, well,
multiply the electron velocity by its kinetic energy, minus the
chemical potential, and that's the heat current. For a phonon do
the same operation and there is a neatly written down heat current.
However, there are more complicated effects that come in if you
have magnetic materials. If you have magnons you know a magnon can
carry heat. And if there are magnetic interactions then energy
can be transferred. And an electron traveling along, interacting
with spins as it goes, is in effect causing the spins to interact
with each other, and there are heat currents involved in that. I
guess people are just at the point of writing down some of the
equations to define this. In superconductors where you get barriers
between superconducting and normal materials I believe there are
questions that have recently been answered concerning the heat current carried. I wasn't trying to say that people have not done
anything; I thought maybe there were still some things to do.

MANY-BODY EFFECT OF THE ELECTRON-PHONON INTERACTION ON THE THERMO-

ELECTRICITY IN METALS*

S. K. Lyo

Department of Physics
University of California
Los Angeles, California 90024   USA

ABSTRACT

The many-body effect of the electron-phonon interaction on
the electron-diffusion thermopower is calculated microscopically.
It is shown that the thermopower is enhanced not by the mass
enhancement but also significantly by a new mechanism, arising from
the electron-phonon modification of the quasi-particle velocity.
The effect of the above result on the high field adiabatic Nernst-
Ettingshausen coefficient, and the magneto-thermopower is discussed.

I.   INTRODUCTION

Recently, Opsal, Thaler, and Bass[1] showed that the electron
diffusion thermoelectric power is enhanced by the electron-phonon
mass renormalization, unlike the dc conductivity and contrary to
previous theoretical calculation.[2] However, it was not clear to
what extent their semiclassical approach[1] treats the effect of
the electron-phonon interaction. Subsequently, Lyo[3] has shown
microscopically that the electron diffusion thermoelectric power
is enhanced not only by the mass enhancement but also significantly
by another mechanism arising from the modification of the quasi-
particle velocity through the electron-phonon interaction.

In this paper we examine the many-body effect of the electron-
phonon interaction on the electron diffusion thermoelectric power.[2]

------

*Work supported in part by the National Science Foundation, Grant
Number DMR 75-19544.

We discuss also how the magneto-thermoelectric power is affected by the electron-phonon interaction. In Section II we present a simple phenomenological quasi-particle description of thermo-electric power. In Section III, a brief microscopic calculation is given. The effect of the electron-phonon interaction on magneto-thermoelectric power is discussed in Section IV.

II. ENHANCEMENT OF THERMOELECTRIC POWER

The thermoelectric power is given by Mott's formula[4]

$$S = \frac{\pi^2}{3e} k_B^2 T \left[\frac{\partial \ln \sigma_{xx}(z)}{\partial z} \right]_{z=\mu} \tag{1}$$

where e, k_B, T, μ, and σ_{xx} are, respectively, electronic charge (negative), Boltzmann's constant, temperature, chemical potential, and dc conductivity. A cubic symmetry will be assumed for simplicity in this paper. The above formula can be derived quite generally from the Landau-Boltzmann equation.[5] Using a relaxation time approximation, the conductivity is given by

$$\sigma_{xx}(\mu) = \frac{e^2}{12\pi^3 h} \int_{\mu=E_{\underset{\sim}{k}}} v_{\underset{\sim}{k}}^{*} \tau_{\underset{\sim}{k}}^{*} \, dS \tag{2}$$

where $\tau_{\underset{\sim}{k}}^{*}$ represents the renormalized relaxation time and $\underset{\sim}{k}$ the crystal momentum. The integral is taken over the Fermi surface.

The quasi-particle energy $E_{\underset{\sim}{k}}$ is given in terms of the bare electron energy $\varepsilon_{\underset{\sim}{k}}$ by

$$E_{\underset{\sim}{k}} = \varepsilon_{\underset{\sim}{k}} + M_{\underset{\sim}{k}}(E_{\underset{\sim}{k}}) \tag{3}$$

where $M_{\underset{\sim}{k}}(z)$ is the real part of the electronic self-energy, arising from a virtual one-phonon process illustrated in Fig. 1. The quasi-particle velocity then defined by

$$v_{\underset{\sim}{k}}^{*} = \frac{1}{\hbar} \nabla_{\underset{\sim}{k}} E_{\underset{\sim}{k}} \quad . \tag{4}$$

Fig. 1. Electron self-energy part.

Inserting (3) in (4), one finds

$$\underset{\sim}{v}_{\underset{\sim}{k}}^{*} = \frac{\underset{\sim}{v}_{\underset{\sim}{k}}}{1 - M_{\underset{\sim}{k}}'(E_{\underset{\sim}{k}})} + \frac{1}{1 - M_{\underset{\sim}{k}}'(E_{\underset{\sim}{k}})} \; \frac{1}{\hbar} \nabla_{\underset{\sim}{k}} M_{\underset{\sim}{k}}(z) \Bigg|_{z=E_{\underset{\sim}{k}}} \quad (5)$$

where $\underset{\sim}{v}_{\underset{\sim}{k}} \equiv \frac{1}{\hbar} \nabla_{\underset{\sim}{k}} \varepsilon_{\underset{\sim}{k}}$ is the undressed electron velocity and prime

means a derivative with respect to the argument. Note that $- M_{\underset{\sim}{k}}'(\mu)$ (>0) is of order unity. The second term of (5) is of

order unity. The second term of (5) is of order $|M_{\underset{\sim}{k}}/\mu|$ (<<1) smaller

than the first term and can be neglected for the purpose of cal-culating the conductivity. Assuming $\lambda \equiv - M_{\underset{\sim}{k}}'(\mu)$ to be isotropic,

one then has $\underset{\sim}{v}_{\underset{\sim}{k}}^{*} = \underset{\sim}{v}_{\underset{\sim}{k}}/(1+\lambda)$. It is well known[2,6] that there is no

effect of mass renormalization on dc conductivity. This means that

$\tau_{\underset{\sim}{k}}^{*}(\mu) = \tau_{\underset{\sim}{k}}^{(o)}(\mu)(1+\lambda)$ where $\tau_{\underset{\sim}{k}}^{(o)}$ is the bare relaxation time.

Using these properties, Opsal et al.[1] have shown that the first term of (5), inserted in (2) and (1), leads to enhanced electron thermopower

$$S_1 = (1+\lambda) \frac{\pi^2}{3e} k_B^2 T \frac{\partial \ln \sigma_{xx}^{(o)}(z)}{\partial z} \Bigg|_{z=\mu} \quad (6)$$

where $\sigma_{xx}^{(o)}$ is the bare conductivity,

$$\sigma_{xx}^{(o)} = \frac{e^2}{12\pi^3\hbar} \int_{z=\varepsilon_{\underset{\sim}{k}}} v_{\underset{\sim}{k}} \tau_{\underset{\sim}{k}}^{(o)} \, dS \quad . \tag{7}$$

Turning to (5), one observes that the second term varies much more rapidly in energy than the first term. Namely, the second term $[\curlywedge M_{\underset{\sim}{k}}(z)]$ varies over the scale of the phonon energy, while the first term over the electron energy. This means that both terms contribute significantly to the thermoelectric power through $\partial v_{\underset{\sim}{k}}^{*}/\partial E_{\underset{\sim}{k}}$. The second term of (5), then, gives an additional contribution to the thermopower

$$S_2 = \frac{\pi^2}{3e} k_B^2 \, T[\sigma_{xx}^{(o)}(\mu)]^{-1} \frac{e^2}{12\pi^3\hbar} \int_{\mu=\varepsilon_{\underset{\sim}{k}}} \frac{[v_{\underset{\sim}{k}} \cdot \frac{1}{\hbar} \nabla_{\underset{\sim}{k}} M_{\underset{\sim}{k}}'(\mu)] \tau_{\underset{\sim}{k}}^{(o)}(\mu)}{v_{\underset{\sim}{k}}} \, dS \tag{8}$$

where use is made of the fact that $z=\mu$ is an inflection point for $M_{\underset{\sim}{k}}(z)$ [i.e., $M_{\underset{\sim}{k}}''(\mu) =0$]. Defining $\xi = \{ \frac{1}{\hbar v_{\underset{\sim}{k}_{xF}}} \frac{\partial}{\partial k_{\underset{\sim}{xF}}} M'_{\underset{\sim}{k}_F}(z) \}_{z=\mu}$

and assuming ξ to be isotropic, (8) becomes

$$S_2 = \frac{\pi^2 \xi}{3e} k_B^2 \, T \quad . \tag{9}$$

In effective mass and Debye approximations, one finds

$$\xi = \lambda/2\mu \quad . \tag{10}$$

Approximating $\partial \ln \sigma_{xx}^{(o)}/\partial z \sim 1/\mu$ in (6), S_2 is then comparable to S_1.

III. MICROSCOPIC ANALYSIS

For convenience, we study a system of Bloch electrons interacting with the lattice and a low concentration of impurities at low temperature ($T \ll \theta_D$, Debye temperature). The thermopower is given by[7]

$$S = \frac{<<JK>>}{eT<<JJ>>} \tag{11}$$

where

$$<<JK>> = \frac{1}{i} \frac{\partial}{\partial \omega} F_{JK}(\hbar\omega + i0) \Big|_{\omega=0} \tag{12}$$

The correlation function is given by

$$F_{JK}(\hbar\omega_r) = \int_0^\beta <J(u)K> \exp(\hbar\omega_r u)du$$

where the angular brackets denote the grand canonical and thermo-dynamic average, $\hbar\omega_r = k_B T 2\pi r i$, $\beta^{-1} = k_B T$, and r is an integer. J(u) is in the imaginary time Heisenberg representation. Finally, using the Frölich Hamiltonian,[6] the heat current operator K/e is given in terms of the charge current operator J and the electronic energy current operator Q by

$$K_{\underset{\sim}{k}'\underset{\sim}{k}} \equiv (eQ-\mu J)_{\underset{\sim}{k}'\underset{\sim}{k}} = ev_{\underset{\sim}{k}x}(\epsilon_{\underset{\sim}{k}}-\mu)\delta_{\underset{\sim}{k}',\underset{\sim}{k}} + e \frac{v_{\underset{\sim}{k}'x} + v_{\underset{\sim}{k}x}}{2} v_{\underset{\sim}{q}}^{(o)} x$$

$$(b_{\underset{\sim}{q}} + b_{-\underset{\sim}{q}}^\dagger) \delta_{\underset{\sim}{k}',\underset{\sim}{k}+\underset{\sim}{q}} \tag{13}$$

where $v_{\underset{\sim}{q}}^{(o)}$ is the bare electron-phonon interaction strength, assumed

to be a function of the momentum transfer only, and $b_{\underset{\sim}{q}}(b_{-\underset{\sim}{q}}^\dagger)$ is a

Boson destruction (creation) operator. The temperature gradient is assumed to be in the x-direction.

The correlation function in (12) is found in terms of the upper charge current vertex correction (defined as EF-vertex part following Holstein's work[6]), which consists of ladders of irreduc-ible scattering part. The latter contains single impurity lines and phonon lines. The lower vertex contains not only the first term of (13) but also terms arising from the second member of (13) and illustrated in Fig. 2. Here the curvy, solid, and incoming

Fig. 2. Effective electronic energy current vertex due to the
electron-phonon interaction.

wiggly lines represent, respectively, phonon propagators, full
electron propagators [to be denoted as $S_k(z)$], and the external
line. The electron-phonon vertices $(\tilde{V}_q)$ and the phonon frequency
$(\tilde{\omega}_q)$ are dressed.[6]

One then finds

$$<<JK>> = e^2 \sum_k v_{kx} \int_{-\infty}^{\infty} dz \; \{[\varepsilon_k + M_k(z) + m_k(z) - \mu] \; [-f^{(-)'}(z)] \; x$$

$$\phi_k(z) \; \delta[z - \varepsilon_k - M_k(z)] + \frac{1}{2\pi} \sum_{\pm} \pm (\varepsilon_k - \mu) f^{(-)}(z) \frac{S'_k(z \pm i0)}{1 - G'_k(z \pm i0)} \; x$$

$$\frac{\partial}{\partial \omega} \Lambda_k(z \pm i0, \; z + \hbar\omega \pm i0)_{\omega=0} \} \qquad (14)$$

where $f^{(-)}(z)$, ω are the Fermi function and the external frequency.
The electronic self-energy part is given in terms of its real and
imaginary parts by $G_k(z \pm i0) = M_k(z) \mp i\Gamma_k(z)$. The self-energy-
like quantity $m_k(z)$ is defined by

$$m_k(z) = v_{kx}^{-1} \sum_{k'q} \sum_{\pm} v_{k'x} |V_q|^2 \; f^{(\mp)}(\varepsilon_{k'}) \delta_{k',k+q} \; P\frac{1}{z - \varepsilon_{k'} \mp \hbar\omega_q} \qquad (15)$$

where $f^{(+)} \equiv 1 - f^{(-)}$ and P indicates the principal part. The
real part of the self-energy $M_k(z)$ is obtained by replacing $v_{k'z}$
in (15) by v_{kx}. The distribution $\phi_k(z)$ in (14) is related to the

EF-vertex part $\Lambda_{\underset{\sim}{k}}$ by[6] $\phi_{\underset{\sim}{k}}(z) = \hbar \Lambda_{\underset{\sim}{k}}(z-i0, z+i0)/2\Gamma_{\underset{\sim}{k}}(z)$ and to the (bare) transport relaxation time by $\phi_{\underset{\sim}{k}}(z) = v_{\underset{\sim}{kx}}\tau_{\underset{\sim}{k}}(z)$. The quantity $\partial\Lambda_{\underset{\sim}{k}}/\partial\omega$ in (14) is given by

$$\frac{\partial}{\partial\omega}\Lambda_{\underset{\sim}{k}}(z+i\eta0,\ z+\hbar\omega+i\eta0)\bigg|_{\omega=0} = \frac{1}{i\hbar}\int_{-\infty}^{\infty}dx\ \sum_{\underset{\sim}{k}'\underset{\sim}{q}}\sum_{\pm}\mp f^{(-)'}(x)\left|V_{\underset{\sim}{q}}\right|^2\ x$$

$$\phi_{\underset{\sim}{k}'}(x)\delta_{\underset{\sim}{k}',\underset{\sim}{k}+\underset{\sim}{q}}\ \delta[x-\varepsilon_{\underset{\sim}{k}'}-M_{\underset{\sim}{k}'}(x)][P\frac{1}{z-x\pm\hbar\omega_{\underset{\sim}{q}}} - \pi i\eta\delta(z-x\pm\hbar\omega_{\underset{\sim}{q}})],$$

$(\eta = \pm 1)$. (16)

Although the phonon ladders do not contribute significantly to the scattering at low temperature, they are directly responsible for (16). Noting that $\phi_{k}(z)$ is inversely proportional to the concentration, (14) and (16) are given to the lowest-order in the latter quantity. Another smallness parameter in the present theory is the ratio of the sound velocity (c_s) to the Fermi velocity (v_f), or equivalently that of the Debye energy to the Fermi energy. The two terms $M_{\underset{\sim}{k}}(z) + m_{\underset{\sim}{k}}(z)$ in the parenthesis of (14) arise from the processes shown in Fig. 2 and account for additional electronic energy current arising from the electron-phonon interaction. Although these quantities are small (i.e., of order of Debye energy), they vary rapidly over the range of the Debye energy (e.g., $\partial M_{\underset{\sim}{k}}(z)/\partial z \sim 1$) and lead to an important contribution. The term proportional to $\partial\Lambda_{\underset{\sim}{k}}/\partial\omega$ in (14) is apparently of higher order in electron-phonon interaction in view of (16). It is known[6] that the contribution from terms of this type of EF-part is negligible for the charge conduction. However, in the present problem, the quantity $\partial\Lambda_{\underset{\sim}{k}}/\partial\omega$ leads to a significant contribution. It turns out that the term proportional to $m_{\underset{\sim}{k}}(z)$ in (14) cancels part of the above mentioned contribution arising from $\partial\Lambda_{\underset{\sim}{k}}/\partial\omega$. One then finds to the lowest order in the smalless parameters[3]

$$S = S_1 + \frac{\pi^2 k_B^2 T}{3e}[-m''_{\underset{\sim}{k}_F}(\mu)] \quad . \tag{17}$$

Using (15), one can readily identify the second term of (17) with S_2 in (9).

IV. MAGNETO-THERMOELECTRICITY

Let us assume that an electric field ε is produced by a temperature gradient $\partial T/\partial x$ in the presence of an external magnetic field H ($//z$) and in the absence of the electric current. Defining the magneto-thermoelectric power $S(H) \equiv \varepsilon_x/(\partial T/\partial x)$ and the Nernst-Ettingshausen coefficient $S_{N.E.}(H) \equiv \varepsilon_y/(\partial T/\partial x)$, one finds[5,8] semiclassicaly, for a large magnetic field (i.e., $\omega_c \tau \gg 1$, ω_c is the cyclotron resonance) and in the absence of the heat flow in the y-direction

$$S_{N.E.}(H) = eL_oT \left(\frac{\sigma_{xy}^{(o)}(H)}{\sigma_{yy}^{(o)}(H)} \right) \frac{d}{dz} \ln \sigma_{xy}(H) \Bigg|_{z=\mu} \tag{18a}$$

and

$$\Delta S \equiv S(H) - S(H\equiv 0) = eL_oT \left\{ 2 \frac{d}{dz} \ln \sigma_{xy}(H) - \frac{d}{dz} \ln[\sigma_{yy}(H) \right.$$

$$\left. \sigma_{xx}(H=0)]_{z=\mu} \right\} \tag{18b}$$

where $L_o = \frac{1}{3} (\pi k_B/e)^2$.

For uncompensated metals without open orbits, one has[9] $\sigma_{xy}(z) = ec [N_e(z) - N_h(z)]/H$ where c, $N_e(z)$ and $N_h(z)$ are speed of light, the number of electrons and holes contained within the energy surface $E_k = z$, respectively. The quasi-particle velocity does not appear in σ_{xy}. Therefore (18a) does not contain a contribution due to the electron-phonon modification of the quasi-particle velocity [cf. (8)]. Then, as pointed out by Opsal et al.,[1] replacing d/dz in (18a) by $\partial/\partial E_k = (1+\lambda) (\partial/\partial \varepsilon_k)$, the Nernst-Ettingshausen coefficient becomes enhanced by a factor $1 + \lambda$. This effect has been observed recently by Fletcher[10] and by Thaler, Fletcher, and Bass.[11]

The diagonal element of the conductivity is given, for a large magnetic field, by[12]

$$\sigma_{yy}(\underset{\sim}{H}) = \frac{1}{H^2} \int_{E_{\underset{\sim}{k}}=\mu} \frac{F(\underset{\sim}{k})}{v_{\underset{\sim}{k}}^* \tau_H^*} \, dS \tag{19}$$

where $F(\underset{\sim}{k})$ and τ_H^* are a certain function of $\underset{\sim}{k}$ and a field-dependent relaxation time,[12] respectively. In a spherical model, one can factor out $v_{\underset{\sim}{k}}^*$ from the integrals in (2) and (19). This means that the quasi-particle velocity does not appear in (18b) and ΔS becomes enhanced simply by a factor $1 + \lambda$.[1] However, for a general band structure, ΔS contains contributions arising from the second term of (5).

ACKNOWLEDGEMENTS

The author wishes to acknowledge valuable conversations with T. Holstein. He is grateful to J. Bass for bringing this problem to his attention.

REFERENCES

1. J.L. Opsal, B.J. Thaler, and J. Bass, Phys. Rev. Lett. <u>36</u>, 1211 (1976).
2. R.E. Prange and L.P. Kadanoff, Phys. Rev. <u>134</u>, A566 (1964).
3. S.K. Lyo, Phys. Rev. Lett. (1977), to be published.
4. N.F. Mott and J. Jones, <u>Theory of Metals and Alloys</u> (Clarendon Press, Oxford, 1936).
5. R.S. Averback and D.K. Wagner, Sol. St. Comm. <u>11</u>, 1109 (1972).
6. T. Holstein, Ann. Phys. (N.Y.) <u>29</u>, 410 (1964).
7. R. Kubo, M. Yokota, and S. Nakajima, J. Phys. Soc. Japan <u>12</u>, 1203 (1957).
8. S.K. Lyo, unpublished.
9. I.M. Lifshitz, M.Ia. Azbel, and M.K. Kaganov, Zh. Eksp. Teor. Fiz. <u>30</u>, 220 (1955) [Soviet Phys. JETP <u>3</u>, 143 (1956)].
10. R. Fletcher, Phys. Rev. B <u>14</u>, 4329 (1976).
11. B.J. Thaler, R. Fletcher, and J. Bass (1977), preprint.
12. D.K. Wagner, Phys. Rev. B <u>5</u>, 336 (1972).

MEASUREMENT OF ELECTRON-PHONON MASS ENHANCEMENT IN THERMO-

ELECTRICITY*

J. Bass, R. Fletcher[†], J. L. Opsal, and B. J. Thaler[††]

Department of Physics

Michigan State University
East Lansing, Michigan 48824 USA

ABSTRACT

Measurements of the high magnetic field limit of the Nernst-Ettingshausen coefficient of both compensated and uncompensated metals can be used to isolate the effect of electron-phonon mass enhancement on thermoelectricity. We describe what is measured in each case and why, and review the experimental results which have been obtained.

I. INTRODUCTION

Measurements of electronic specific heat yield an electronic density of states $N(\mu)$ at the Fermi energy μ which is enhanced by many-body effects, with the electron-phonon mass enhancement usually dominant. Until recently, it was generally believed that d.c. transport properties would not be similarly enhanced.[1] However, measurements of the difference between the high magnetic field and zero field limits of the diffusion thermopower of

---

* Supported in part by the NSF under grants DMR-75-14138, DMR-77-04680 and DMR-75-01584.

† Permanent Address, Queen's University, Kingston, Ontario, Canada.

†† Present Address, Physics Department, Northwestern University, Evanston, Illinois, USA.

Al^2 revealed that this belief was in error as regards thermoelectricity. With evidence that enhancement is present in thermoelectricity, it is of interest to inquire whether the magnitude
of this enhancement is the same as that observed in electronic
specific heat. This question can be answered experimentally by
measuring the adiabatic Nernst-Ettingshausen coefficients of both
compensated and uncompensated metals in the limit of high magnetic
field $\vec{B}$. In this paper we: Define the most convenient Nernst-
Ettingshausen coefficients to measure for compensated and uncompensated metals, Q^a and P^a respectively; show how each yields a
"transport specific heat" γ^t which can be compared with the independently measured electronic specific heat γ^c; describe how each
coefficient is measured; discuss the limitations on the accuracy
with which γ^t can be extracted; review published results; and,
finally, suggest some experiments which still need to be done.

II. TRANSPORT RELATIONS AND DEFINITIONS OF Q^a AND P^a

We start with the transport equations relating the electrical
and thermal current densities, $\vec{J}$ and $\vec{U}$, to the electric field $\vec{E}$ and
the temperature gradient $\vec{\nabla}T$:^{3}

$$\vec{J} = \sigma(\vec{B})\vec{E} + \varepsilon''(\vec{B})\vec{\nabla}T \tag{1a}$$

$$\vec{U} = -T\tilde{\varepsilon}''(-\vec{B})\vec{E} - \lambda''(\vec{B})\vec{\nabla}T \tag{1b}$$

Here σ, ε'' ($\tilde{\varepsilon}''$ is the transpose of ε''), and λ'' are, respectively,
the electrical conductivity, the thermoelectric, and the thermal
conductivity tensors at temperature T and magnetic field $\vec{B}$. For $\vec{B}$
directed along a 3-fold or higher symmetry axis (defined as the
z-axis) of a single crystal, there are only three independent
components for each tensor, two of which determine the behavior of
the sample in transverse magnetic fields: e.g. for ε'' these are
ε''_{xx} and ε''_{yx}. The thermoelectric tensor of a metal consists of two
components, an electron diffusion component $(\varepsilon'')_d$ and a phonon-
drag component $(\varepsilon'')_g$, with only the diffusion component being subject to effects of mass enhancement^4. In this paper we will be
interested solely in the high magnetic field limit of $(\varepsilon''_{yx})_d$. It
it this quantity which yields γ^t.

The electronic specific heat coefficient of a metal has the
form $\gamma^c = \pi^2 k^2 N(\mu)/3$, where k is Boltzmann's constant. For a metal
with no open orbits in the plane perpendicular to $\vec{B}$, the high
magnetic field limit of $(\varepsilon''_{yx})_d$ is $\pi^2 k^2 N(\mu)T/3B$.[3] We write $(\varepsilon''_{yx})_d$
as $\gamma^t T/B$, thereby defining a transport specific heat γ^t which can
be compared with γ^c. The important feature of the high field limit
of $(\varepsilon''_{yx})_d$ which makes it suitable for accurately establishing the
presence of mass enhancement in thermoelectricity is that it is
completely independent of scattering. This eliminates from the
problem the only source of theoretical uncertainty. Moreover, it
also eliminates other predicted many-body effects which are associated with scattering.[5,6,7]

Since at high fields, both $(\varepsilon''_{yx})_d$ and $(\varepsilon''_{yx})_g$ vary as $1/B$,[4]
measurement of the high field limit of ε''_{yx} at any temperature T
yields the sum of these two quantities. They are separated by
using the fact that $(\varepsilon''_{yx})_d$ varies linearly with T, whereas $(\varepsilon''_{yx})_g$ is
expected to vary approximately as T^3.

A sample for measurement of the Nernst-Ettingshausen coefficient
is typically in the form of a strip with limbs as shown in Fig. 1.
The sample is maintained in vacuum, with one end connected to a cold-

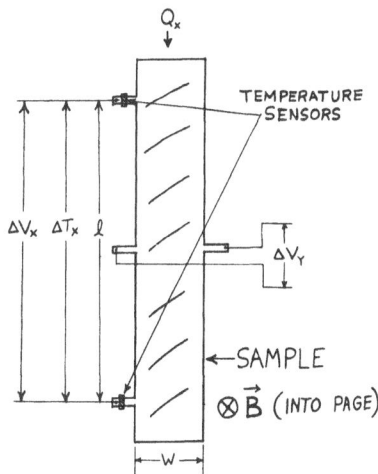

Fig. 1. A typical sample for Nernst-Ettingshausen Coefficient
measurements, with various experimental quantities indicated.

sink. A heater is attached to the other end and used to supply heat current Q_x which flows down the sample along the x-axis.

The most convenient quantities to measure for the extraction of γ^t are not the same for compensated and uncompensated metals. In the former case one measures Q^a, defined as

$$Q^a = -E_y/(\partial T/\partial x) = (\Delta V_y \ell)/(\Delta T_x w), \qquad (2a)$$

and for the latter P^a, defined as

$$P^a = E_y/U_x = -(\Delta V_y t)/Q_x . \qquad (2b)$$

Both are measured with boundary conditions $\vec{J} = 0$ and $U_y = 0 = U_z$. In eqs. (2a) and (2b), t is the sample thickness and the other quantities are as defined in Fig. 1. In the high field limit, the diffusion portions of Q^a and P^a yield γ^t according to the relations[8]

$$\gamma^t = Q_d^a B/\rho_{xx} T \qquad (3a)$$

$$\gamma^t = -P_d^a [\pi^2 k^2 (n_e - n_h)^2]/3B \qquad (3b)$$

Here ρ_{xx} is the diagonal component of the resistivity tensor for the compensated metal and $(n_e - n_h)$ is the difference between the number of electrons and holes per unit volume in the uncompensated metal.

III. LIMITATIONS ON THE ACCURACY OF γ^t

(a) _Phonon-drag_. To isolate $(\varepsilon''_{yx})_d$ it is necessary to choose experimental conditions under which it dominates $(\varepsilon''_{yx})_g$. This means going to low temperatures since $(\varepsilon''_{yx})_g$ decreases much more rapidly than $(\varepsilon''_{yx})_d$ as T approaches zero temperature. In practice it seems necessary to go below about $\Theta_D/100$ (Θ_D = Debye temperature of the metal) to achieve accuracies in γ^t of several percent, and several times lower to achieve 1% or better.

(b) _Other Limitations_. Even if effects of phonon-drag could be entirely eliminated, there are still residual sources of uncertainty which make 1% accuracy difficult to achieve. We briefly indicate the most important of these for Q^a and P^a.

A. Compensated Metal, Q^a

Examination of Eqs. (2a) and (3a) indicates that the determination of γ^t for a compensated metal requires measurement of several quantities. Most of these can be determined with an uncertainty of less than 1% using existing technology and reasonable care. The most difficult to determine accurately appear to be the geometrical factors in the ratio $\ell^2/(w^2 t)$ and, to a lesser extent, the temperature difference ΔT_x. Taken together, these make an absolute accuracy of better than a few percent hard to achieve.

B. Uncompensated Metal, P^a

In uncompensated metals, on the other hand, greater accuracy appears feasible, both because temperature differences do not have to be measured and because the determination of geometrical factors is more tractible; all other sources of error being no more difficult than for compensated metals. As indicated by Eq. (2b), the only geometrical factor which must be measured is the thickness t, and there exists an excellent experimental check on its value -- namely the Hall coefficient, $R_H = \rho_{yx}/B = \dfrac{-\Delta V'_y \cdot t}{IB}$, where I is the electric current and $\Delta V'_y$ is the Hall voltage. In the high field limit, R_H approaches the value $1/[(n_e - n_h)e]$, where e is the negative electronic charge. R_H can be calculated with high accuracy from known properties of the metal of interest. Under conditions where phonon-drag is negligibly small, there appears to be no fundamental reason why γ^t cannot be determined with an accuracy of better than 1% in an uncompensated metal, provided that adequate voltage sensitivity is available.

IV. DATA AND RESULTS

Measurements of Q^a have been made on Cd,[8,9] Sb,[10] W[8,11] and Mo.[12] While none of these measurements contradict the presence of mass enhancement, the only one which appears to allow a quantitative estimate of its value is the last, for which the ratio $\gamma^t/\gamma^c = 1.06 \pm 0.06$ was obtained. The major problem with most of the measurements seems to have been the difficulty in adequately correcting for effects of phonon-drag.

Measurement of P^a has, so far, been made only on Al,[13] which produced the ratio $\gamma^t/\gamma^c = 1.00 \pm 0.03$. Fig. 2 shows a plot of $-P^a/B$ versus T^2, with data for the temperature range 1.8K to 5K. γ^t is determined from the T = 0K intercept of the data. As can been seen, the major source of uncertainty in γ^t is due to the uncertainty in the extrapolation to T = 0K from data taken only above 1.8K.

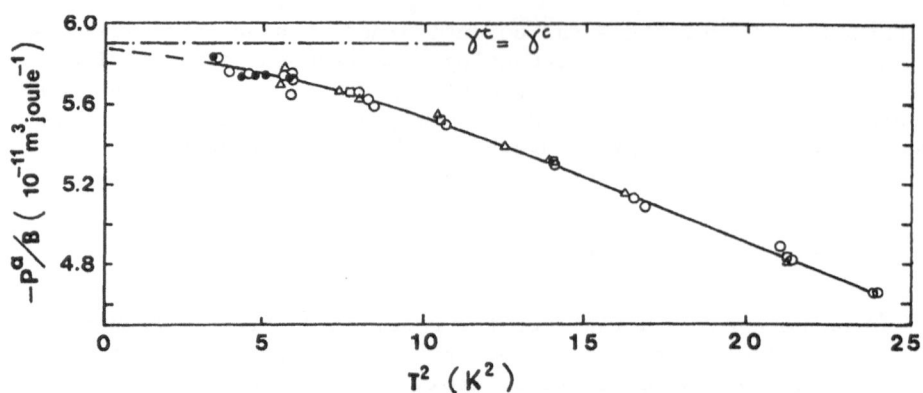

Fig. 2. $-P^a/B$ as a function of T^2 for a polycrystalline Al sample for magnetic fields of 1.5T($\triangle$), 1.8T($\square$), 2.0T(o), and 2.1T($\bullet$). The broken line indicates the zero temperature value predicted for $\gamma^t = \gamma^c$.

V. CONCLUSIONS AND SUGGESTIONS FOR FURTHER WORK

        Measurements of Q^a for Mo and of P^a for Al demonstrate convincingly that, as expected from theory, electron-phonon mass enhancement is present in thermoelectricity and that in $(\varepsilon''_{yx})_d$ it has the value expected to within at least a few percent. The main interest in further measurements appears to lie at lower temperatures, so as to completely eliminate effects of phonon-drag. Worthwhile measurements well below 1K include the following: (1) P^a for Al and alkali metals such as Na and K, to see whether sufficiently accurate data can be obtained to establish that electron-electron mass enhancement is also present in thermoelectricity (e.g. such enhancement is expected to produce roughly a 1% contribution in Al[14] and larger contributions in the alkali metals. (2) Q^a for W and Cd, to see whether the expected values of γ^t are finally obtained. Such measurements are necessary in order to make sure that inadequate treatment of phonon-drag is indeed the reason for disagreement between published data and expectation.

REFERENCES

1. R.E. Prange and L.P. Kadanoff, Phys. Rev. A134, 566 (1964).
2. J.L. Opsal, B.J. Thaler and J. Bass, Phys. Rev. Lett. 36, 1211 (1976).
3. R.S. Averback and D.K. Wagner Solid St. Commun. 11, 1109 (1972).
4. J.L. Opsal, J. Phys. F: Metal Phys. 11, 2349 (1977).

5. P.E. Nielsen and P.L. Taylor, Phys. Rev. B10, 4061 (1974).

6. S.K. Lyo, This volume and Phys. Rev. Lett. 39, 363 (1977).

7. A. Hasegawa, Solid St. Commun. 15, 1361 (1974); J. Phys. F 4, 2164 (1975).

8. R.J. Douglas and R. Fletcher, Phil. Mag. 32, 73 (1975).

9. C.G. Grenier, K.R. Efferson and J.M. Reynolds, Phys. Rev. 143, 406 (1966).

10. R.S. Blewer, N.H. Zebouni and C.G. Grenier, Phys. Rev. 174, 700 (1968).

11. J.R. Long, Phys. Rev. B3, 1197 (1971).

12. R. Fletcher, Phys. Rev. B14, 4329 (1976).

13. B.J. Thaler, R. Fletcher and J. Bass, J. Phys. F: Metal Phys. (In Press).

14. T.M. Rice, Ann. Phys. 31, 100 (1965).

DISCUSSION

K. Böning: In the last part, the second slide, I think, you showed two curves of the high field Nernst-Ettingshausen coefficient, and the two curves showed different saturation values. Is this effect understood?

J. Bass: That effect is exactly this: the 2.3K value I show here; 2.3 squared is about 5, and gives this number, here. 4.6K squared is about 25, and gives the value up there. So the effect is understood if you accept my argument that it is phonon-drag. If you do not accept the argument that it is phonon drag, then it is not understood.

J. Garland: Your Hall coefficient of aluminum, although it looks nominally constant, appears to be decreasing slightly. This reminds me of other data I have seen on simple metals which show this slow decrease in the high field Hall coefficient. Is there a possible decrease of the Hall coefficient in your experiment?

J. Bass: Let's say it is not impossible. We are not sure why that is occurring; whether it represents a remnant of an inability to perfectly average forward and reverse fields, whether it represents a possible small magnetoresistance of the standard resistor against which we are making these measurements - we think it does not - or whether it represents a small but real effect. At the moment, and for present purposes, in this field range the agreement is within one percent and I don't want to push our data more than that. (Our present uncertainty is of order half a percent.) However, the question raised is an interesting one, and we hope to investigate it further when we get more accurate data to see if it could possibly be a real effect.

P.B. Allen: I'm not sure I understand what you mean by the electron-electron enhancement effect. Are you suggesting there should be, because of coulomb scattering, a breakdown of the observed equality between thermal specific heat and transport specific heat?

J. Bass: I have every reason to expect that when we make more accurate measurements we will get exactly the same results as one gets for electronic specific heat. On the other hand, I want to see this result experimentally before I accept it as gospel. If, on the contrary, I see a difference between the two, I would not be completely astonished. I have learned in this field not to trust the theorists too far, and not to trust my own intuition either if it can be checked experimentally. I am merely raising here a possibility which I think should be investigated.

SEMI-CLASSICAL THEORY OF PHONON-DRAG MAGNETOTHERMOELECTRICITY

IN METALS

Jon L. Opsal

Physics Department
Michigan State University
East Lansing, Michigan 48824, USA

ABSTRACT

Using semi-classical transport theory, exact expressions are obtained for the low temperature phonon-drag contribution to the thermoelectric tensor of a metal in a high transverse magnetic field, H. It is shown that the off diagonal elements (ε_{xy} and ε_{yx}) of the thermoelectric tensor to leading order go as $\frac{1}{H}$ and, under certain symmetry conditions, are independent of the impurity scattering. It is also shown that under the same conditions the diagonal elements to leading order go as $\frac{1}{H^2}$. It is especially significant that ε_{xy} does not contain any of the complex many-body effects predicted by Nielsen and Taylor[1] and by Hasegawa[2]. Since ε_{xy} is also amenable to measurement, the real possibility exists, for the first time, of testing the theory of phonon-drag in metals.

When studying thermoelectricity in metals at low temperatures one is confronted with the problem of separating the electron-diffusion and phonon-drag contributions. Most of the work thus far has been done in zero magnetic field on the thermopower, S, which is assumed to be the sum of a diffusion part S^d and a phonon-drag part S^g. Using simple models, one expects S^d to depend linearly on the temperature T and S^g to vary as T^3. Consequently, the temperature dependence of the experimentally observed S is normally used when attempting to isolate the two contributions S^d and S^g.

Recently, however, serious doubt has been cast on the validity of this approach by the theoretical work of Nielsen and Taylor[1] and Hasegawa[2] who predict contributions to S^d which are expected to have a temperature dependence similar to S^g. These additional contributions to S^d come from electron-phonon many-body corrections to the scattering of electrons by impurities and phonons. While no calculations have yet been done for real metals, these effects are expected to be significant based on free electron model calculations. It is therefore clear that any further understanding of zero field thermopower will rely on our ability to calculate accurately either S^d or S^g.

The aim of this paper is to show that by considering thermo-electric transport in a high transverse magnetic field for a metal having only closed orbits, one finds a thermoelectric transport coefficient which is independent of the complicated many-body effects mentioned above, and for which the phonon-drag can experimentally be isolated accurately. In order to make the physical significance of the results to be presented more transparent, it is convenient to introduce the phenomenological linear transport equation,

$$\vec{J} = \overset{\leftrightarrow}{\sigma} \cdot \vec{E} - \overset{\leftrightarrow}{\varepsilon} \cdot \vec{\nabla}T \quad , \tag{1}$$

which relates the electric current density $\vec{J}$ to the electric field $\vec{E}$ and temperature gradient $\vec{\nabla}T$. For cubic systems in zero magnetic field, the conductivity $\overset{\leftrightarrow}{\sigma}$ and thermoelectric tensor $\overset{\leftrightarrow}{\varepsilon}$ become scalar quantities and with the boundary condition $\vec{J} = 0$, the measured thermopower is simply $S = E_x/\nabla_x T = \varepsilon/\sigma$. However, as discussed above there are many contributions to S which are difficult to calculate and apparently impossible to separate experimentally. In the presence of a magnetic field, $\vec{H}$ (with the direction of $\vec{H}$ defining the z axis), $\overset{\leftrightarrow}{\sigma}$ and $\overset{\leftrightarrow}{\varepsilon}$ become tensors and in general, several independent measurements are required to determine their individual components. At first sight this appears to have made the problem much more complicated. However, from the semi-classical transport theory of Lifshitz, Azbel, and Kaganov[3], we already know that the leading term (in 1/H) for σ_{xy} can be readily calculated in the high field limit ($\omega\tau \gg 1$) for a metal with only closed orbits and the result is completely independent of scattering. Furthermore, it is found that the diffusion part of ε_{xy} is also independent of scattering[4] under the same conditions. In the present paper it will be shown that the phonon-drag part, ε_{xy}^g, in addition to being proportional to $1/H$, is also, under certain conditions, independent of scattering in the high field limit. This means that for any particular metal, ε_{xy}^g has a specific value which might depend on the orientation of $\vec{H}$ relative to the crystal, but unlike phonon-drag in S[5-7] it is not sensitive to the nature of the impurity scattering. It is of particular importance that ε_{xy} does not contain any of the complex many-body effects predicted by Nielsen and Taylor[1] and by Hasegawa[2]. Since ε_{xy} is also amenable to measurement[8], we now have, for the first time, a real possibility of testing the theory of phonon-drag in metals.

Although not strictly necessary, it is simpler to analyze the thermoelectric transport problem in terms of the Peltier tensor $\overleftrightarrow{\pi}$ (with zero temperature gradient) rather than $\overleftrightarrow{\varepsilon}$ and we shall proceed in this manner. The heat current density $\vec{Q}$ when linearly related to $\vec{E}$ defines $\overleftrightarrow{\pi}$ by

$$\vec{Q} = \overleftrightarrow{\pi} \cdot \vec{E} \quad , \tag{2}$$

and the elements of $\overleftrightarrow{\pi}$ are related to those of $\overleftrightarrow{\varepsilon}$ by the Onsager relation $\pi_{ij}(\vec{H}) = T\,\varepsilon_{ji}(-\vec{H})$. In the presence of the electric field, both the electron and phonon system will be driven out of equilibrium; the electrons because of their direct interaction with $\vec{E}$ and the phonons because of their interactions with the electrons. The electron-diffusion and phonon-drag contributions to $\overleftrightarrow{\pi}$ are then obtained, respectively, from the electron and phonon contributions to the heat current.

Let us consider first the electrons which we will characterize by a single particle distribution function f_k. We can then use the dynamics of a single electron to find f_k. An electron in the presence of $\vec{E}$ will experience a force $e\vec{E}$ which changes its energy ε_k by an amount $d\varepsilon_k$, $d\varepsilon_k = e\vec{E} \cdot \vec{v}_k dt$, in the time interval dt, where $\vec{v}_k$ is the electron's velocity. If we assume a relaxation time τ_k, then the average change in energy is given by

$$\delta \epsilon_k = e\vec{E} \cdot \int_0^\infty dt \ \vec{v}_k \ e^{-t/\tau_k} \tag{3}$$

where e^{-t/τ_k} is the probability that the electron has not under-
gone a collision in time t from its most recent collision. Since
we are in effect assuming $\delta\epsilon_k$ to be small, f_k can be obtained by
simply displacing the equilibrium distribution f_k° by $\delta\epsilon_k$; that is,

$$f_k = f_k^\circ - \frac{\partial f_k^\circ}{\partial \epsilon_k} \ \delta\epsilon_k \quad . \tag{4}$$

We expect a similar result for the phonon distribution n_q,

$$n_q = n_q^\circ - \frac{\partial n_q^\circ}{\partial \epsilon_q} \ \delta\epsilon_q \quad , \tag{5}$$

except that the displacement $\delta\epsilon_q$, while still linear in E, depends
also on the electron-phonon interaction. We can obtain an expression for $\delta\epsilon_q$ by calculating that fraction of the electron's
additional energy $\delta\epsilon_k$ which is ultimately transferred to the
phonon system. The net probability, relative to all phonon processes which can occur, for creating a phonon of wavevector $\vec{q}$ due
to the interaction with an electron of wavevector $\vec{k}$, when
multiplied by $\delta\epsilon_k$ and the result summed over all initial and final
electron states, should give $\delta\epsilon_q$. We can write this explicitly as

$\delta\epsilon_q = r_q^{-1} \sum_{kk'} (p_k^{k'q} - p_{kq}^{k'}) \ \delta\epsilon_k$, where $p_k^{k'q}$ and $p_{kq}^{k'}$ are, respectively,

equilibrium transition rates for emission or absorption of a phonon
of wavevector $\vec{q}$ in which an electron is scattered from state $\vec{k}$ to
$\vec{k}'$,

$r_q = \sum_{kk'} p_{kq}^{k'} + n_q^\circ(n_q^\circ+1)/\tau_q$, and τ_q is a phonon relaxation time

corresponding to all phonon scattering processes other than electron-
phonon scattering. Since $p_{kq}^{k'} = p_{k'}^{kq}$, by interchanging $\vec{k}$ and $\vec{k}'$ in

the first sum we have

$$\delta\varepsilon_q = r_q^{-1} \sum_{kk'} P_{kq}^{k'}(\delta\varepsilon_{k'} - \delta\varepsilon_k) \quad . \tag{6}$$

The problem has now been reduced to finding $\delta\varepsilon_k$, which is simply another way of saying that the phonon distribution is completely determined by the electron distribution. In zero magnetic field $\delta\varepsilon_k = e\vec{E} \cdot \vec{v}_k \tau_k$, and the results for $\overset{\leftrightarrow}{\pi}{}^g$ are exactly those obtained previously by Bailyn[7]. When $\vec{H} \neq 0$ the integral in Eq. (3) is more complicated since $\vec{v}_k$ must be obtained as a function of time according to the equation of motion $\hbar \frac{d\vec{k}}{dt} = \frac{e}{c} \vec{v}_k x \vec{H}$, which in general is difficult to solve.

As an example, consider the free electron case which can be solved exactly with the result,

$$\delta\varepsilon_k = \frac{e\vec{E} \tau \cdot [\vec{v}_k - \vec{v}_k x \vec{\omega} \tau]}{1 + (\omega\tau)^2} \quad , \tag{7}$$

where $\vec{\omega} = \frac{e\vec{H}}{mc}$ and $\vec{H}$ is perpendicular to $\vec{E}$. Recalling that the direction of $\vec{H}$ defines the z axis, let's focus our attention for the moment on the off diagonal element of the phonon-drag contribution, π_{xy}^g. Neglecting Umklapp processes in Eq. (6) and assuming full phonon-drag (i.e., $r_q = \sum_{kk'} P_{kq}^{k'}$) we obtain $\pi_{xy}^g(H) = \frac{\omega\tau}{1+(\omega\tau)^2}$ $\pi_{xx}^g(0)$, where $\pi_{xx}^g(0) = \frac{e\tau}{m} \frac{1}{3} C_g T$ and C_g is the lattice specific heat. In the high field limit H→∞,

$$\pi_{xy}^g = \frac{cT}{H} \frac{1}{3} C_g \quad , \tag{8}$$

the result obtained previously by Zebouni, Blewer, and Grenier[8]. The important features of this last result are that π_{xy}^g is (like π_{xy}^d [4]) independent of impurity scattering and proportional to 1/H.

In addition, it is independent of the sign of the charge carrier.
Interestingly enough, these features are retained for Bloch
electrons in a metal having only closed orbits provided $\vec{H}$ is along
an axis of two-fold or higher symmetry and the orbits possess
this same symmetry. A rigorous proof of this last statement is
given elsewhere[9], but the precise form for π^g_{xy} can be obtained
using the more intuitive approach of the present paper. In the
high field limit, the integral in Eq. (3) can be done exactly and
after some manipulation we find $\delta\epsilon_k = \frac{\hbar c}{H}(E_y\tilde{k}_x - E_x\tilde{k}_y)$, where $\tilde{k}_x$ and
$\tilde{k}_y$ are components of the wavevector measured from the center of
symmetry of the orbit, $\tilde{k}_x = k_x - \frac{1}{2\pi}\oint d\phi k_x$ and $\tilde{k}_y = k_y - \frac{1}{2\pi}\oint d\phi k_y$,
and ϕ is the orbit variable describing the position of the electron
along its orbit. Using this result in Eq. (6) and then calculating
the phonon heat current, we obtain for π^g_{xy},

$$\pi^g_{xy} = \frac{\hbar cT}{H} \sum_q \frac{\partial n^o_q}{\partial T} v_{qx}\, r^{-1}_q \sum_{kk'} p^{k'}_{kq}(\tilde{k}'_x - \tilde{k}_x), \qquad (9)$$

where $\vec{v}_q$ is the phonon velocity. Under the symmetry conditions
specified above, this result is shown in Ref. (9) to be exact by
using the known high field solution of the Boltzmann equation.
Although Eq. (9) is simple in appearance, and it is true that when
Umklapp processes can be neglected it reduces to the simple result
in Eq. (8), the fact that in most metals Umklapp processes per-
sist down to the lowest attainable temperatures implies that any
quantitative statements about π^g_{xy} must come from a detailed
numerical calculation.

Since the diagonal elements π^d_{xx} and π^d_{yy} to leading order go
as $1/H^2$,[3] we want to consider the possibility of $1/H$ contributions
to π^g_{xx} and π^g_{yy}. In exactly the same manner used to obtain Eq. (9),
we find for π^g_{yy} to leading order in $1/H$, $\pi^g_{yy} = \frac{\hbar cT}{H}\sum_q \frac{\partial n^o_q}{\partial T} v_{qy}\, r^{-1}_q \sum_{kk'}$
$p^{k'}_{kq}(\tilde{k}'_x - \tilde{k}_x)$, which will not vanish. However, when $\vec{H}$ is along a
three-fold or higher symmetry direction, or when $\vec{H}$ is along a two-
fold symmetry direction and the x or y axis is also a two-fold symmetry direction, the $1/H$ contribution to π^g_{yy} will be zero. Therefore, π^g_{yy} and similarly π^g_{xx} are, in this case, proportional to

$1/H^2$. It is important to note that even if conditions are such that the $1/H$ contributions to π^g_{xx} and π^g_{yy} do not vanish, it is still possible to experimentally isolate π_{xy} and π_{yx} through the adiabatic Nernst-Ettingshausen effect[8]. However, in a measurement of the adiabatic magnetothermopower (defined as $S_a = E_y / \nabla_x T$ with $\vec{J}=0$ and $Q_y = 0 = Q_z$) any $1/H$ contribution to π_{xx} and π_{yy} would result in a contribution to S_a proportional to H which changes sign when the direction of H is reversed.

In conclusion, the real possibility now exists for testing the theory of phonon-drag thermoelectricity in metals. All of the parameters needed to calculate π^g_{xy} as given by Eq. (9) are in many cases known quite accurately and the experimental determination of π^g_{xy} is relatively straightforward. A favorable test of this theory, it seems, would then justify using the calculated phonon-drag thermopower in zero field to isolate the diffusion thermopower involving the many-body effects predicted by Nielsen and Taylor[1] and by Hasegawa[2].

REFERENCES

1. P.E. Nielsen and P.L. Taylor, Phys. Rev. B10, 4061 (1974).
2. A. Hasegawa, Solid St. Commun. 15, 1361 (1974); J. Phys. F 4 2164 (1974).
3. I.M. Lifshitz, M.I. Azbel, and M.I. Kaganov, Zh. Eksp. Teor. Fiz. 31, 63 (1956) (Eng. transl.: Soviet Phys. JETP 4, 41 (1957).
4. M.I. Azbel, M.I. Kaganov, and I.M. Lifshitz, Zh. Eksp. Teor. Fiz. 32, 1188 (1957) (Eng. transl.: Soviet Phys. JETP 5, 967 (1957)).
5. R. Fletcher and J.S. Dugdale, published in conference proceedings, LT-10, Moscow, p. 246 (1966).
6. C. Van Baarle, Physica 33, 424 (1967).
7. M. Bailyn, Phys. Rev. 157, 480 (1967).
8. R.S. Blewer, N.H. Zebouni, and C.G. Grenier, Phys. Rev. 174, 700 (1968).
9. J.L. Opsal, J. Phys. F7, 2349 (1977).

THE MAGNETOTHERMOPOWER OF SINGLE CRYSTAL INDIUM*

B. J. Thaler† and R. Fletcher††

Physics Department
Michigan State University
East Lansing, Michigan 48824 USA

ABSTRACT

It has been previously shown that the application of a magnetic field produces large changes in the thermopower S^a of In and Al (as well as other metals) at temperatures of order $\theta_D/10$, and initial attempts to explain these results have not been completely successful. By measuring a sufficient number of transverse magnetotransport properties, six being the minimum, it is possible to isolate the contributions of the two components of the thermoelectric tensor ε'', i.e. ε''_{xx} and ε''_{yx}, to the magnetothermopower. We have used this approach to investigate the origin of the magnetothermopower of a high purity In single crystal with fields of 0-2T parallel to the tetrad axis, at temperatures of 2.5-8 K. For fields greater than a few hundred gauss, we find that the magnetothermopower is basically due to ε''_{yx}, and in fact for most temperatures and fields $S^a \simeq \rho_{yx}\varepsilon''_{yx}$ where ρ_{yx} is the Hall resistivity. This result enables us to prove that phonon drag is mainly responsible for the observed effects at these intermediate temperatures.

Until recently, it was generally assumed, without much experimental evidence, that a magnetic field would not have any significant effect on the thermopower of metals, particularly in the case of the phonon drag. It is now known that the thermopower of a variety of

---

*Work supported by N.S.F. Grants DMR-77-04680 and DMR-75-01584.
†Present address: Physics Department, Northwestern University, Evanston, Illinois 60201.
††Permanent address: Physics Department, Queen's University, Kingston, Ontario, Canada K7L 3N6.

metals does in fact exhibit large changes in the presence of moderate fields.[1] To proceed with the discussion in more detail, it is convenient to introduce the thermoelectric tensor ε'' according to the equation

$$\tilde{J} = \rho^{-1} \tilde{E} + \varepsilon'' \nabla T,$$

where ρ is the isothermal resistivity tensor, $\tilde{J}$ the electric current density, and $\tilde{E}$ and ∇T the electric field and temperature gradient, respectively. Using this equation, and defining the thermal resistivity tensor γ by $\nabla T = -\gamma \tilde{U}$ (with $\tilde{J} = 0$), $\tilde{U}$ being the thermal current density, one finds that the adiabatic thermopower S^a [$=E_x/(\partial T/\partial x)$ with $\tilde{J} = 0 = U_y$] is given by

$$S^a = -\varepsilon''_{xx} [\rho_{xx} - \frac{\gamma_{ys}\rho_{yx}}{\gamma_{xx}}] + \varepsilon''_{yx} [\rho_{yx} + \frac{\gamma_{yx}\rho_{xx}}{\gamma_{xx}}] \tag{1a}$$

which, for convenience, we will rewrite as

$$S^a = -\varepsilon''_{xx} A + \varepsilon''_{yx} C. \tag{1b}$$

We have assumed the field B to be along z and that this direction is a high symmetry crystal axis (at least 3 fold). At zero field the above expression reduces to $-\varepsilon''_{xx}\rho_{xx}$, and at high fields the only term that we are able to eliminate on theoretical grounds, at least for uncompensated metals with no open orbits, is this same term, i.e. $-\varepsilon''_{xx}\rho_{xx}$ (because $\rho_{yx}\gamma_{yx} \gg \rho_{xx}\gamma_{xx}$ in this limit).

At very low temperatures, where diffusion effects predominate and one can assume elastic scattering, the change in thermopower of Al that is induced by a magnetic field has received a satisfactory explanation.[2] In this case the change in thermopower is predominantly associated with the term $\varepsilon''_{yx} C$, which in the low temperature limit reduces to $2\rho_{yx}\varepsilon_{yx}$.

The origin of the large thermopower changes that occur near $\theta_D/10$, due to the application of a magnetic field, is not understood (θ_D is the Debye temperature). A calculation in this range is a far more ambitious undertaking because of the predominance of inelastic scattering and phonon drag. The aim of the present work is to attack the problem from an experimental standpoint. We can directly measure all the resistivities ρ_{xx}, ρ_{yx}, γ_{xx} and γ_{yx} that appear in Eq. 1a for a particular sample; furthermore, we can extract the components ε''_{yx} and ε''_{xx} from measurements on S^a and the Nernst-Ettingshausen coefficient P^a ($=E_y/U_x$ with $\tilde{J} = 0 = U_y$), this latter being given by an expression similar to Eq. 1b, i.e.

$$(P^a/\gamma_{xx}) = \varepsilon''_{xx} C + \varepsilon''_{yx} A, \tag{2}$$

where A and C have the same meanings as in Eq. 1b.

We have previously carried out such a program on a pure single crystal of Mo at temperatures below 4.2 K where diffusion effects were dominant;[3] this enabled us to pinpoint the source of the unexpected behavior of the magnetothermopower of that material. Our present investigation is aimed at a temperature region where phonon drag should play a significant role, and In proves to be a convenient material because of its low Debye temperature. This choice enables us to reach a temperature of about $\theta_D/14$ by 8 K, a limit imposed by the superconducting - normal transition of our Nb-Ti potential leads at fields of about 2T.

The In single crystal was in the form of a strip with protruding limbs to which the potential probes and thermometers were attached. It was grown in a high purity graphite mold from Cominco 69 grade material and had its long axis along [100]. The measured residual resistivity ratio (R_{293}/R_0) was about 50,200, and we estimate a value for the bulk material, after correcting for surface scattering effects, of 91,000. By 5.5 K, 90% of the electrical resistivity and 96% of the thermal resistivity are due to phonon scattering. The magnetic field $\vec{B}$ was perpendicular to the current and directed along [001], the tetrad axis.

Space is too limited to enable us to present all our data. However, it is worth pointing out that an overall check on the experimental accuracy is provided by the limiting high field values of ρ_{yx} and γ_{yx} which have well known theoretical values;[4,5] the experimental results yield these values to within our estimated uncertainties of a few percent.

Our experimental results on S^a are shown in Fig. 1. It is seen that S^a always becomes more negative in the presence of the magnetic field, in agreement with previous data on polycrystalline In.[6,7] We also notice that S^a saturates at sufficiently high fields. This is in accordance with Eq. 1a because theory shows[4,5,8] that, for an uncompensated metal with no open orbits, the various coefficients have high field limits as follows: $\rho_{xx} \sim$ constant, $\gamma_{xx} \sim$ constant, $\rho_{yx} \sim B$, $\gamma_{yx} \sim B$, $\varepsilon''_{xx} \sim B^{-2}$, $\varepsilon''_{yx} \sim B^{-1}$ (these last two being correct even with phonon drag). To attain the high field limit, the electrons are required to make many orbital revolutions before being scattered; it follows that as the temperature increases and electronic scattering becomes more frequent, we need higher fields to achieve saturation, and this is what we observe.

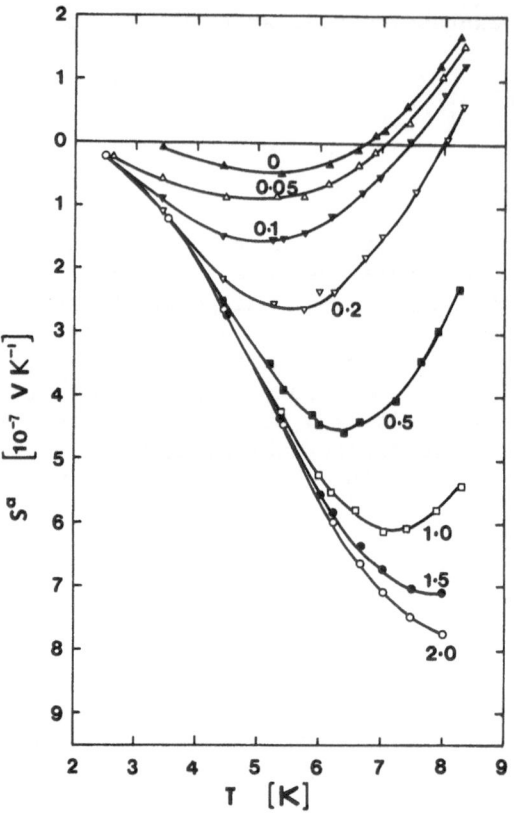

Fig. 1. The measured thermopower S^a as a function of temperature. The magnetic field (in Tesla) is given on each curve. For clarity, many of the low temperature points are omitted.

    Using all our data, we are able to evaluate ε''_{xx} and ε''_{yx}. It turns out that ε''_{yx} plays the major role in determining the behavior of S^a and we illustrate this by plotting, in Fig. 2 $\varepsilon''_{yx}C$ (see Eq. 1b) for various field values. It is seen that the general shape and magnitude of S^a and $\varepsilon''_{yx}C$ are very similar. The other contribution, namely $-\varepsilon''_{xx}A$, is of oppostie sign and can be obtained from Figs. 1 and 2 as the difference, $S^a - \varepsilon''_{yx}C$. For $T \gtrsim 5$ K, we find that $|\varepsilon''_{xx}A| \lesssim |\varepsilon''_{yx}C/2|$ and indeed Eq. 1b shows that when $A = 0$ (i.e. when $\rho_{xx}\gamma_{xx} = \rho_{yx}\gamma_{yx}$), then S^a is determined purely by $\varepsilon''_{yx}C$. One can also show that the factor C is itself dominated by ρ_{yx} (in Eq. 1a, the ratio $(\rho_{yx}\gamma_{xx}/\gamma_{yx}\rho_{xx})$ is typically 4-5 for $B \gtrsim$ few kG) which leads us to conclude that the observed magnetothermopower is, to a large extent, a measure of $\rho_{yx}\varepsilon''_{yx}$.

    We know that ε''_{yx} is a linear sum[8] of a diffusion part $(\varepsilon''_{yx})_d$ and a phonon drag part $(\varepsilon''_{yx})_g$. It is possible to accurately

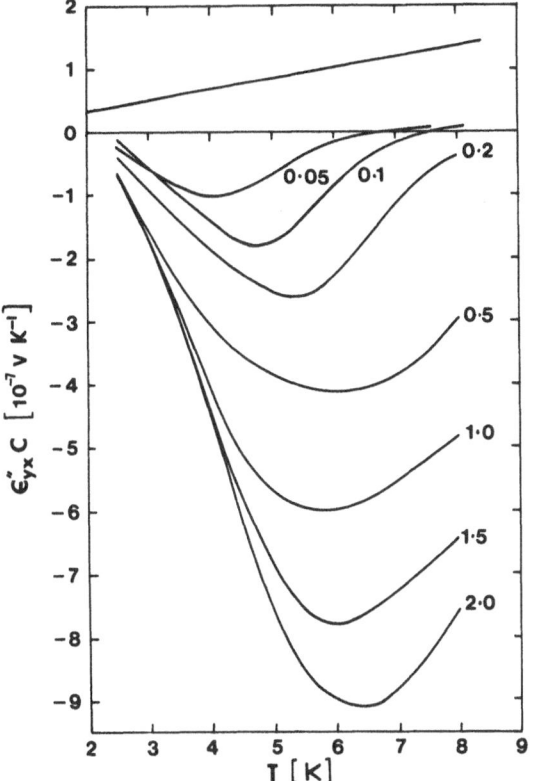

Fig. 2. The contribution $\varepsilon''_{yx}C$ to the thermopower. The straight line is the calculated high field value of $(\varepsilon''_{yx})_d\rho_{yx}$ and is given by $1.7\times10^{-8}T$ VK^{-1}.

calculate the high field limit of $\rho_{yx}(\varepsilon''_{yx})_d$ and it is found to have a value of $+1.7\times10^{-8}T$ VK^{-1}, independent of scattering. This has the wrong sign and is an order of magnitude smaller than S^a (or $\rho_{yx}\varepsilon''_{yx}$) at higher temperatures. Thus the observed effects must result from $(\varepsilon''_{yx})_g\rho_{yx}$ and although this term is amenable to calculation, at least in the high field limit, such a calculation has not yet been attempted.

We can summarize our findings as follows:

1. S^a is a simple quantity only at zero field where it yields $-\varepsilon''_{xx}\rho_{xx}$; in the presence of a field the situation is far more complex, though experimentally the dominant contribution is $\varepsilon''_{yx}\rho_{yx}$ in the case of In. The zero and high field limits of S^a are not related in any simple manner. Previous experiments[2] on Al in the diffusion regime have indicated that the <u>change</u> in thermopower

between the zero and high field limits is approximately $C\varepsilon''_{yx}$ (or $2\varepsilon''_{yx}\rho_{yx}$) and in fact the component arising from $-\varepsilon''_{xx}A$ is relatively constant (and not small in the high field limit). In the present case, the term $C\varepsilon''_{yx}$ (now approximately $\varepsilon''_{yx}\rho_{yx}$) actually dominates the high field results, but it would still be correct to say that the change in thermopower in the two cases is due essentially to the same term.

2.   The dominance of ε''_{yx} enables us to prove that phonon drag is responsible for the observed effects at intermediate temperatures. Caplin et al.[4] have, in fact, previously suggested that phonon drag plays a major role, but they were not able to show this with any certainty. In retrospect, it is clear that the attempt of Blatt et al.[9] to fit all the experimental data in terms of a diffusion model is incorrect, though of course their general arguments with respect to the diffusion term may still be valid.

3.   As Eq. 1a shows, the magnetothermopower is a rather complex quantity - far more so than any other transport coefficient in the high field limit. In general, an experimental coefficient is far closer to those quantities that are actually calculated than is the case for S^a. Even P^a reduced to the relatively simple result $P^a = -\varepsilon''_{yx}\gamma_{yx}\rho_{yx}$ in the high field limit. As we have seen, the magnetothermopower contains six independent coefficients and in the high field limit, the only new information that S^a provides (i.e. which cannot be obtained more accurately and easily from other measurements) is ε''_{xx}. However ε''_{xx} cannot be extracted with high accuracy because of the presence of all the other quantities, and because it tends to be overwhelmed by the contribution due to ε''_{yx}. It is possible to show that ε''_{xx} can be obtained far more accurately from experimental measurements on samples with a Corbino disc geometry[10] (assuming that the field is oriented along a crystal axis of at least threefold symmetry and that there are no open orbits). Thus we could argue that if the present results are indicative of the general situation, the measurement of magnetothermopower is possibly not quite as useful nor as interesting as we might have presupposed.

REFERENCES

1. For a survey, see F.J. Blatt, P.A. Schroeder, C.L. Foiles and
    D. Greig, Thermoelectric Power of Metals (Plenum Press, New
    York, 1976), p. 217.
2. J.L. Opsal and D.K. Wagner, J. Phys. $\underline{F6}$, 2323 (1976).
3. R. Fletcher, J.L. Opsal and B.J. Thaler, J. Phys. $\underline{F}$ (in press).
4. I.M. Lifshitz, M. Ia Azabel' and M.I. Kaganov, Zh. Eksp. Teor.
    Fiz. $\underline{31}$, 63 (1956) [Sov. Phys. JETP $\underline{4}$, 41 (1957)].
5. M. Ia Azabel', M.I. Kaganov and I.M. Lifshitz, Zh. Eksp. Teor.
    Fiz. $\underline{32}$, 1188 (1957) [Sov. Phys. JETP $\underline{5}$, 967 (1957)].
6. A.D. Caplin, C.K. Chiang, J. Tracy and P.A. Schroeder, Phys.
    Stat. Sol. $\underline{26A}$, 497 (1974).
7. R.S. Averback and J. Bass, Phys. Rev. Lett. $\underline{26}$, 882 (1971).
    Errata, Phys. Rev. Lett. $\underline{34}$, 631 (1975).
8. J.L. Opsal, J. Phys. $\underline{F}$ (in press).
9. F.J. Blatt, C.K. Chiang and L. Smrcka, Phys. Stat. Sol. $\underline{24A}$,
    621 (1974).
10. The Corbino disc geometry refers to a sample in the form of a
    disc with radial current flow. The field $\vec{B}$ is normal to the
    disc.

GIANT QUANTUM OSCILLATIONS IN THE THERMOELECTRIC PROPERTIES OF

ALUMINUM

J. B. Sampsell and J. C. Garland

The Ohio State University

Columbus, Ohio   USA

ABSTRACT

We have measured the two thermoelectric coefficients S(T, H) and G(T, H) for polycrystalline and single crystal aluminum samples in magnetic fields to 5T from 3 - 6K. All measurements were made in a longitudinal geometry with the electrical current and heat current in the samples parallel to the applied magnetic field. An analysis of the oscillations is given in terms of magnetic breakdown and current distortion effects.

Measurements of the thermoelectric power of aluminum in a transverse magnetic field have been reported by Thaler and Bass[1] and by Kesternich and Papastaikoudis.[2] In each case the field dependence of the thermoelectric power of single crystal aluminum was dominated by giant quantum oscillations (GQO), whose origin was attributed to magnetic breakdown of the third zone β-orbits. The oscillations are expected to be pronounced only in carefully aligned single crystal specimens with the magnetic field aligned transverse to the direction of heat flow. In this paper we report measurements of the thermoelectric power and of the thermoelectric function G of aluminum in a longitudinal magnetic field. In contrast to our expectations, we have found both of these longitudinal thermoelectric coefficients to exhibit large quantum oscillations, both in oriented single crystal specimens and in polycrystalline specimens. We attribute the presence of these oscillations to macroscopic distortions of the heat current in the specimens induced by specimen inhomogeneities. In strong magnetic fields, inhomogeneities whose size are comparable to a mean free path are believed to induce large irregularities in the electric and heat current flow pattern of metals and have been suggested as a cause for the anomalous linear magnetoresistance frequently observed in aluminum and the other simple metals.

The experimental apparatus, shown schematically in Fig. 1, is similar in design to that described previously by Garland and Van Harlingen.[3] An rf biased SQUID was used as a galvanometer in a null-detecting potentiometric configuration. With careful attention to shielding and to minimizing vibration the circuit was capable of 10^{-13} V resolution in fields of 6.0 T. Temperature differences across the samples were measured with field-independent Lakeshore capacitance thermometers which were independently calibrated during each experimental run against a germanium standard thermometer.

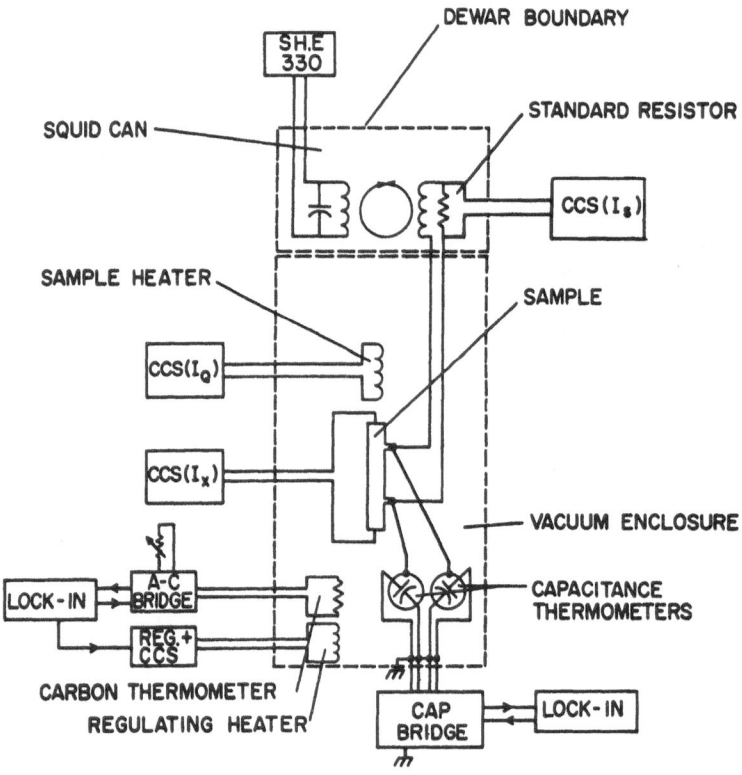

Fig. 1. A schematic diagram of the apparatus used in this experiment, showing the SQUID detector, constant current supplies, capacitance thermometers, and temperature regulator. The four-terminal germanium reference thermometer is not shown in the diagram.

The results of measurements made on a polycrystalline sample
and on a single crystal sample whose (100) axis was oriented along
the magnetic field direction are shown in Figures 2 and 3; part (a)
of each figure shows data for the polycrystal while part (b) shows
the single crystal results. Figure 2 shows the magnetic field
dependence of the longitudinal thermoelectric function G, defined
by

$$G = (\frac{J_z}{U_z})_{\vec{E}=0} \tag{1}$$

where J_z and U_z are the electric and thermal currents oriented along
the magnetic field direction. The relationship between G and the
transport coefficients of the metal may be seen by noting that in a
longitudinal geometry the general transport equations simplify to

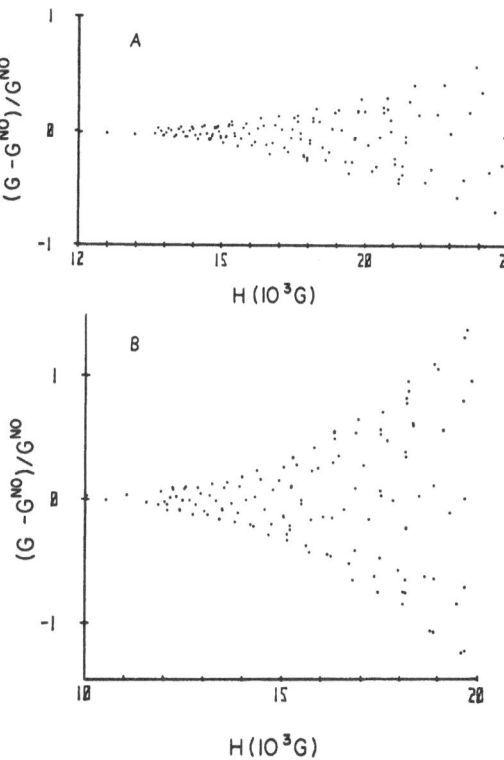

Fig. 2. The field dependence of the oscillatory component of the
longitudinal thermoelectric function G, (a) for polycrystalline
aluminum, and (b) for single crystal aluminum.

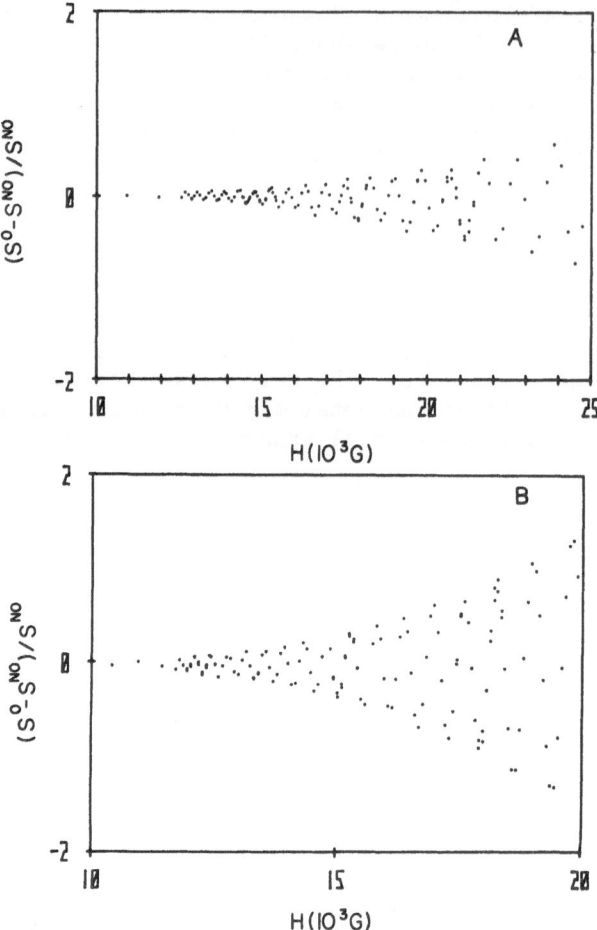

Fig. 3. The field dependence of the oscillatory component of the longitudinal thermoelectric power S, (a) for polycrystalline aluminum, and (b) for single crystal aluminum.

$$J_z = \sigma_{zz} E_z - \varepsilon_{zz} \frac{\partial T}{\partial z} \qquad (2)$$

$$U_z = T\varepsilon_{zz} E_z - \kappa_{zz} \frac{\partial T}{\partial z} \qquad (3)$$

where σ_{zz} and κ_{zz} are the electrical and thermal conductivity tensor components respectively, and ε_{zz} is the z-component of the thermo-electric transport tensor. Combining Eq. (1) and Eq. (2) we see that

$$G = \frac{\varepsilon_{zz}}{\kappa_{zz}} \tag{4}$$

measured in the absence of an electric field.

Figure 3 shows the magnetic field dependence of the longitudinal thermoelectric power S, defined by

$$S = \frac{E_z}{\partial T/\partial z} = \frac{\varepsilon_{zz}}{\sigma_{zz}} \quad . \tag{5}$$

Data for both G and S are shown in a reduced form since the observed oscillations are superimposed upon a slowly varying component; the quantities shown in the figures are $(G-G^{NO})/G^{NO}$ and $(S-S^{NO})/S^{NO}$ where G^{NO} and S^{NO} are the non-oscillating components of G and S. The quantum oscillations were found to be periodic in 1/H with a frequency of $2.93 \pm .01 \times 10^6$G. This frequency compares with a value of $2.96 \pm .03 \times 10^6$G measured by Larson and Gordon[4] for the β-orbit of aluminum using the de Haas-van Alphen effect. The amplitude of the oscillations was fit to an expression of the form

$$\exp - \left(\frac{2\pi^2 m^* c\ k_B T}{ehH}\right) \tag{6}$$

in order to determine the value of the electronic effective mass. We obtain a value of $m^* = 1.1\ m_e$ which is about 20 percent larger than the phonon enhanced mass calculated by Thaler and Bass.[1] Despite the discrepancy, we believe our data show convincing evidence for thermoelectric mass enhancement due to many-body effects.

The data of Figs. 2 and 3 show that magnetic breakdown effects are smaller in polycrystalline samples than in single crystals by approximately a factor of three. By assuming a random crystallite orientation, we would expect the oscillations on the polycrystalline sample to be at least an order of magnitude smaller than those actually observed. We have considered the possibility that our anomalous results might be due to a preferential ordering of the crystallites, but careful inspection of the samples using both visual and x-ray techniques have revealed no such ordering. Furthermore, we have found the relative magnitudes of the GQO in polycrystalline samples to be independent of the details of sample preparation. We believe that our results may be understood by taking into account distortions of the current flow in the samples introduced by inhomogeneities. We have not yet tried to account for the effects of inhomogeneities on thermoelectric coefficients explicitly. However, an analysis based on this model, to be published elsewhere, is able

to account for similar anomalous oscillations observed in the longitudinal magnetoresistance by assuming that the samples contain about
0.4% randomly distributed non-conducting inclusions.

ACKNOWLEDGMENTS

This research was supported in part by a grant from the National
Science Foundation.

REFERENCES

1. B.J. Thaler and J. Bass, J. Phys. F 5, 1554 (1975).
2. W. Kesternich and C. Papastaikoudis, Phys. Stat. Solidi B 64,
    K41 (1974).
3. J.C. Garland and D.J. Van Harlingen, Phys. Rev. 13 10, 4825
    (1974).
4. C.O. Larson and W.L. Gordon, Phys. Rev. 156, 703 (1967).
5. A.V. Gold, Solid State Physics, Vol. I: Electrons in Metals
    (ed. by J.F. Cochran and R.R. Haering) Gordon and Breach, N.Y.,
    1968.

DISCUSSION

R.R. Bourassa: I am somewhat confused about the meaning of the
frequencies you get in the polycrystalline sample.

J.B. Sampsell: If you get some expression for some transport property, the magnetic breakdown part of it will be expressed in terms
of the "transmission-reflection coefficients" across the little
breakdown orbit in the third zone. Those transmission-reflection
coefficients will oscillate with the phase length around that orbit.
Hence they will oscillate as the area of that orbit, so we can
determine the area by looking at them. Looking at these frequencies
and comparing them to de Haas-van Alphen measurements, they agree.
So that's a measurement of the area of the orbit.

R.R. Bourassa: Which orbit?

J.B. Sampsell: The beta orbit, the small orbit in the third zone.

W.R. Datars: Does the cross section of that orbit change very much
if you rotate the field?

J.B. Sampsell: The cross section changes some, but it is not a
remarkable change. I would say that it changes in the neighborhood
of 15% within the region that you can tilt the crystal away from

the on-axis direction and still expect to see a large contribution
to this effect.

W.R. Datars: Haven't you answered your question then, why you see
the oscillations?

J.B. Sampsell: No, I don't think so. It changes by enough so that
I would expect that effect to wash out.

J. Bass: I want to get clear one thing you said. Did you say that
you had been able to explain similar oscillations in the electrical
resistivity using the current distortion argument? That is, oscillations with periods like this? Or did I mishear what you said.

J.B. Sampsell: Yes, that is what I said.

J. Bass: So you can, in some fashion, obtain periods of the same
value as the beta orbits by current distortion?

J.B. Sampsell: Yes, let me explain. Look at the data of Parker
for aluminum; they made calculations for magnetoresistance and they
measured magnetoresistance and obtained pretty good agreement between
theory and experiment. The same theory predicts nearly unseeable
oscillations in the longitudinal magnetoresistivity; two orders of
magnitude down from the transverse. Yet, when they measure it they
see that they are not over one order of magnitude down. At the
time they ascribed this to possible misalignment or any number of
sample geometry effects; they had not really intended to make longitudinal measurements, they just had a chance to do this and made
the measurements. Our experiment was designed to make these measurements. We are very sure of our orientation, and we saw the
effects. We can place our data on top of theirs, taken 5 years
apart and on different samples, and they nearly overlap. Using our
current distortion theory and just fitting empirically the oscillations in ρ_{xx} and assuming a 0.4% void concentration - I am not
saying there are 0.4% voids but something that causes distortion
like voids - we get good agreement. The periods fit, obviously.
There is also agreement between oscillation size and threshold, and
not only that, there is a fit to the gentle curvature in the non-
oscillatory part of the resistivity.

J. Garland: As I sat in back of the room and looked at these points
scattered on the graph it occurred to me that there might be some
doubting Thomases here who are not quite sure that there are indeed
oscillations. I'm inviting any of you skeptics to hold the view
graphs at an angle and squint along them, and you will see the
oscillations are really there.

THERMOELECTRIC MEASUREMENTS AT MILLIKELVIN TEMPERATURES[*]

W. P. Pratt, Jr., C. Uher, P. A. Schroeder, and J. Bass

Michigan State University

East Lansing, Michigan USA

I. INTRODUCTION

In the late 1950s and early 1960s physicists at the National Research Council (NRC) in Canada made a pioneering series of thermoelectric measurements on metals to temperatures below 0.1 K.[1-7] A fitting tribute to their ultralow temperature work is the fact that until recently not a single measurement they made below 0.3 K had been superceded.

Recent developments have combined to make possible more precise thermoelectric measurements of ultrapure metals and their alloys at ultralow temperatures: 1) The ^{3}He-^{4}He dilution refrigerator provides continuous refrigeration down to 10 mK. 2) The commercially available Superconducting Quantum Interference Device (SQUID) makes possible voltage measurements which are limited primarily by the Johnson noise developed in the samples. 3) There have been improvements in the techniques for purifying metals.

We wish to discuss here some of the thermoelectric measurements that we have performed at ultralow temperatures and in zero magnetic field. We begin by briefly reviewing in Section II the thermoelectric parameters which are best suited for measurement at ultralow temperatures. In Section III we discuss how Items 1) and 2) above make possible precision thermoelectric measurements at these temperatures. Finally, in Section IV we present some highlights of our work.

---

[*]Work supported by the U. S. National Science Foundation under grants DMR-75-14138 and DMR-75-01584.

II.   TRANSPORT PARAMETERS

We begin with the one dimensional transport equations which describe the flow of electric current (j = current density) and of heat ($\dot{q}$ = heat flow density) under the influence of an electric field E and temperature gradient ∇T.

$$j = \sigma E \quad - \varepsilon \nabla T \tag{1a}$$

$$\dot{q} = \varepsilon TE \quad - \kappa \nabla P \tag{1b}$$

The equations hold for a cubic metal in the absence of a magnetic field whenever E and ∇T are parallel to each other. For a normal metal we shall make use of the fact that $S^2/L \ll 1$ where S is the thermoelectric power and L is the Wiedemann-Franz ratio (see Eqs. 2c and 3). We obtain from Eqs. (1a) and (1b) some of the measurable transport parameters:

$$\sigma = \rho^{-1} \quad = \left. \frac{j}{E} \right|_{\nabla T=0} \tag{2a}$$

$$\kappa = \left. \frac{-\dot{q}}{\nabla T} \right|_{j=0} \tag{2b}$$

$$S = \left. \frac{E}{\nabla T} \right|_{j=0} \quad = \frac{\varepsilon}{\sigma} \tag{2c}$$

$$G = \left. \frac{j}{\dot{q}} \right|_{E=0} \quad = \frac{\varepsilon}{\kappa} \tag{2d}$$

$$\Gamma = - \left. \frac{\dot{q}}{E} \right|_{j=0} \quad = \frac{\sigma \kappa}{\varepsilon} \tag{2e}$$

In terms of the Wiedemann-Franz ratio we have

$$L = \frac{\kappa}{\sigma T} \quad = \frac{S}{GT} \quad = \frac{\Gamma S}{\sigma T} \tag{3}$$

At ultralow temperatures the dominant scatterers of electrons in a metal will be either impurities or the sample surface. Both types of scatterers are normally assumed to scatter electrons elastically. Under these conditions L will take on the Sommerfeld value of $L_o = 2.44 \times 10^{-8} \ V^2/K^2$. We shall neglect for the time being the possibility of electron-phonon scattering and phonon drag effects at ultralow temperatures. Since L presumably equals L_0, one need only make two measurements to completely determine the transport coefficients in Eqs. (1a) and (1b). At ultralow temperatures the following pairs of parameters can be most precisely measured:

$$G, \rho \quad \text{and} \quad \Gamma, \rho$$

The measurements of the resistivity are usually limited primarily by inaccuracies in the determination of the cross-sectional-area-to-length ratio for the samples (of order $\pm$ 1%). The parameters G and Γ have the advantage over the more conventionally measured quantity S that no temperature difference ΔT across the sample need be measured; such ΔT measurements at ultralow temperatures can be rather inaccurate. This advantage of G over S measurements was pointed out originally by Garland.[8]

In the ultralow temperature limit of elastic scattering we expect the following behaviors for the parameters defined by Eqs. (2a)-(2e):

$$\rho = \rho_0 \tag{4a}$$

$$\kappa = \alpha T \tag{4b}$$

$$S = \beta T \tag{4c}$$

$$G = G_0 \tag{4d}$$

$$\Gamma = \Gamma_0 \tag{4e}$$

Note that ρ, G, and Γ have the advantage of attaining constant values in the low temperature limit. The parameter G has a slight advantage over Γ in that S can be determined directly using Eq. (3) without a measurement of ρ.

The experimental procedure at ultralow temperatures is straightforward. One begins by measuring either the ρ, G or the ρ, Γ pairs of parameters. If any of the Eqs. (4a), (4d), or (4e) are not satisfied, then measurements of S and/or κ may be required in order to determine L. If the past is any guide, any anomalous behavior of these parameters will most likely be due to the presence of magnetic impurities. It has been observed that such scattering results in $L \neq L_0$.[9]

However, in interpreting the deviations of L from L_0 one must realize that the precision of the calibration of the thermometers and the reliability of the temperature scale used to do the calibration have a bearing on the significance of such deviations. In terms of directly measured quantities, we can express L in the form:

$$L = \frac{R\dot{Q}}{T\Delta T} \tag{5}$$

where R is the electrical resistance of the sample and $\dot{Q}$ is the heat flow rate. The quantities R and $\dot{Q}$ can usually be measured much more precisely than the product $T\Delta T$. Deviations of L from L_0 on the scale of $\pm 1/2\%$ at ultralow temperatures are probably not significant.

III. THERMOELECTRIC MEASUREMENTS USING DILUTION REFRIGERATORS AND SQUIDS

Our experimental setup is depicted in Fig. 1 along with certain transport parameters which are defined in terms of directly measurable quantities. The reference resistance R_r, the reference and sample currents I_r and I_s, and the sample heat flow rate $\dot{Q}$ can usually be measured to an accuracy of better than 0.1%. The parameters R_s, K, and Γ' are related to their counterparts in Eqs. (2a), (2b), and (2e) by the standard cross-sectional-area-to-length ratio.

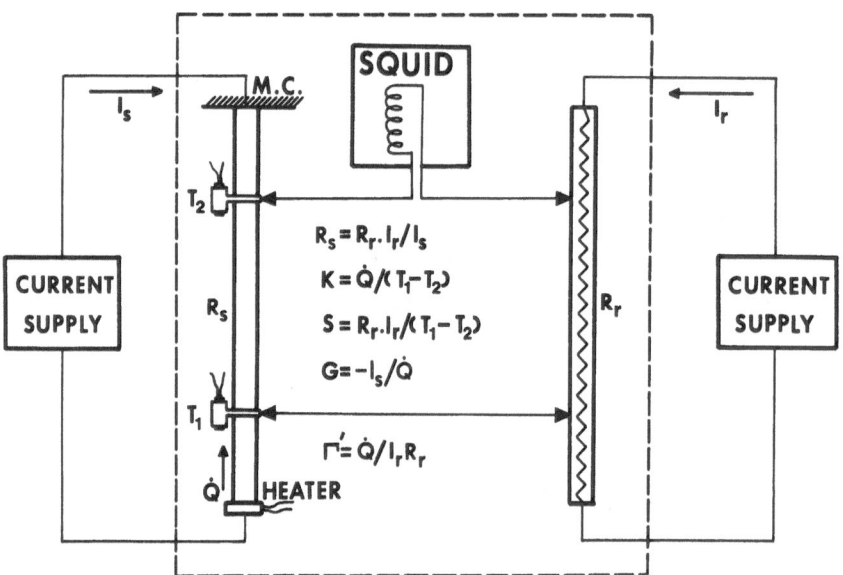

Fig. 1. Schematic of apparatus for measuring transport parameters. All components which are enclosed by the dashed lines are at or below 4.2 K.

The wires which end with arrows are superconducting. The SQUID acts as a null detector in that a current will flow through the super-conducting coil, which is magnetically coupled to the SQUID, if the potential differences across R_r and the sample resistance R_s are not equal. Temperatures T_1 and T_2 are measured by calibrated germanium resistance thermometers (cryocal CR-50). Thermal contact between the sample and the copper body of the dilution refrigerator mixing chamber ("M.C.") is made through a series of copper-to-copper and copper-to-sample spot-welds. (For a complete discussion of ultra-low temperature techniques including the use of SQUIDS, consult the book by O. V. Lounasmaa.[10]) Similar circuits with a r.f. biased SQUID null detector have been used in thermoelectric measurements above 1 K by Dee, Guénault, and Pickett,[11] by Fujita and Ohtsuka,[12] and by Tracy.[13]

We might add that the high sensitivity of the SQUID circuit makes possible rather precise measurements of the ratio R_s/R_r. We have attained precisions approaching 10 ppm for this ratio,[14,15] and Rowlands and Woods[16] have attained precisions of 0.1 ppm with a SQUID null detector. Such high precision measurements require that the temperature of R_r be closely regulated, usually at 4.2 K, and that $R_s \cong R_r$. Satisfying these requirements results in less precise volt-age measurements for the thermoelectric parameters at ultralow tem-peratures. This is because the voltage noise in the circuit is due primarily to Johnson noise across R_s and R_r, and keeping R_r non zero and at 4.2 K produces more Johnson noise than if R_r were shorted or kept at the temperature of the sample. For a given sample at ultra-low temperatures, the refrigeration capacity fixes the maximum value of $T_1 - T_2$ which in turn establishes a maximum value for the thermo-electric voltage across the sample.

For the discussions which follow we shall choose values of R and S which are representative of very pure metals. We use $R_s = R_r = 1 \times 10^{-7}$ Ω which is the resistance of a sample of 1-mm diam-eter and 5-cm length, made either from Ag with a residual resistance ratio (RRR) of 10,000 or W with RRR = 35,000. Typical ultralow tem-perature values of S for "ultrapure" samples of these metals are of the order of $1 \times 10^{-8} T$ (V/K).

A. Refrigerator Power and Thermometry

In Fig. 2 we present the refrigeration powers for two commercial-ly available machines, manufactured by S.H. E. Corporation,[17] and for our local refrigerator ("MSU-II"). The two commercial machines rep-resent the largest and smallest cooling capacities purchasable at present. For each refrigerator these cooling capacities were deter-mined by applying heat to the copper body of the mixing chamber and measuring the temperature of this copper body. The commercial DRI-337

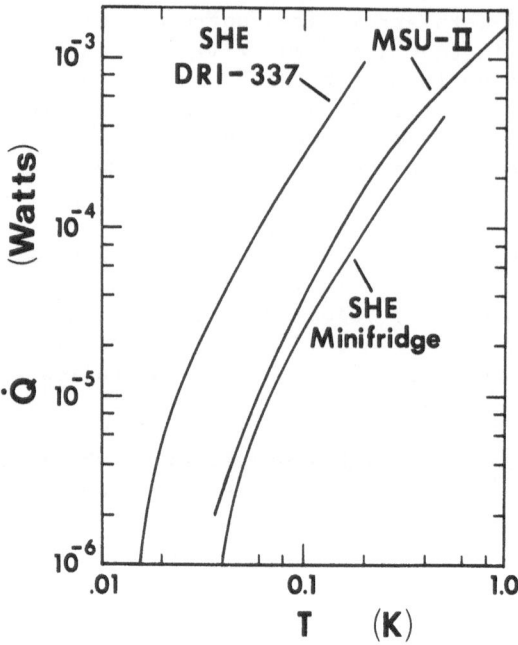

Fig. 2.   Three examples of dilution refrigerator powers below 1 K.

machine is new and has an unusually large ^{3}He circulation rate of
4.5×10^{-4} moles/sec which accounts for its high cooling power.   Our
circulation rates are a factor of 5 to 10 smaller.   An earlier mixing chamber on our refrigerator, which was not designed to cool
metallic samples, routinely maintained temperatures as low as 8 mK.
A new mixing chamber was constructed (MSU-II) which made the mount-
ing of the samples much easier and had considerably larger cooling
capacity.   Unfortunately, for a variety of reasons this new design
did not attain as low a temperature as before.   However, the versa-
tility of this new design, we feel, outweighs this disadvantage.

Using Eq. (5) with $L = L_O$ and $R_S = 1 \times 10^{-7} \Omega$, we can compute ΔT
as a function of T from the cooling-power curves of Fig. 2.   We
present in Fig. 3 a plot of $\Delta T/T$ against T for the high capacity
S.H.E. machine and for our MSU-II machine.   We have also added a bit
of realism to these calculations by placing an electrical resistance
of $\approx 0.5\mu\Omega$ between the sample and the copper body of the mixing chamber to represent the various copper bars and spot welds between them.

The dashed curve in Fig. 3 shows approximately what values of
$\Delta T/T$ would be measured with an uncertainty of $\pm 100\%$ for our germanium
resistance thermometers.   Our temperature measuring procedure consists of simply noting the change in T_1 as the heat flow $\dot{Q}$ is applied

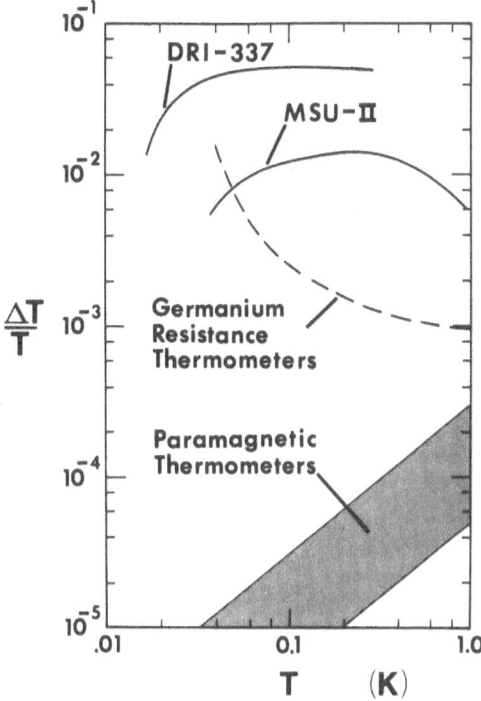

Fig. 3. The relative temperature difference, $\Delta T/T$, across a metallic sample with $R_s = 10^{-7}\Omega$ (solid curves). The dashed curve and the shaded region are explained in the text.

for fixed T_2. The measuring precision could probably be much improved by a more careful regulation of the refrigerator temperature and by the use of matched carbon resistors to measure directly $T_1 - T_2$. This dashed curve in Fig. 3 combines a variety of factors including: the sensitivity to electromagnetic interference at the lowest temperatures, the temperature stability of the refrigerator, and the R vs T characteristics of the thermometers.

The shaded region represents possible values of $\Delta T/T$ with $\pm 100\%$ uncertainty when magnetic thermometers are employed. These thermometers could be constructed along the lines suggested by Giffard et al.,[18] who compressed a few mg. of cerrous magnesium nitrate (CMN) together with an equal volume of gold powder into a hole in a piece of high purity copper. This method of construction ensures good thermal contact between the magnetic thermometer (CMN) and the surrounding copper. The susceptibility of CMN follows well Curie's law above 20 mK. A SQUID magetometer would be employed to measure the magnetization of the CMN. Such a system would be ideally suited for direct measurements of $T_1 - T_2$.

Clearly, precise measurements of $\Delta T/T$ with $R_s = 1 \times 10^{-7}\Omega$, on our system are not possible at the present time. However, for samples with $R \gg 1 \times 10^{-7}\Omega$, $\Delta T/T$ can be made much larger, and for such samples we have made measurements of κ to a precision of approximately 2%. As Fig. 3 shows a combination of increased cooling capacity and improved differential thermometry techniques should result in significant improvements in the precision of ΔT measurements.

B. Voltage Measurements with SQUIDs

The equivalent circuit for the SQUID null detector is given in Fig. 4 where ΔV is the thermoelectric voltage across R_s. The current I, through the signal coil of self inductance L', can induce a flux ϕ in the SQUID sensor via the effective mutual inductance M. Typical values are $M \sim 2 \times 10^{-8}$ H and $L' \sim 2 \times 10^{-6}$ H. Noise in the SQUID sensor and in its electronics is usually expressed in terms of an effective mean-square flux noise as referred to the SQUID sensor. This effective flux noise thus contains contributions from the superconducting device proper and from its associated electronics. Expressed in terms of an effective current noise through the signal coil, one has a root-mean-square device noise current typically in the range $(\overline{I_D^2})^{\frac{1}{2}} = 10^{-11}B_o^{\frac{1}{2}}$ to $10^{-10}B_o^{\frac{1}{2}}$ amps, where B_o is the bandwidth of the output of the SQUID electronics and is usually ≈ 1 Hz.[10,17] Noise due to electromagnetic interference and vibration in the earth's magnetic field can be eliminated by proper system design.

There remains to be considered the Johnson voltage noise across R_s and R_r. The mean-square voltage noise across the series combination of R_s and R_r is given by:

$$\overline{V_J^2} = 4k_B(R_r T_r + R_s T_s)B \tag{6}$$

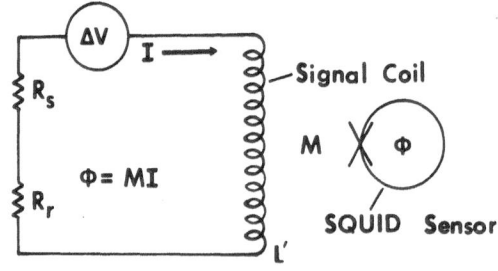

Fig. 4. Equivalent circuit for SQUID null detector.

where k_B is Boltzmann's constant, the T's refer to the temperatures of R_r and R_s, and B is the bandwidth. The bandwidth B_i of the input circuit is defined as $B_i = (R_r + R_s)/4L'$. For the chosen values of R_r, R_s, and L' we have $B_i = 0.025$ Hz. If $B_i \ll B_o$, then we substitute $B = B_i$ in Eq. (6). This is because the SQUID senses current, and the broad-band voltage noise produces a current noise only in the bandwidth B_i. If $B_i \gg B_o$, then we have $B = B_o$. The former condition applies for our case, and thus we have

$$\overline{v_J^2} = k_B(R_r T_r + R_s T_s)(R_r + R_s)/L' \tag{7}$$

For $R_r = R_s = 1 \times 10^{-7} \Omega$, we obtain

$$\overline{(v_J^2)} = 3.7 \times 10^{-16}(T_r + T_s)^{\frac{1}{2}} \text{ volts} \tag{8}$$

For the equivalent device voltage noise, we have

$$\overline{(v_D^2)}^{\frac{1}{2}} = (R_r + R_s)(10^{-11} \text{ to } 10^{-10})B_o^{\frac{1}{2}}$$

$$= 2 \times 10^{-18} \text{ to } 2 \times 10^{-17} \text{ volts} \tag{9}$$

with $B_o = 1$ Hz. Equations (8) and (9) are plotted in Fig. 5. The device noise is clearly negligible in comparison to the Johnson noise for this case.

Also plotted in Fig. 5 are the thermoelectric voltages ΔV across R_s, which were obtained from the two $\Delta T/T$ curves in Fig. 3 with $S = 1 \times 10^{-8}T$ (V/K). As can be seen, measurements of ΔV to $\pm 1\%$ are possible for the S.H.E. refrigerator at 0.02K and for ours at 0.04 K, with $T_r = 4.2$ K which is the condition of largest Johnson noise. If we have $T_r = T_s$ or make $R_r = 0$, then a measurement of ΔV would have a precision of approximately 0.1 % at these two temperatures.* A comparison of Figs. 3 and 5 shows why at the present time

---

*If we choose instead $R_s \gg 10^{-7}\Omega$, then we have $B_i \gg B_o \cong 1$ Hz, and Eq. (6) becomes $\overline{v_J^2} = 4k_B(R_r T_r + R_s T_s)$. For example, let us assume that $R_s = R_r = 4 \times 10^{-5}\Omega$, $T_r = T_s = 0.02$K, and $S = 1 \times 10^{-8}T$ (V/K). At 0.02 K only 2.5×10^{-8} watts of cooling power are required to attain $\Delta T/T = 0.1$. With $\Delta T/T = 0.1$ at 0.02 K we obtain $\left[\overline{v_J^2}\right]^{\frac{1}{2}}\big/\Delta V = 0.02$. Thus reasonably precise measurements of ΔV are also possible at these larger values of R_s.

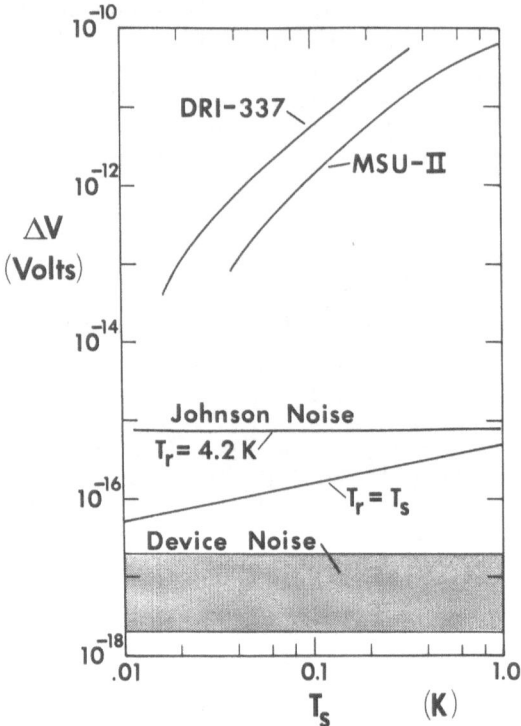

Fig. 5. Thermoelectric voltage across a sample with $R_s = 10^{-7}\Omega$ and $S = 10^{-8}T$ (V/K), upper curves. The Johnson noise and the device noise limit the voltage resolution of the SQUID and are discussed in the text.

our measurements of G or Γ are so much more precise than a measurement of the assumed S for very pure metals. An obvious improvement would be to reduce the Johnson noise by shorting out R_r when one is not making measurements of R_s. For this reason measurements of G will be preferrable to measurements of Γ.

We wish to emphasize that the discussion in this section, with its explicit presentation of less than ideal refrigeration capacity and thermometry for our system, is meant to be viewed optimistically That is, in spite of these difficulties interesting results have been obtained.

IV. EXAMPLES OF THERMOELECTRIC
MEASUREMENTS AT ULTRALOW TEMPERATURES

We wish to present some examples of our ultralow temperature thermoelectric measurements by discussing three metals: W, Pd(Ru) alloys, and the semimetal Bi. J. E. Graebner and colleagues at Bell Laboratories have also combined the use of SQUIDs with a dilution refrigerator to make ultralow temperature transport studies. They performed these measurements on Rh(Fe) alloys down to 20 mK.[19]

A. Tungsten

These studies were motivated by the work of Garland and Van Harlingen[8] (hereafter referred to as GVH). They had observed for five zone-refined tungsten samples between 1.2 and 4 K that an empirical fit to the temperature dependence of G contained an anomalous term which diverged positively as $T^{-1/2}$. This divergence was largest for the sample with the smallest RRR. This behavior is, of course, contrary to the predictions of theory for simple elastic scattering of electrons by impurities which requires $G \rightarrow G_0$ as $T \rightarrow 0$. We undertook these measurements on W to see if this anomalous behavior persisted to lower temperatures. Samples of W can be made ultrapure by zone refining which makes them ideally suited for ultralow temperature studies. Since W has the highest melting point of any metal, most impurities are evaporated in the zone refining process.

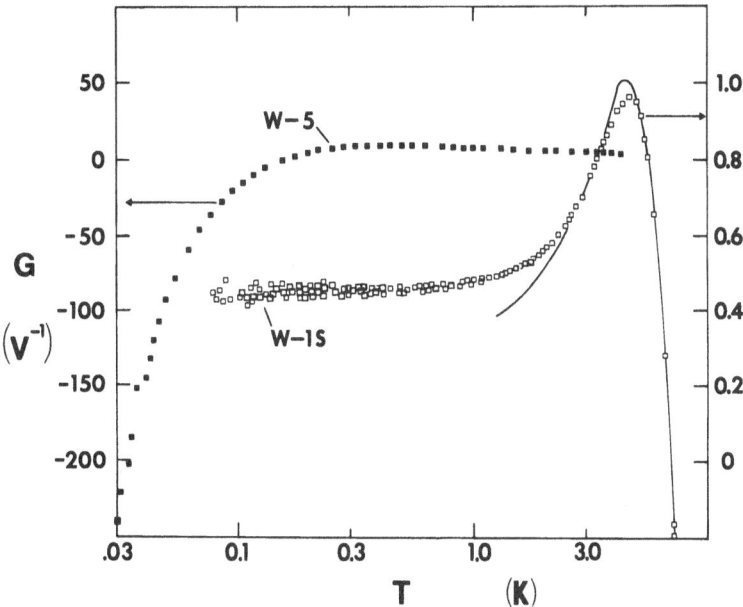

Fig. 6. Thermoelectric ratio measurements on W to 0.03 K. Note that the arrows point to the corresponding ordinates for the displayed data.

In Fig. 6 we present our G measurements on a three-pass zone refined single crystal (W-1S) having a 1.4-mm diameter and RRR = 44,000. From 0.5 K down to 0.07 K, the normally expected $G = G_0$ behavior is obtained. Also shown as a solid curve are the results of GVH for their RRR = 63,000 sample.[8] The sets of data are encouragingly similar over their common temperature region. We believe that our ultralow temperature results are indeed representative of very pure tungsten for which the electron scattering is dominated by simple potential scattering by impurities and boundaries. We have verified that an "ultrapure" sample (RRR=77,000), provided by GVH, behaves at ultralow temperatures in a manner similar to sample W-1S in Fig. 6. Also shown in Fig. 6 are our measurements on sample W-5 of GVH which has RRR=9,500. In contrast to sample W-1S, this sample has a G which becomes increasingly more negative with decreasing temperature below 0.2 K and a resistivity which shows a minimum at 0.6 K. Such behaviors would seem to indicate that dissolved magnetic impurities are present in this sample. We have also verified that similar "less pure" samples exhibit such large negative values of G at the lowest temperatures, along with in some cases resistance minima.

Sample W-5 also exhibits a weak positive maximum in G near 0.4 K. Other samples loaned to us by GVH show that the magnitudes of the $T^{-1/2}$ divergence seen above 1.2 K and of this positive maximum are correlated. Furthermore, the magnitudes of the large negative value of G at the lowest temperatures and of the positive maximum at higher temperatures are also correlated. We tentatively attribute these positive and negative deviations from the $G = G_0$ behavior to dissolved magnetic impurities of unknown origin.

B.   Palladium-Ruthenium Alloys

The purpose of this study is quite simple: to test the conventional theory for electron transport to ultralow temperature in a regime where the electron scattering should be completely elastic over the whole temperature range below 4 K.[22] This choice of alloy was made in order to satisfy two criteria: 1) Fairly concentrated non-magnetic alloys will be required in order to suppress the effects of the always-present magnetic impurities. 2) Such alloys must have negligible phonon drag contributions to S at 4 K. Prior work has established that the latter condition is satisfied for Pd(Ru) alloys.[23]

In Fig. 7 we present plots of S against T for four alloys: 0.1%, 0.5%, 1%, and 5% Ru in Pd. The expected linear behavior of S upon T is confirmed. In fact deviations from this linear behavior only appear at temperature below approximately 0.15 K. These deviations are in part due to inaccuracies in the calibration of our germanium resistance thermometers below 0.1 K. However, in this same

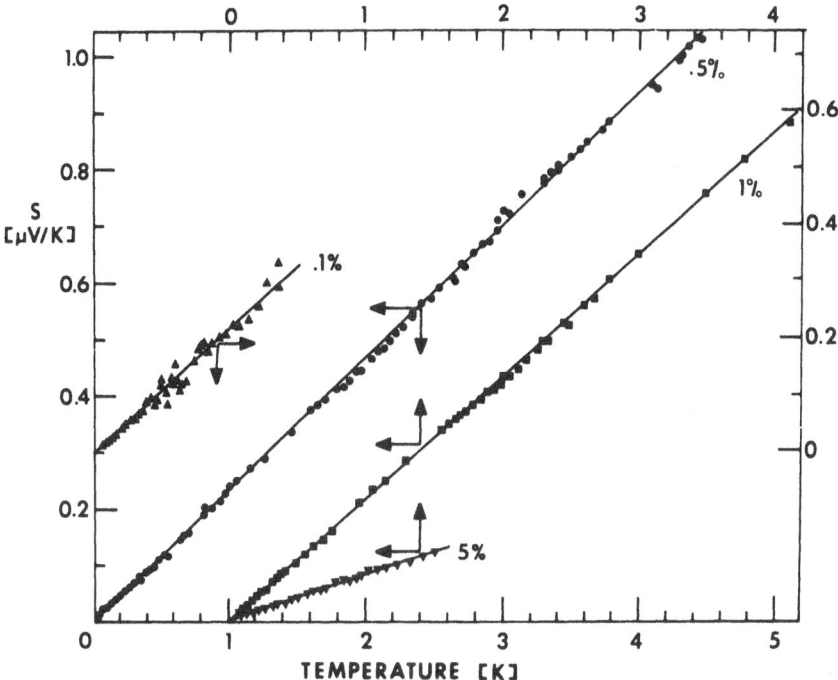

Fig. 7. Thermopower of Pd(Ru) alloys to ultralow temperatures. The arrows indicate which ordinate and abscissa should be used.

temperature region, G exhibits significant departures from being temperature independent. The cause of this unexpected behavior is not known, although magnetic impurities are suspected.

C. The Semimetal Bismuth

Bismuth is a semimetal with a carrier density which is approximately 10^{-5} of the density of a typical metal. The electrons and holes have Fermi energies of $E_F^- = 0.026$ and $E_F^+ = 0.011$ eV, respectively, and the carrier spectrum is degenerate at helium temperatures. The partial diffusion thermopower of electrons and holes for an assumed isotropic carrier spectrum and a simple impurity scattering is given approximately as

$$S_d^- = 0.024T/E_F^- \approx -0.9\,T \qquad (\mu V/K) \qquad (6a)$$

$$S_d^+ = 0.24T/E_F^+ \approx 2T \qquad (\mu V/K). \qquad (6b)$$

The total diffusion thermopower is then

$$S_d = \frac{\sigma^- S_d^- + \sigma^+ S_d^+}{\sigma^- + \sigma^+} \tag{7}$$

where σ^- and σ^+ are the partial electrical conductivities of electrons and holes. Since the electron conductivity is 3 to 4 times the hole conductivity, one expects a negative sign for the diffusion thermopower and a magnitude not exceeding approximately 1 μV/K at liquid helium temperatures. Contrary to this expectation, experiments show huge positive maxima in the thermopower reaching several tens of μV/K. Such maxima are believed to be associated with dragging of carriers by phonons. A large phonon drag is favorable in Bi because the carrier density is very low and the umklapp phonon-phonon scattering is negligible at these temperatures. It is naturally of interest to investigate what happens to the phonon drag at lower temperatures. So far, the positive peaks have been ascribed to the phonon drag of holes. Would perhaps a similar phonon drag acting on electrons be observed at lower temperatures?

In Fig. 8 we present two graphs of S against T. This Bi single crystal sample had its trigonal axis at 49° to the long axis of the sample, a diameter of 3 mm, and a RRR=190. In the upper graphs of Fig. 8, the rapid drop of S at temperatures below the positive peak is evident. Note that S changes sign near 0.45 K and passes through a minimum of about -0.4 μV/K near 0.28 K. Below approximately 0.075 K on the lower graph, it would seem that the linear temperature dependence for the expected diffusion thermopower has been attained. The narrow line, a linear extrapolation of S to T=0, has a negative slope of ≈1 μV/K^2 which is in agreement with naive expectations regarding Eq. (7). It is tempting to ascribe the negative minimum to the phonon drag of electrons. However, further experiments with crystals of different orientations will be required before any definite conclusions can be made concerning this possibility.

V. CONCLUSIONS

We have shown that reasonably precise thermoelectric measurements on a variety of metals can be made to temperatures as low as 20 mK with commercially available refrigerators and SQUID null detectors. In the future our program will have two main components: First, there are no intrinsic limitations which would keep us from making reasonably precise thermoelectric measurements to temperatures below 5 mK. We are presently developing a second stage of refrigeration by adiabatic demagnetization[10] to attain these very low

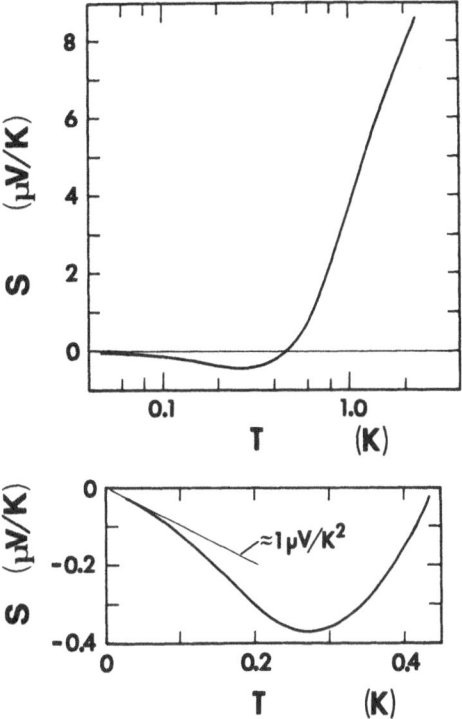

Fig. 8. Two graphs of the thermopower for the same single crystal sample of Bi. The lower graph is a linear plot of S against T, which emphasizes the lower temperature region.

temperatures. Second, the area of thermoelectric measurements to ultralow temperatures in high magnetic fields is totally unexplored. We are building a second dilution refrigerator system which will be a dedicated facility for such studies. Some of our first work in this new area will be to apply the discovery by Opsal, Thaler, and Bass[24] of phonon mass enhancement effects in thermoelectricity to metals other than aluminum where the lower Debye temperatures will require ultralow temperatures in order to clearly separate diffusion terms from phonon drag terms.

REFERENCES

1. D.K.C. MacDonald, W.B. Pearson, and I.M. Templeton, Proc. Roy. Soc. (London) A256, 334 (1960).

2. A.M. Guénault and D.K.C. MacDonald, Proc. Roy. Soc. (London) A264, 41 (1961).

3. A.M. Guénault and D.K.C. MacDonald, Proc. Roy. Soc. (London) A274, 154 (1963).

4. D.K.C. MacDonald, W.B. Pearson, and I.M. Templeton, Phil. Mag.
    3, 657 (1958).

5. D.K.C. MacDonald, W.B. Pearson, and I.M. Templeton, Proc. Roy.
    Soc. (London) A266, 161 (1962).

6. D.K.C. MacDonald, W.B. Pearson, and I.M. Templeton, Phil. Mag.
    5, 867 (1960).

7. D.K.C. MacDonald, W.B. Pearson, and I.M. Templeton, Phil. Mag.
    4 380 (1959).

8. J.C. Garland and D.J. Van Harlingen, Phys. Rev. B10, 4825 (1974).

9. R.G. Sharma and M.S.R. Chari, J. Low Temp. Phys. 15, 79 (1974);
    G.S. Poo, Phys. Rev. B13, 451 (1976).

10. O.V. Lounasmaa, Experimental Principles and Methods Below 1 K
    (London and New York: Academic Press, 1974).

11. R.H. Dee, A.M. Guenault, and G.R. Pickett, J. of Phys. E 9, 807
    (1976).

12. T. Fujita and T. Ohtsuka, Japanese J. of Appl. Phys 15, 881
    (1976).

13. J. Tracy, Phys. Lett. 48A, 219 (1974).

14. E.L. Stone, M.D. Ewbank, J. Bass, and W.P. Pratt, Jr., Phys.
    Lett. 59A, 168 (1976).

15. C. Uher and W.P. Pratt, Jr., Phys. Rev. Lett. 39, 491 (1977).

16. J.A. Rowlands and S.B. Woods, Rev. Sci. Instr. 47, 795 (1976).

17. Catalogue from S.H.E. Corporation, 4174 Sorrento Valley Blvd.,
    San Diego, California 92121.

18. R.P. Giffard, R.A. Webb, and J.C. Wheatley, J. Low Temp. Phys.
    6, 533 (1972).

19. J.E. Graebner, J.J. Rubin, R.J. Schultz, F.S.L. Hsu, W.A. Reed,
    and R.J. Higgins, Amer. Inst. Physics Conf. Proc. 24, 445
    (1974).

20. E.L. Stone, M.D. Ewband, W.P. Pratt, Jr., and J. Bass. Phys.
    Lett 58A, 239 (1976).

21. C. Uher, D. Mundinger, and J. Bass, J. of Physics F 7, 1691
    (1977).

22. P.A. Schroeder and C. Uher, J. Low Temp. Phys. 29, 487 (1977).

23. D. Greig, T.K. Brunck, and P.A. Schroeder, Phil. Mag. 25, 1009
    (1972).

24. J. Opsal, B.J. Thaler, and J. Bass, Phys. Rev. Lett. 36, 1211
    (1976).

THERMOPOWER NEAR MAGNETIC AND ORDER-DISORDER CRITICAL POINTS

R. D. Parks*

University of Rochester, Rochester, New York 14627 USA

and

R. Orbach[†]

University of California, Los Angeles, California 90024 USA

ABSTRACT

The thermopower anomalies observed at ferromagnetic, antiferromagnetic, and order-disorder critical points are characterized by specific heat like singularities in the temperature derivatives of the thermopower S. The nature of the anomaly is prescribed by the temperature dependence of the zero time localized spin-spin (or concentration-concentration) correlation functions $g_{\vec{k}}(t=0)$ for large momentum transfers ($|\vec{k}| \sim 2k_F$). Inelastic scattering is important as a consequence leading to an expression for the critical thermopower which contains convoluted functions of the frequency dependent localized spin-spin correlation functions $g_{\vec{k}}(\omega)$. The observed temperature dependence of S is thought to reside primarily in the behavior of the resistivity ρ near the critical temperature, the variation of the thermopower integrals with temperature being rather smooth in this regime. When the momentum transfer is small (e.g. ferromagnetic semiconductors) so that the localized spin system's fluctuations critically slow down, elastic scattering will dominate the thermopower. In such a case, S is directly proportional to the localized spin-spin correlation function $g_{2k_F}(\omega=0)$.

---

*Work supported in part by the Army Research Office and the Office of Naval Research.

†Work supported in part by the National Science Foundation and the Office of Naval Research.

The purpose of this review will be to summarize our current
understanding of thermopower anomalies at second order phase transitions in metallic systems. We shall focus on ferromagnetic,
antiferromagnetic, and order-disorder transitions, the last being
formally equivalent to the second. However, the general results
obtained can be equally well applied to other types of second order
transitions, in particular, those in which the dominant contribution to the thermopower anomaly arises from critical fluctuations
of the scattering centers.

The thermopower (TEP) of Nickel (Figs. 1 and 2) is typical of
that of metallic ferromagnets and is qualitatively similar to the
critical resistivity in that it monotonically increases with temperature through the critical point and its temperature derivative
closely resumbles the λ-shaped specific heat anomaly. This behavior
has been explained in terms of a single model which considers the
effect of critical spin fluctuations on the scattering of conduction
electrons.[2] The coupling between the localized spins and the conduction electrons is taken as the exchange interaction, and is assumed to be isotropic. The same model was considered previously,[3-5]
but the treatment was limited to elastic scattering (absence of

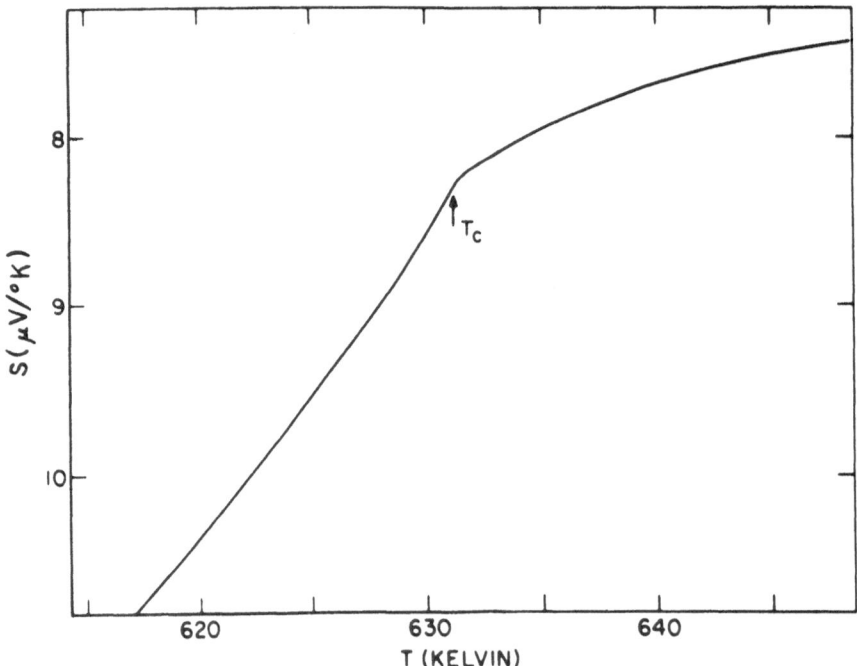

Fig. 1. Thermopower of Nickel (with respect to platinum reference)
in the vicinity of the Curie point (after Tang, et al).[1]

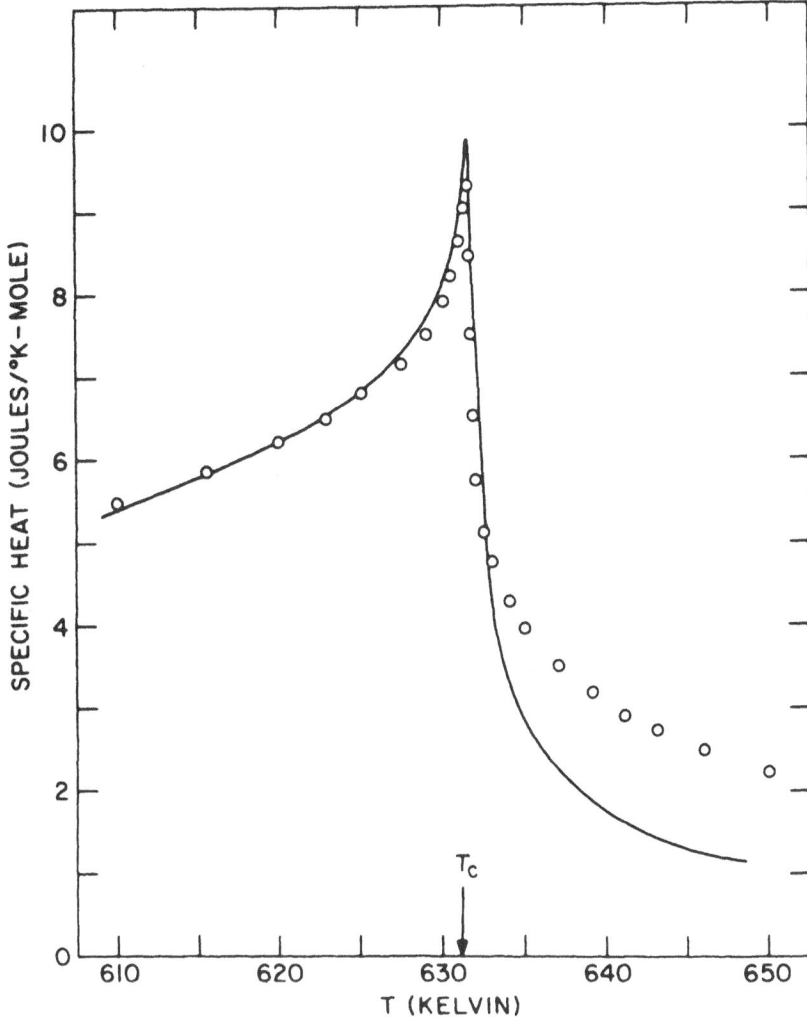

Fig. 2. dS/dT for Nickel (Ref. 1) compared to specific heat results (circles) of Connelly, et al [D. L. Connelly, L. S. Loomis and D. E. Mapother, Phys. Rev. B $\underline{3}$, 922 (1971)].

energy transfer between the conduction electrons and the localized spin system). As we shall argue later, the conduction electrons move so rapidly through the metal that the interaction with the localized spin excitations is essentially instantaneous. This leads to the dominance of the thermodynamic limit of the localized spin-spin correlation function, i.e., the value at equal times. As a consequence, the conduction electrons exchange energy with the localized spins over the full frequency width of the latter's response function.

It is shown in Ref. 2 that the thermopower can be written as:

$$S/T \approx \text{const} + \frac{C}{\rho}\,(1/\tilde{\tau}_c) \tag{1}$$

where C is a constant, ρ is the resistivity, and $1/\tilde{\tau}_c$ is the transport integral appropriate to the thermopower. It has the form:

$$\frac{1}{\tilde{\tau}_c} = \frac{2V_c m k_F I^2}{3\pi^2 N\hbar^3} \int_{-\infty}^{\infty} d\omega\,(\beta\hbar\omega)\,(e^{\beta\hbar\omega}-1)^{-1}\;\times$$

$$\left(g_{2k_F}(\omega) + \frac{3}{8\pi^2 k_F^2}(\beta\hbar\omega)^2 \int_0^{2k_F} dk\, k g_k(\omega)\right). \tag{2}$$

where V_c is the volume of the unit cell, k_F the Fermi momentum, I the localized-conduction electron spin exchange coupling constant, N the number of electrons, $\beta = 1/k_B T$, and $g_k(\omega)$ the frequency dependent localized spin-spin correlation function defined by the Fourier transform of

$$g_{\vec{k}}(t) = \langle S_{-\vec{k}}(0)\cdot S_{\vec{k}}(t)\rangle \tag{3}$$

Only if one takes $g_{\vec{k}}(\omega) = g_{\vec{k}}(0)\cdot\delta(\omega)$ does (2), in combination with (1), reduce to the result[3-5] for elastic scattering (in the limit $\rho_c \ll \rho$, ρ_c being the critical resistivity):

$$S/T = \text{const} + C g_{2k_F}(T)/\rho \tag{4}$$

where the symbol T implies the spin-spin correlation function at $2k_F$ is evaluated at temperature T, appropriate to zero frequency. In fact, as can be seen immediately from (2), the actual correlation function (first integral) is the t = 0 function evaluated at $2k_F$ at temperature T. The second integral is a small correction to the first at small frequencies.

The reason for the importance of the t = 0 correlation function in the case of metals is that the conduction electrons, to first approximation, will always see spin fluctuations as if they are instantaneously frozen. To see this, we have to show that the time spent by an electron in the vicinity of a fluctuation, which is given by $(kv_F)^{-1}$, v_F being the Fermi velocity, is much shorter than the fluctuation lifetime ω_k^{-1} regardless of the wavelength of the fluctuation k^{-1}. The most severe test of this criterion is posed by the shortest wavelength fluctuations, viz., $k^{-1} \sim a$ (an atomic spacing). Using $v_F \sim 10^8$ cm/sec, we have $(kv_F)^{-1} \sim 10^{-16}$ sec, whereas the maximum energy of the fluctuations, given by $k_B T_c$, corresponds to $1/\omega_k \sim 10^{-13}$ sec for $T_c \sim 100$ K, hence the inequality is easily satisfied. Critical slowing down near T_c of the longer wavelength fluctuations increases the ratio kv_F/ω_k for these fluctuations. Therefore, in the case of a metal the instantaneous (equal time correlation function) picture would appear to be a good approximation.

It would be very informative to use the results of modern dynamic scaling theory to predict the critical behavior of the equal time correlation function. Unfortunately, scaling theory only applies to long wavelengths in the vicinity of T_c.[6] The large momentum transfer, coupled with the high frequency components contained in (2), makes the thermopower integral impossible to simplify. An interesting result for the electrical resistivity (where a similar problem involving large momentum transfers is present) has very recently been exhibited by Geldart.[7] He showed that the static ($\omega = 0$) contributions to the correlation function dominate the electrical resistivity near T_c, the leading inelastic corrections amounting to only 10 to 20%. No such examination of the thermopower integral has yet been performed, but it would clearly be of great interest.

Since the equal time spin-spin correlation function contained in the resistivity contributes the dominant temperature effect near the transition temperature, we shall summarize what is known about it. Until the late sixties there existed only the Ornstein-Zernike approximation, in which $g_{\vec{k}}(t=0)$ has the Lorentzian form shown in Fig. 3a. This leads to a peak in the resistivity[8] or TEP near T_c. Ising model studies by Fisher and Burford,[9] together with neutron scattering studies, established the correct behavior for $g_{\vec{k}}(t=0)$ as shown in Fig. 3b. The rapid plummeting of $g_{\vec{k}}(t=0)$ for $\vec{k} = 0$ as one approaches T_c from above can be understood physically by noting that it is expensive to create short wavelength fluctuations in the presence of long wavelength fluctuations, the latter becoming increasingly numerous as T_c is approached. The fact that $dg_{\vec{k}}(t=0)/dT \to \infty$ at T_c for $k \neq 0$, leading to specific heat-like anomalies in $dS(T)/dT$ and $d\rho/dT$, has not been explicitly shown in Ising model calculations, but has been argued for, indirectly but convincingly, by Fisher and Langer.[10] Note that the TEP of Nickel (Fig. 1) looks qualitatively similar to $g_{(1/a)}(0)$ in Fig. 3b, which is consistent with the predictions of Eq. (4) if we note that $a^{-1} \sim k_F$. The cross-over to Ornstein-Zernicke behavior at higher temperatures is not apparent in the

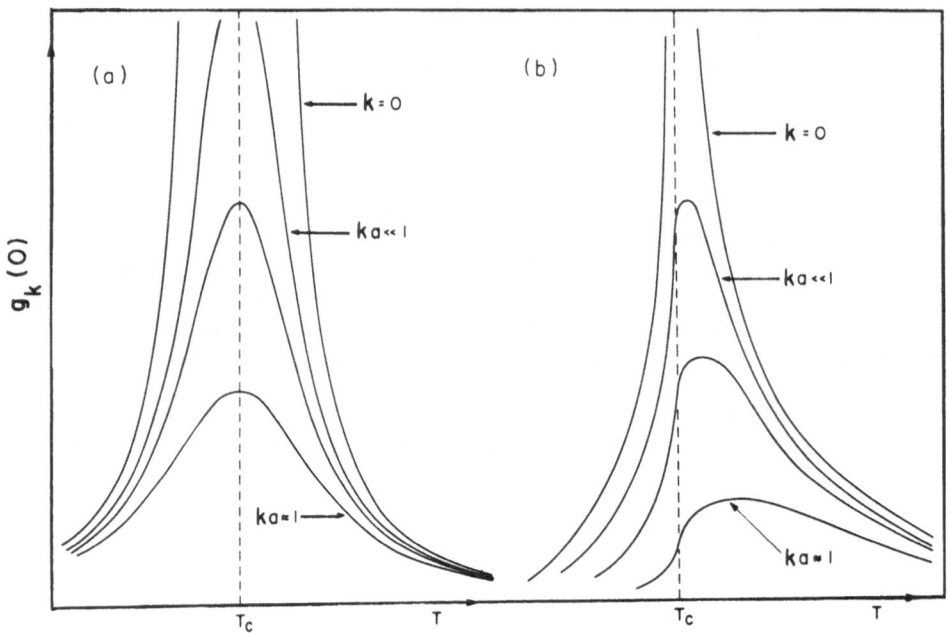

Fig. 3. Zero time correlation function calculated in the Ornstein-Zernicke approximation (a) and after Fisher and Burford (b).[9]

Nickel data but can be seen in the TEP results[4] for GdNi$_2$. This cross-over is manifested particularly strikingly in the temperature derivative (Fig. 4), since it changes the entire character of the otherwise specific heat like lambda.

At this point we digress to discuss the correspondence between the critical thermopower and the critical resistivity. In every system discussed in this review the resistive anomaly is qualitatively similar to the observed thermopower anomaly. This can be understood by noting the relation between the result for the critical resistivity[10] (ignoring inelastic scattering):

$$\rho_c = \text{const} + \text{const} \int_0^{2k_F} g_k(T) \, k^3 dk \qquad (5)$$

and Eq. 1. The resistivity enters into the denominator of the expression for S, while the numerator contains integrals over the frequency dependent localized spin-spin correlation function for large momentum transfer. We cannot see any reason why these integrals should have significant variations with temperature in the vicinity of the critical temperature. Hence, the dominant source

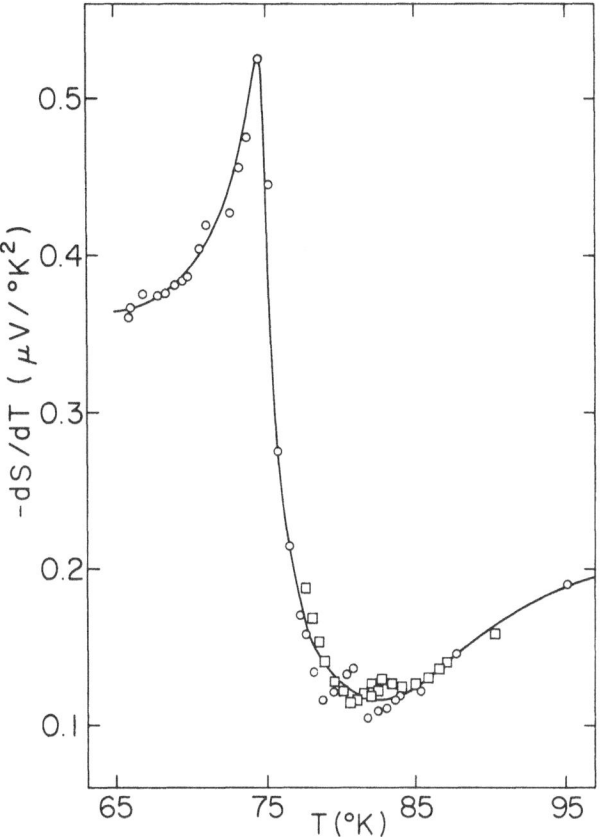

Fig. 4. Temperature derivative of absolute thermopower of $GdNi_2$ (after Thomas et al).[4] Different symbols correspond to different averaging periods in computer differentiation (see Ref. 4).

of the thermopower temperature anomaly appears to be the resistivity which appears explicitly in the expression for S. It is for this reason that the temperature variation of the thermopower in ferromagnetic metals appears to "track" that of the resistivity.

We consider next the cases of an antiferromagnetic transition and formally equivalent (as an Ising model prototype) order-disorder transition. In these cases a new element enters the picture, viz., the wave vector $\vec{K}$ associated with the periodicity of the ordered phase. The behavior of $\vec{K}$ for this case is shown in Fig. 3b if one relabels the curves $\vec{k} \to \vec{k} + \vec{K}$; hence $\vec{k} = \vec{K}$ is the diverging mode. Two quite different behaviors for the resistivity result, depending upon whether the vector $\vec{K}$ spans the Fermi surface, i.e., connects different portions of the Fermi surface. If it does not, as in the case of beta-brass,[3,11] the problem becomes formally similar to that

of a ferromagnet and one observes a ferromagnetic anomaly[3] both in
ρ and S as seen in Nickel. However, if $\vec{K}$ spans the Fermi surface,
the nature of the anomaly is qualitatively different. Recent calculations[11-13] which corrected earlier work have shown that the
expected resistance anomaly for this case is the mirror image of
the ferromagnetic anomaly, with $d\rho/dT$ being negative at T_c, yet
with $d\rho/dT$ being specific heat like within a region which embraces
T_c. In the region below but not too close to T_c, yet another phenomenon comes into play. If the new Brillouin zones associated with
the wave vector of the order parameter cut the Fermi surface, this
produces energy gaps in the conduction band, which in turn leads to
a decrease in carrier density and hence increased resistance (see,
e.g., Ref. 14). However, sufficiently close to T_c short range fluctuations should dominate, leading to specific heat like behavior in
$d\rho/dT$ a' la Refs. 11-13. While, to these authors' knowledge, the
critical TEP has not been formally worked out for an antiferromagnet,
we might expect the anomaly to be qualitatively similar to the resistivity anomaly as in the case of ferromagnets. This correspondence is seen in the case of Chromium,[15] for which we show the TEP
results in Fig. 5.

We show in Figs. 6 and 7 preliminary results[16] for the TEP of
the archetypal order-disorder system Fe_3Al. This system exhibits
Ising model exponents in the critical behavior of the X-ray scatter-

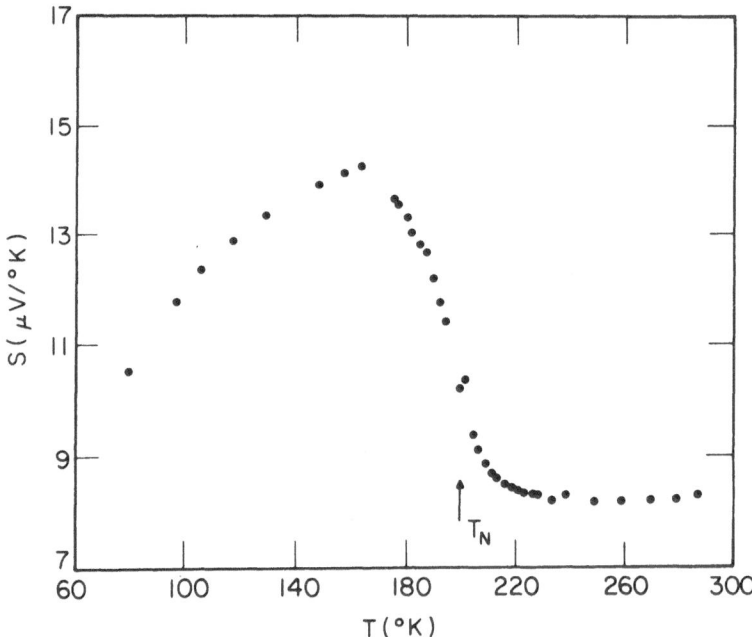

Fig. 5. Absolute thermopower of Chromium (with 1.2 at.% Al) in the
neighborhood of the Néel temperature (after Arajs et al).[15]

ing observed by Guttman et al.,[17] the transition is second order, the system is isotropic and the ordered state is simple. The ordering vector associated with the low temperature phase cuts the fermi surface as in Chromium, unlike beta brass. The TEP results in Figs. 6 and 7 are qualitatively similar to those observed for Chromium and the thermal derivative shows both the classic λ-shaped specific heat anomaly and the suggestion of a cross-over to Ornstein-Zernicke behavior at higher temperatures. An exponent analysis of the results has not yet been made.

We have exhibited above the experimental results for the critical thermopower for a number of metallic systems which might be classified as archetypal ferromagnetic, antiferromagnet and order-disorder systems. In the case of Nickel and Chromium, for which the spins are believed to be at least partially itinerant, the TEP results are qualitatively similar to those for $GdNi_2$, beta brass and Fe_3Al, for which the ordering elements are localized. This is not particularly surprising since all the information about the critical

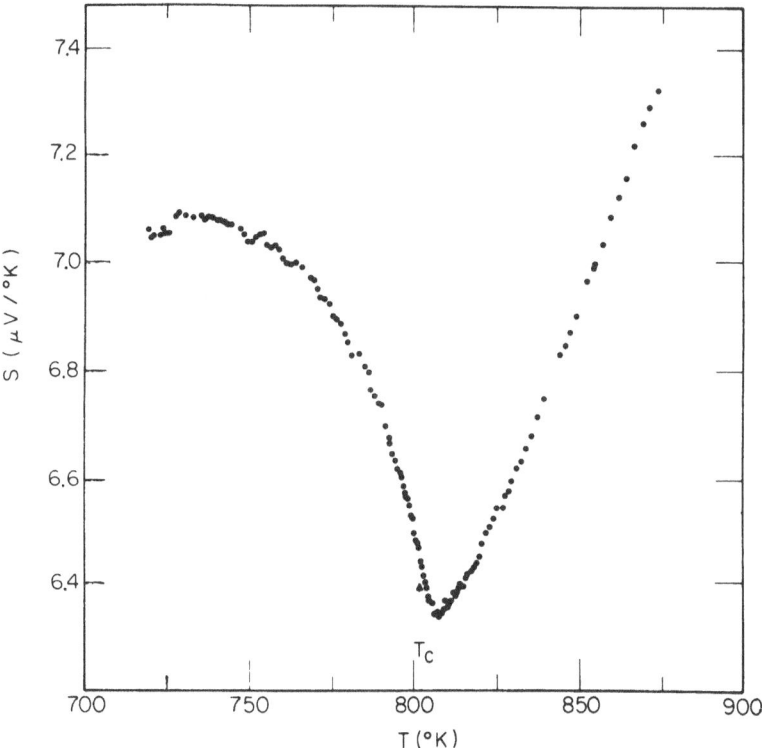

Fig. 6. Preliminary thermopower results for Fe_3Al (nominal composition and with respect to copper reference) in the vicinity of the order-disorder transition.[16]

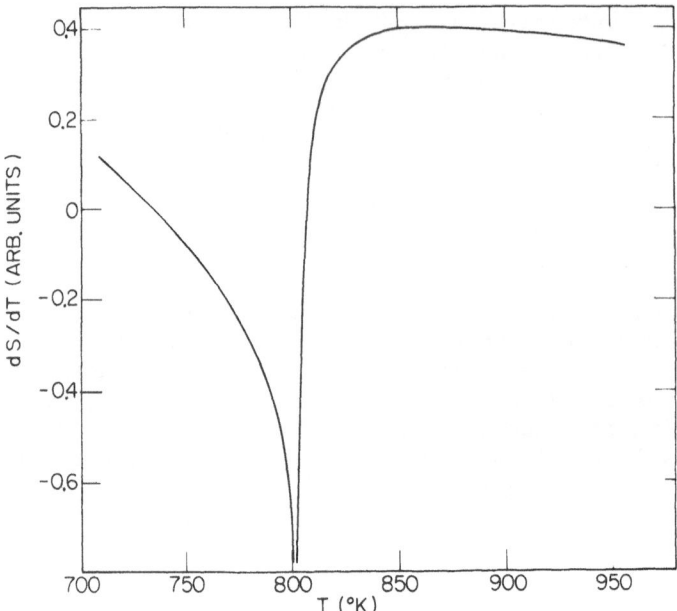

Fig. 7. Thermal derivative of curve shown in Fig. 6. Note disappearance of quasi-linear background in S(T) above T_c.

scattering centers [e.g., Eqs. (2) and (4)] is contained in the correlation functions $g_{\vec{k}}$. It is known from numerous neutron scattering studies that the behavior of the $g_{\vec{k}}$'s can be qualitatively similar in itinerant and localized systems. However, a serious problem would be encountered in the case of the transition metals if the d-electrons were to contribute significantly to the conduction process (as carriers). This complication has not been considered in theories of either the critical resistivity or the critical thermopower.

In ending, we believe that detailed studies of the thermopower and resistivity near critical points can in principle supply important information about both the statics and dynamics of short wavelength critical fluctuations. The most useful information will come not necessarily from the immediate vicinity of the critical point, but probably further from T_c in the region of the cross-over to Ornstein-Zernike behavior. The high resolution possible in transport experiments relative to that obtainable in neutron scattering experiments compensates, in a sense, for the fact that the quantities of interest are more convoluted in the former types of experiments. Studies of ferromagnetic or antiferromagnetic degenerate semiconductors with small, and better yet variable, Fermi velocities, would be

particularly interesting since they would allow the study of longer
wavelength fluctuations. For example, in the case of ferromagnetic
semiconductors, the momentum transfers contained in (1) would per-
force be small. In such a case, dynamic scaling leads to critical
slowing down of the response frequencies. It is easily seen that
in such a case, (1) reduces to the elastic limit (4). Now a thermopower measurement would yield directly the temperature dependent
localized spin-spin correlation function at zero frequency. This
would be a very interesting measurement.

A new theoretical direction would be an attempt to duplicate
the Geldart[7] analysis of the resistivity, assessing the importance
of inelasticity for the magnitude of S in the vicinity of the critical temperature. It would also be of some interest to calculate
the thermopower in the critical regime for an antiferromagnet, especially in light of the analogy with order-disorder transitions.

REFERENCES

1. S.H. Tang, R.P. Craig and T.A. Kitchens, Phys. Rev. Lett. 27, 593 (1971).
2. Ora Entin-Wohlman, Guy Deutscher, and R. Orbach, Phys. Rev. B14, 4015 (1976).
3. G.A. Thomas, K. Levin and R.D. Parks, Phys. Rev. Lett. 29, 1321 (1972).
4. I. Zoric, G.A. Thomas and R.D. Parks, Phys. Rev. Lett. 30, 22 (1973).
5. R.D. Parks, A.I.P. Conf. Proc. 5, 630 (1972).
6. B.I. Halperin and P.C. Hohenberg, Phys. Rev. 177, 952 (1969).
7. D.J.W. Geldart, Phys. Rev. B15, 3455 (1977).
8. P.G. De Gennes and J. Friedel, J. Phys. Chem. Solids 4, 71 (1958).
9. M.E. Fisher and R.J. Burford, Phys. Rev. 156, 583 (1967).
10. M.E. Fisher and J.S. Langer, Phys. Rev. Lett. 20, 655 (1968).
11. T.G. Richard and D.J.W. Geldart, Phys. Rev. B15, 1502 (1977).
12. S. Alexander, J.S. Helman and I. Balberg, Phys. Rev. B13, 304 (1976).
13. T. Kasuya and A. Kondo, Solid State Comm. 14, 249 (1972).
14. W.J. Nillis and S. Legvold, Phys. Rev. 180, 581 (1969).
15. Sigurds Arajs, Norman L. Reeves and Elmer E. Anderson, J. Appl. Phys. 42, 1691 (1971).
16. D. Charkaborty and R.D. Parks, to be published.
17. L. Guttman and H.C. Schnyders, Phys. Rev. Letters 22, 520 (1969).

DISCUSSION

K.H. Fischer: Have you considered what might happen to the thermo-power or to the resistivity in a spin glass near the freezing temperature?

R.D. Parks: No; the resistivity is hard enough. Thermopower is like an order of magnitude more difficult usually, especially in complex situations like that.

B.R. Coles: I was glad to hear Dr. Parks emphasize the importance of remapping of the Fermi surface in the antiferromagnetic case. I should emphasize that the change in the effective number of carriers, of course, enters into his normal term in the resistivity as well as into the critical scattering term. That means, where phonon scattering and impurity scattering are large the resistivity and the thermopower in the antiferromagnet near T_c will almost be dominated totally by those effects. I think it will be rather difficult to separate out experimentally effects due to the criti-cal scattering term only. It is rather interesting for this conference that the first sign of those effects was, in fact, seen by White and Woods in measurements at NRC on alpha manganese. It was a very striking effect and predated, in fact, really reliable indications from neutron scattering of the antiferromagnetism of alpha manganese.

R.D. Parks: Let me comment. Of course, above T_c this effect does not enter into the picture. Now, just below T_c it is generally believed that the effects on the conduction band, the short range effects, dominate. That would lead to a behavior like that of the specific heat, and one way of deconvoluting the two is through scal-ing theory which makes certain predictions about the prefactor of specific heat above and below a transition. If you bring in a new mechanism it would seem that you would affect this constant on the low temperature side. Hence, by looking at the ratio of the pre-factors of the two divergences you might be able to make a separation of the two effects.

D. Greig: I wonder if Dr. Parks could explain what is meant by the second point in his summary, calling for more thorough studies. Is this an experimental matter?

R.D. Parks: It is experimental. I showed, for example, results on $GdNi_2$. Well, it is hard to get band structure people to do much with compounds like that. The point is, you have to find something that gives you a nice, well-defined cross-over effect and, at the same time, you have to know all the parameters you need, all the band parameters, that is.

D. Greig: Where in relation to the transition temperature is the cross-over?

R.D. Parks: The cross-over appears about ten to thirty percent above T_c. That's rather a sensitive function of the sample parameters as it turns out.

H.J. Van Daal: Is there anything clearly established about the Brillouin superzone effect due to the antiferromagnetism; that is, about the relation between the magnitude of the band gap and the magnetic coupling strength. Let's say, if the Néel temperature increases, does that imply, for instance, a higher resistivity?

R.D. Parks: It is very hard, as Brian Coles mentioned, to separate these two components. The suggestion I just made - and we have not tried it yet - is to look at the prefactors of the specific heat. The first step must be to try to make such a separation, and that has not been done, not as yet, I think.

H.J. Van Daal: I mention this because we have observed an interesting relation in chromium-aluminum alloys where with increasing aluminum concentration the Néel temperature increases tremendously to about 1000K and you also see a tremendous increase in resistivity and a correspondingly large decrease of the electronic specific heat coefficient γ. We think that is a Brillouin superzone effect. It could be a relation between these properties.

B.R. Coles: I think Dr. Van Daal is quite right. The increase below the Néel temperature is associated strongly with the superzone effect. But the reason it increases so strongly is, I think, that the modulation of the scattering includes the modulation of the impurity scattering term. Therefore, as your otherwise temperature independent impurity scattering term increases the size of the temperature modulation of that term through the Brillouin zone effect also increases. That is the main reason; this was seen by Arrott and me some years ago in manganese-iron alloys. I think that if you look for example at vanadium-chromium alloys in which the Néel temperature decreases, even so, you will, nevertheless, find an increase in the anomalous resistance term because of the increase in impurity scattering.

THERMOPOWER OF DILUTE MAGNETIC ALLOYS

K. H. Fischer

Institut für Festkörperforschung der Kernforschungsanlage

Jülich

D-5170 Jülich, Germany

ABSTRACT

Calculations and measurements of the thermopower in Kondo and localized spin fluctuation (LSF) systems are critically reviewed. For Kondo systems various theories including a recent one of Keiter for the thermopower in a magnetic field are discussed. A relationship between the scattering t-matrices for LSF systems with simple and transition metals as host (as described by the Anderson and Wolff models) is pointed out. Theoretical results for the thermopower in LSF systems are discussed.

1. INTRODUCTION

Among the various Kondo anomalies the giant thermopower $S_d(T)$ is perhaps the most spectacular one.[1] One observes a broad peak the maximum of which might be as large as 40 µV/K. Fig. 1 shows the thermopower $S_d(T)$ for Au Fe, Au V, and Au Co.[1] The temperature of the maximum (minimum) can be roughly identified with the Kondo temperature T_K. Depending on the system, T_K varies over several decades. The thermopower $S_d(T)$ is positive on the l.h.s. of the 3d transition metal series and negative on the r.h.s., the border line being roughly Cr or Mn. In Cu Cr and Cu Mn one observes[2] $S_d < 0$. However, the effect is rather small, and therefore the magnetic impurity contribution is difficult to separate from other contributions. For non-interacting impurities and for a single scattering mechanism the thermopower should be independent of the

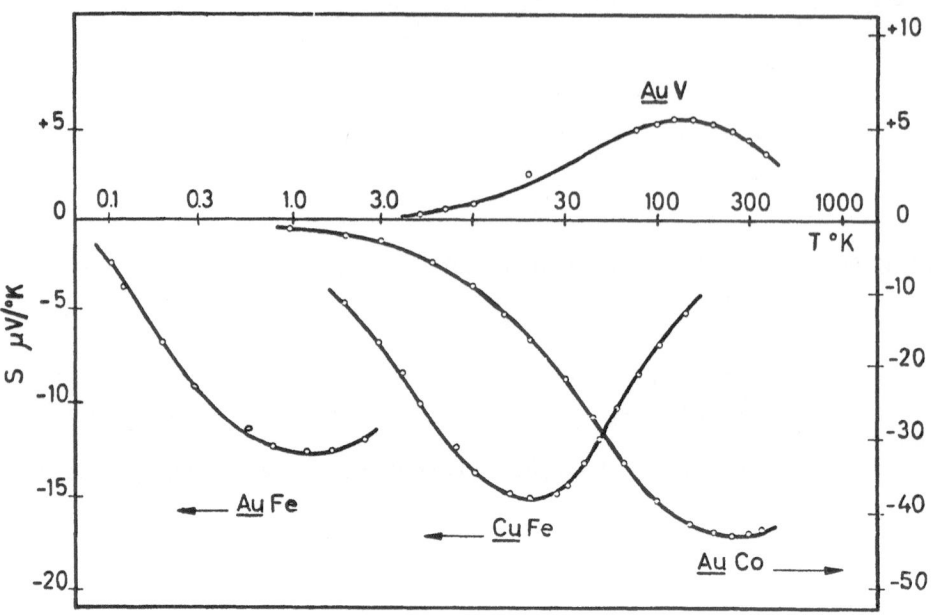

Fig. 1. Representative examples of the temperature dependence of
the thermopower for 3d transition metal impurities in Cu and Au
(from Ref. 1).

impurity concentration c. The dependence on c observed,[1] for instance, in Cu with 300 ppm and 750 ppm Fe should be due to interactions between the magnetic impurities.

A giant peak in the thermopower is observed not only in Kondo
systems. Local spin fluctuation (LSF) systems such as Th U of Al Mn
show very similar effects,[3] and the theory indeed confirms such a
behaviour[4] under certain conditions to be discussed below. This, of
course, raises the question of the relation between Kondo and LSF
systems. In the following, we will discuss both types of dilute
magnetic alloys, considering mostly the thermopower as one of the
most sensitive methods for the investigation of these systems. We
will not discuss crystal field and valence fluctuation effects.

2. KONDO SYSTEMS

From the point of view of theory (and Kondo after all is a
theorist) a Kondo system is defined by a spin S of fixed magnitude

which is coupled to the conduction electrons of the host via the effective exchange interaction

$$H_J = - J \, \underset{\sim}{S} \, \underset{\sim}{S}_e(o) \quad , \tag{1}$$

$\underset{\sim}{S}_e(o)$ being the spin density of the host electrons at the impurity
site. The exchange coupling has to be <u>antiferromagnetic</u> (J < 0).
In addition, one has the energy of independent band electrons.
Physical properties are calculated under the assumption that one has
a small concentration of <u>independent</u> impurities.

For the thermopower no dramatic effects are obtained from the
interaction (1), since H_J preserves the particle-hole symmetry[5]
and therefore leads to a relaxation time $\tau(\varepsilon) = \tau(-\varepsilon)$ [compare Eqs.
(5) and (6); the energy ε is measured with respect to the Fermi
energy ε_F; the band energy is assumed to by symmetric with respect
to ε_F]. This symmetry is broken if one includes in the Hamiltonian
a potential (i.e. spin independent) interaction

$$H_V = V \, \hat{n}_e(o) \quad , \tag{2}$$

where $\hat{n}_e(o)$ is the electron density operator at the impurity site.
The interaction (2) is rather strong in Kondo systems: one has
<u>resonance scattering</u> of the conduction electrons into the partly
filled 3d orbitals of the impurity, as has been discussed first by
Friedel.[6] If the resonance is near to the Fermi surface (F.S.),
this leads to a large spin independent impurity resistivity. The
effective exchange (1) is essentially due to the mixing between band
states and localized impurity 3d states, together with the Coulomb
interaction between the electrons in the impurity 3d states.[7]

The Kondo effect has been reviewed by many authors.[8] As Wilson[9]
stressed, in the limit T → 0, the renormalized exchange interaction
$|J_{eff}|$ becomes very large, leading for $S = \frac{1}{2}$ to the trapping of one
band electron spin by the impurity spin into a singlet state. However, the system can be polarized[10] by additional electrons which
leads to an effective interaction between the remaining N-1 electrons. The system of N-1 interacting electrons can then be treated[10]
in the framework of Landau's Fermi liquid theory. For T <<T_K, all
physical properties obey as a function of temperature simple power
laws: the specific heat due to the magnetic impurities is proportional to T; the resistivity for V = 0 is

$$\rho = \rho_o [1 - \pi^2 (T/T_K^*)^2] \quad , \tag{3}$$

and the impurity susceptibility at $T = 0$ is $\delta\chi = (2\mu_B)^2/\pi T_K^*$. The resistivity ρ_0 corresponds to the unitarity limit for s-scattering (i.e. to an exchange scattering phase shift $\delta_J = \pm\pi/2$)

$$\rho_0 = \frac{2m^* c}{n_0 e^2 \pi N(o)} \quad , \tag{4}$$

where n_0 is the electron density, m^* the effective mass, and $N(\varepsilon)$ the density of states. In contrast to T_K, T_K^* is defined by the <u>low</u> temperature properties. These results can easily be generalized for $V \neq 0$, and one expects the thermopower for $T \to 0$ to become proportional to T. All physical properties are in the whole temperature range universal functions of T_K (respectively T_K^*), the potential scattering phase shift δ_V, and the properties of the host.

The resistivity ρ, the diffusion part of the thermopower S_d, and the thermal conductivity κ are obtained from the transport integrals K_n ($e < 0$)

$$\rho = (e^2 K_0)^{-1} \ , \ S_d = \frac{1}{eT} \frac{K_1}{K_0} \ , \ \kappa = \frac{1}{T}\left(K_2 - \frac{K_1^2}{K_0}\right) \tag{5a-c}$$

with ($f(\varepsilon)$ in the Fermi function)

$$K_n = \frac{n}{m^*} \int d\varepsilon \ \varepsilon^n \ \tau(\varepsilon,T) \ (-df(\varepsilon)/d\varepsilon) \quad . \tag{6}$$

We neglected in (6) the energy dependence of $N(\varepsilon)$ and of the Fermi velocity $v(k)$ in the energy window $(df/d\varepsilon) \neq 0$, since it leads only to small corrections to the Kondo anomalies. The Kondo effect manifests itself in a strongly energy and temperature dependent relaxation time τ. As a consequence the usual Sommerfeld expansion of the integrals (6) breaks down, and the relation[11] $S_d \sim (d \ln \sigma/d\varepsilon)_{\varepsilon=\varepsilon_F}$ does <u>not</u> hold.

The most extended work on the transport properties of Kondo systems has been performed in the framework of the Suhl-Nagaoka theory[12,13] in which only the simplest correlations between impurity spin and conduction electrons are taken into account. This theory becomes exact for $T \gg T_K$. It holds approximately at $T \sim T_K$, and breaks down for $T \ll T_K$. One obtains a rather complicated expression for the scattering t-matrix and hence for the relaxation $\tau(\varepsilon)$ defined by

$$\tau^{-1}(\varepsilon) = - 2c \ Im t(\varepsilon + i\eta) \quad , \quad \eta = 0_+ \ . \tag{7}$$

Unfortunately, τ and not τ^{-1} enters into equation (6). Exact analytic expressions for the transport integrals can be obtained only if the energy dependent scattering is small compared to elastic scattering mechanisms such as the potential scattering from the interaction (2) or from additional non magnetic impurities. As a consequence, the Gorter-Nordheim rule $\rho_{tot} S_d = \sum_i \rho_i S_{di}$ and the Matthiessen rule $\rho_{tot} = \sum_i \rho_i$ are strongly violated (ρ_i are the resistivity contributions due to the various scattering mechanisms, and S_{di} are the corresponding thermopowers). With this restriction, one obtains for the thermopower[14],[15] (e < 0)

$$S_d = \frac{\pi k_B}{4e} \frac{\rho_o}{\rho'(T)} \frac{\pi^2 S(S+1) \sin 2\delta_V}{[\ln^2 T_K/T + \pi^2 S(S+1)]^{3/2}} \quad . \tag{8}$$

Here, the Kondo temperature T_K is defined as (2D is the band width)

$$T_K = 1.13 \, D \, \exp|\gamma^{-1}| \quad , \quad \gamma \equiv J \, N(o) \, \cos^2 \delta_V, \quad |J \, N(o)| \ll 1. \tag{9}$$

Maki[16] replaces the resistivity ρ' due to the elastic scattering by the resistivity ρ_m of the magnetic impurities (the sign in Ref. 16 is in error). Equation (8) for S_d yields as a function of $\ln T$ a broad peak which is symmetric with respect to T_K. The exact (numerical) evaluation[17] of the transport integrals (6) indicates that the peak with increasing δ_V is increasingly shifted from T_K (compare Fig. 2). The width of the peak depends strongly on the magnitude of the impurity spin S. The theory is based on the assumption of integer or half integer spin values, but better fits can be obtained if S is assumed to be a continuous variable[18]. In the Anderson model[19] the impurity spin indeed is no conserved quantity, and its thermal expectation value can assume within certain limits arbitrary values. The Kondo model turns out to be a limiting case of the Anderson model which is obtained by means of the Schrieffer-Wolff transformation[7]. Unfortunately, the theory for this more general model is not yet as advanced as the theory for the Kondo model.

The result (8) yields the correct order of magnitude. The giant thermopower of Kondo systems turns out to be an interference effect between exchange and potential scattering. The latter leads to a shift of the sharp Kondo resonance which for V = 0 is tied to the F.S.

The theory of transport properties of Kondo systems in a magnetic field is much more complicated. Keiter[20] succeeded in solving the Suhl equations for this more general case. The transport

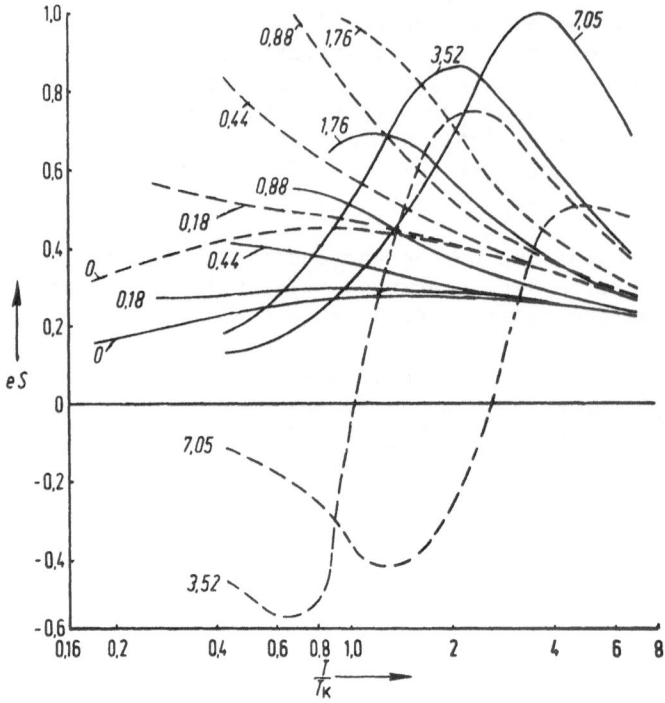

Fig. 2. Diffusion part S_d(T) of the thermopower for Kondo systems
as a function of temperature for various magnetic inductions B/B_K
and for the potential scattering phases $\delta_V = 10°$ (full lines) and
$\delta_V = 20°$ (dashed lines) (from Ref. 20).

coefficients Eq. (6) turn out to be (apart from the phase shift δ_V)
universal functions of T/T_K and B/B_K when B is the magnetic induction
and B_K a "Kondo induction". Fig. 2 shows the thermopower S_d(T,B)
for S = 1/2 for various fields B and for the potential scattering
phase shifts $\delta_V = 10°$ and $\delta_V = 20°$. The zero field thermopower
S_d(T,B = 0) shows a broad maximum which with increasing potential
scattering is shifted to lower temperatures. With increasing field
B the maximum is shifted to higher temperatures and becomes sharper.
This might be the reason for the maximum observed[21] in S_d(B) for
fixed T (Fig. 3): for small fields S_d(B) increases; however, for
sufficiently high fields the shift in the maximum of S_d(T) with B
fixed leads to a decrease of S_d(B) with T fixed. For sufficiently
large phase shifts δ_V and fields B, the thermopower even changes
sign.

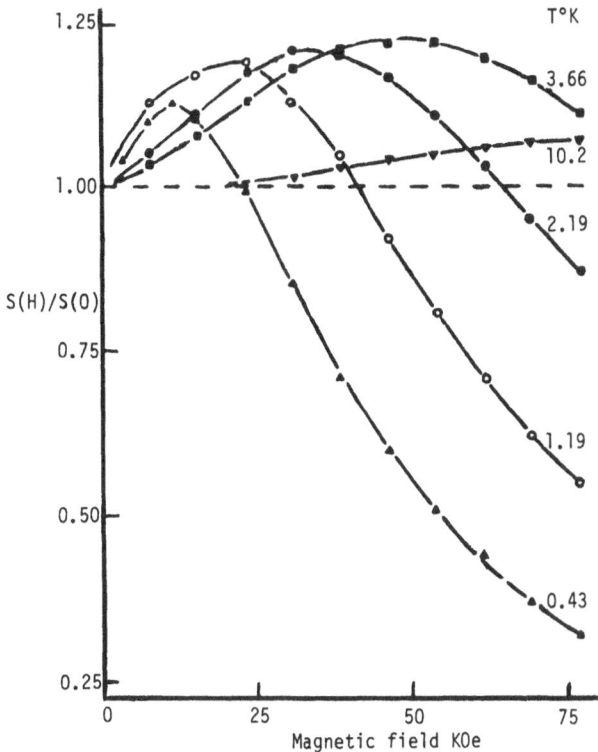

Fig. 3. Change of the thermopower of Au + 0.03% Fe in a magnetic field for fixed temperature T (from Ref. 21).

3. LSF SYSTEMS

The Kondo theory is limited to systems in which the magnitude of the localized spin is fixed. In the Anderson model, this corresponds to the limit in which zero and doubly occupied impurity d-states are forbidden or (in the HF picture) to the case of two Friedel resonances, one of which is well above and one of which is well below the Fermi surface. The LSF theory describes the opposite limit of a single Friedel resonance near the F.S. Experiments on LSF are discussed in considerable detail in Rizzuto's review article.[3]

In the Anderson model the impurity d-states are assumed to be orthogonal to the sp band states of the host which in <u>simple metals</u> are responsible for most properties. In <u>transition</u> metals as host (i.e. Pd, Pt, Rh, Ir, etc.) the Fermi energy is situated between d states. The d-density of states near ε_F is large compared to the sp density of states, and the d-states cannot be assumed to be orthogonal to the impurity states. In the <u>Wolff model</u>[22], impurity and host d-states therefore are assumed to belong to the same system which is described by Wannier states of a single band. The impurity is characterized by a scattering potential V_o, by a screened Coulomb interaction between host d-electrons, and possibly by modified matrix elements for the hopping of an electron from or into the impurity site.

In the theory of Lederer and Mills[23] and Kaiser and Doniach[24] transport properties of systems with transition metals as host are treated in the framework of a two band model. The d-electrons (including those at the impurity site) are coupled to the conduction electrons via exchange. This coupling is treated in Born approximation. This corresponds to considering the influence of the d-electron spin fluctuations on the conduction electrons in linear response. The exchange interaction maintains the particle-hole symmetry. Therefore, the model fails to explain the observed anomalous thermopower, as does the Kondo model with $V = 0$.

In a more realistic model, one would have to take into account mixing between d and conduction (sp) electrons, but this seems to be intractable. One is forced to simplify the problem by dealing with a <u>single band</u> Wolff model in which magnetic and transport properties are due to the same electrons. If the electron-electron interaction U_{1d} <u>within the host</u> can be neglected (which might be a reasonable assumption for all 4d and 5d transition metals <u>besides</u> Pd and Pt), this model is rather similar to the Anderson model. Both models yield indeed the same anomaly in the specific heat. However, the Anderson model has one additional (orthogonal) orbital. This makes both models in the following sense "complementary": the t-matrices (which in both models contain s-scattering only) fulfill the relation[25]

$$t^{Wolff}(z) = t_{u.\ell.} - t^{Anderson}(z), \quad z = \omega + i\eta, \quad |\omega| \ll D, \quad (10)$$

$t_{u.\ell.} = i(\pi N(o))^{-1}$ being the s-scattering matrix element in the unitarity limit. Thus a resonance at the F.S. leads in the Anderson model in HFA to maximum scattering $t = t_{u.\ell.}$. In the Wolff model the (broadened) impurity level fits perfectly into the host, and there is no scattering at all. In the opposite case of a resonance far from the F.S. in the Anderson model the scattering near $\omega = 0$ is not affected at all, whereas in the Wolff model one has a strong scattering potential. The Kondo effect for $V = 0$ yields at $T = 0$ a sharp resonance at the F.S. which in Kondo systems leads to the resistivity $\rho(o) = \rho_{u.\ell.}$ and in the Wolff systems to $\rho(o) = 0$. An example for the second case might be <u>Rh</u> Fe[26]. The relation between the thermopower in both models (assuming again the potential scattering to be large compared to the exchange scattering) reads

$$S_d^W(T) \ (\rho^W + \rho') = - \ S_d^A(\rho^A + \rho') \tag{11}$$

Fig. 4 shows the thermopower $S_d(T)$ calculated[4] for the Wolff model with $U_{id} = 0$ in LSF approximation. Its shape and magnitude are remarkably similar to those for Kondo systems. The same holds true for other physical quantities such as resistivity and thermal resistivity. The parameters which enter into the calculation are the width Δ of the resonance, the LSF temperature T_s as defined by the dynamic susceptibility at the impurity site $\chi(\omega) \sim (k_B T_f - i\omega)^{-1}$, the band width 2D, and the HF scattering potential V^{HF}.

4.   CONCLUSIONS

We stressed the point that very similar anomalies are obtained in Kondo systems and LSF systems of both the Anderson and the Wolff

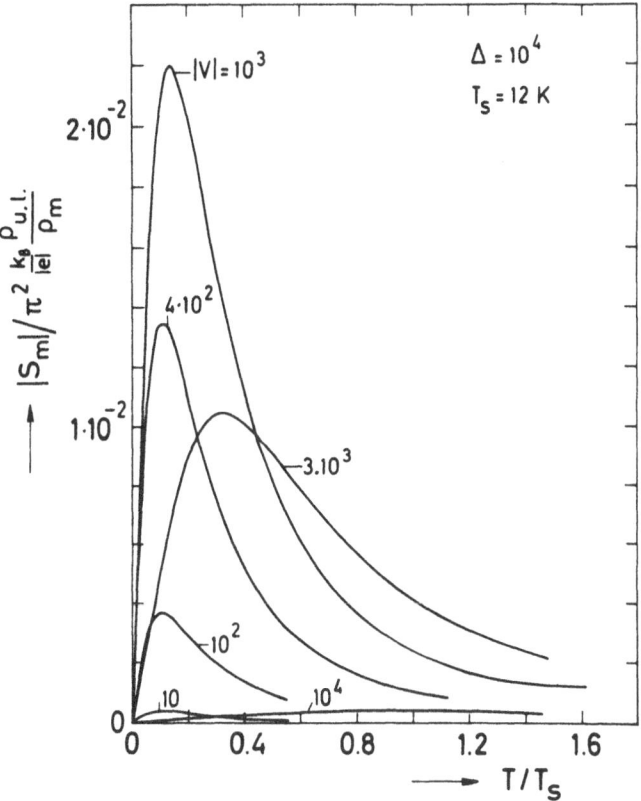

Fig. 4.   Diffusive part $S_d(T)$ of the thermopower of LSF (Wolff) systems as a function of temperature with the energy shift V^{HF} of the impurity level in HFA as parameter (from Ref. 4).

type. Unfortunately, a unified theory which covers both the Kondo
and the LSF limit is still missing. The existing models are extremely simplified. They neglect most of the internal structure of the
impurity by considering a single d-orbital with a spin degree of
freedom. One might expect that the impurity orbital angular momentum is not entirely quenched and manifests itself in certain experiments. If this is the case, spin orbit coupling also should be
included. Further simplifications arise in the explicity computa-
tions in which nearly all details of the band structure are neglected. In ideal Kondo systems with low temperatures this is justified
because of the sharpness of the Kondo resonance. In systems with
higher characteristic temperature T_K or T_s this approximation might
be questionable.

ACKNOWLEDGMENT

I would like to thank Professor D. Wagner for a critical reading of the manuscript.

REFERENCES

1. M.D. Daybell in _Magnetism_, Vol. _5_, edited by H. Suhl (Academic
   Press, New York, 1973) p. 138.
2. J.J. De Jong, thesis, University of Leiden, 1974 unpublished.
3. C. Rizzuto, Rep. Prog. Phys. _37_, 147 (1974).
4. K.H. Fischer, J. Low Temp. Phys. _17_, 87 (1974).
5. H. Keiter, E. Müller-Hartmann and J. Zittartz, Z. Phys. _223_,
   48 (1969).
6. J. Friedel, Can. J. Phys. _34_, 1190 (1956), Nuovo Cim. Suppl.
   _7_, 287 (1958).
7. J.R. Schrieffer and P.A. Wolff, Phys. Rev. _149_ 491 (1966).
8. Compare, for instance, the articles in _Magnetism_, Vol. _5_,
   edited by H. Suhl (Academic Press, New York, 1973).
9. K.G. Wilson, _Collective Properties of Physical Systems_, Nobel
   Symposia Vol. _24_ (Academic Press, New York, 1974).
10. P. Noziéres, J. Low Temp. Phys. _17_, 31 (1974); Phys. Bull.
    (G.B.) _25_, 457 (1974).
11. J.M. Ziman, _Electrons and Phonons_, (Oxford University Press,
    London, 1962) p. 397.
12. H. Suhl, Phys. Rev. _138_, A 515 (1965); Physics (USA) _2_, 39
    (1965); Phys. Rev. _141_, 483 (1966); compare also Refs. 8, 14.
13. Y. Nagaoka, Phys. Rev. _138_, A 1112 (1965).
14. Compare, for instance, K.H. Fischer, phys. stat. sol. _46_, 11
    (1971).
15. The resistivity ρ_0 introduced in Eq. (2.61) of Ref. 14 is _half_
    of the unitarity limit.
16. K. Maki, Progr. Theor. Phys. (Kyoto) _41_, 586 (1969).

17. E. Müller-Hartmann, unpublished.

18. F. Steglich, Z. Physik B $\underline{23}$, 331 (1976).

19. P.W. Anderson, Phys. Rev. $\underline{124}$, 41 (1961).

20. H. Keiter, Z. Physik B $\underline{23}$, 37 (1976); B $\underline{26}$, 169 (1977).

21. R. Berman, J. Kopp and C.T. Walker, Proc. LT 11, St. Andrews 1968, Vol. 2, p. 1238.

22. P.A. Wolff, Phys. Rev. $\underline{124}$, 1030 (1961).

23. P. Lederer and D.L. Mills, Phys. Rev. $\underline{165}$, 837 (1968); compare also Ref. 8.

24. A.B. Kaiser and S. Doniach, Int. J. Magnetism $\underline{1}$, 11 (1970).

25. K.H. Fischer, Proc. ICM 73, Moscow ("Nauka", Moscow, 1974) Vol. $\underline{1}$, p. 288, and to be published.

26. B.R. Coles, Phys. Lett. $\underline{8}$, 243 (1962).

DISCUSSION

I.M. Templeton: I'd like to draw attention to the extra information one can obtain on systems like this from de Haas-van Alphen studies. You find that it is possible to separate the scattering of spin-up and spin-down electrons. In that case, for instance, you can show in the copper-chromium system that, while you can measure the exchange interaction, there is no anisotropy of spin scattering. In the case of copper-iron there is a quite marked anisotropy of spin scattering, and while, of course, we cannot get any information about energy derivatives, it seems most likely that the explanation for the incredibly small thermopower in copper-chromium - it is about three orders of magnitude down compared with copper-iron - is this symmetry effect, so that the derivatives for spin-up and spin-down simply cancel out in that particular case.

K.H. Fischer: I am not quite sure that I can agree with that statement. In the case of copper-chromium one has, in the resistivity, a very big step, like the step I have shown for spin fluctuation systems. If the step is very big, one can estimate that the resistivity nearly goes to the unitarity limit in this model. However, this means that the potential phase shift is very small, and if the potential phase shift is small one gets a very small thermopower. Of course, one can also make more physical argument; looking at the periodic table the thermopower changes from positive to negative. Somewhere in between it has to go to small values and perhaps to zero. This seems to be the case when you come to chromium.

THERMOPOWER AND MAGNETOTHERMOPOWER OF PtCo[*]

C. W. Lee, J. R. Kuhn[†], C. L. Foiles, and J. Bass

Department of Physics

Michigan State University
East Lansing, Michigan 48824 USA

ABSTRACT

Measurements of the thermopower of dilute PtCo alloys containing from 0.31 to 3.74 at.% Co have been made over the temperature range 1.35 to 300K. The thermopowers of the alloys contain pronounced negative peaks which decrease in magnitude and move rapidly to higher temperatures as the Co concentration increases. Magnetothermopower measurements have been made on the same alloys for the temperature range 1.35K to 5K and magnetic fields up to 20kG. Application of a magnetic field is observed to shift the extrema of the negative thermopower peaks to higher temperatures.

PtCo is a giant moment alloy, in which the average magnetic moment per Co atom is roughly twice as large as that for pure Co.[1] As the temperature is lowered, PtCo appears to undergo magnetic ordering, characterized by: (a) a maximum in the low field magnetic susceptibility;[1,2,3] (b) a Schottky anomaly in the electronic specific heat;[4,5] and (c) either an abrupt change in slope or a maximum in the temperature derivative of the electrical resistivity, depending upon the Co concentration and perhaps also upon the method of sample preparation.[3,6,7,8] We denote as characteristic tempera-

---

* Supported in part by the NSF under grants DMR-75-14138 and DMR-77-04680.

† Present Address: Physics Dept., Princeton University, Princeton, New Jersey, USA.

tures T_0 the temperatures of the susceptibility maximum, of the maximum in the Schottky anomaly, and of either the abrupt change in
slope or the derivative maximum in the electrical resistivity. These
temperatures are indicated in Fig. 1 as open symbols for a range of
Co concentrations. We see that above about 1 at.% Co, all three prop
erties give a single curve for T_0. Below 1% Co, however, the characteristic temperatures diverge somewhat for the different properties.

In a recent letter[8] we reported that measurements from 4.2K to
300K of the thermopower of PtCo alloys containing from 0.31 to 3.74
at.% Co revealed pronounced negative peaks which decreased in magnitude and moved rapidly to higher temperatures with increasing Co
concentration. The extrema of these peaks were present in this
temperature range only for the three most concentrated alloys, containing 0.56, 1.65, and 3.74% Co. To complete the picture of the
thermoelectric behavior of these alloys, we have now extended our
thermopower measurements to 1.35K, thereby observing the extremum of
the peak for the 0.31% alloy, and we have also studied the magneto-
thermopowers of these alloys from 1.35K to 4.5K. This paper contains the results of these measurements.

Fig. 2 shows the thermopowers of the four alloys in zero magnetic field for the temperature range 1.35 to 90K, along with that

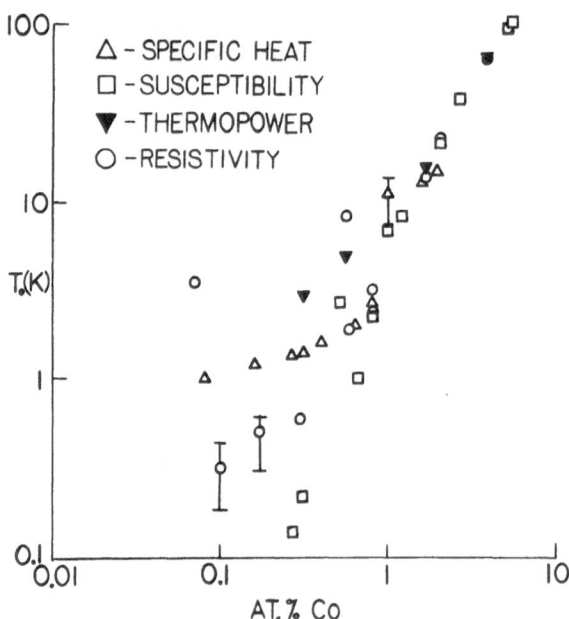

Fig. 1. The variation with Co concentration of the characteristic
temperatures T_0 determined from measurements of electrical resistivity, magnetic susceptibility, electronic heat capacity, and therm(
power.

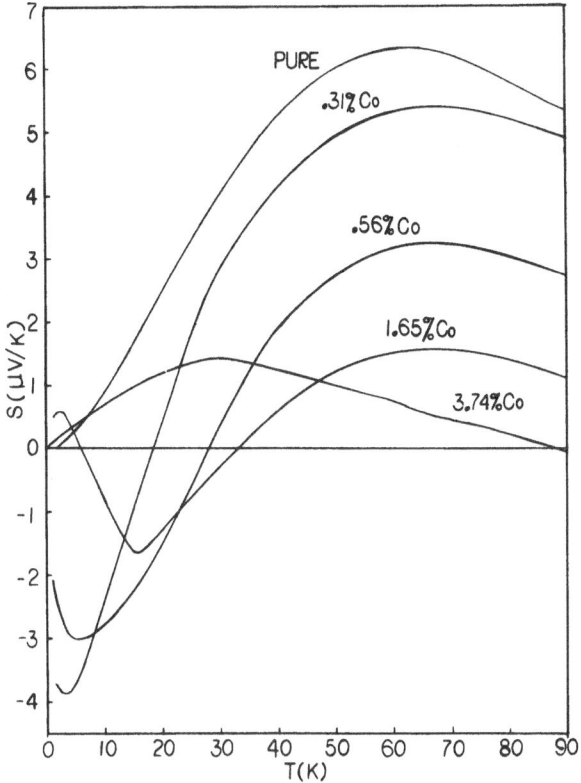

Fig. 2. The thermopower of pure Pt and four dilute P̲t̲Co alloys from
1.35K to 90K. The concentration of Co in each alloy is indicated
in atomic percent.

for a pure Pt control sample prepared identically to the alloys except that no Co was deliberately added. In this temperature range,
the thermopower of pure Pt is dominated by a large, positive phonon-
drag peak, with a maximum at about 63K. For each of the alloys,
there is an additional negative peak, with the peak extrema located
approximately at 3K, 5K, 16K, and 66K* for the 0.31, 0.56, 1.65,
and 3.74% alloys respectively. The temperatures of these peak extrema are indicated in Fig. 1 by filled triangles. We see that for
Co concentrations greater than 1% these temperatures are in excellent agreement with the characteristic temperatures for all three
other properties shown. For Co concentrations less than 1%, the

_____

* As indicated in the original paper[8], the peak for the 3.74%
  alloy is comparable to experimental uncertainty, and therefore
  must be considered less definitively established than for the
  other concentrations.

thermoelectric temperatures appear to agree best with the characteristic temperatures for specific heat.

It has been shown[9] that application of a magnetic field shifts
the locations of the maxima of specific heat peaks in PtCo to higher
temperatures. This is qualitatively the same behavior as is induced
by increasing the impurity concentration. For 0.5% Co in Pt, the
peak maximum was located at about 2K in zero magnetic field and increased approximately linearly with increasing field up to about 6K
in 27kG. This corresponds to a rate of increase of about 0.15K/kG.
If we assume, by analogy with the effect of increasing impurity
concentration, that the magnetic-field-induced shifts in the loca-
tion of the specific heat maxima represent shifts in effective magnetic ordering temperatures, and assume further, based upon the
agreement of the zero field thermopower negative extrema temperatures with the specific heat characteristic temperatures, that the
thermopower negative peak extrema also occur at or near magnetic
ordering temperatures, then we might expect application of a magnetic field to shift the thermopower peak extrema to higher temperatures too. We therefore decided to test this hypothesis.

Our magneto-thermopower apparatus was designed for measurements
to only a little above 4.2K. We therefore began our investigations
with the Pt-0.31% alloy, since it is the only one of our alloys
whose thermopower peak extremum occurs below 4K in zero magnetic
field. Fig. 3 shows the thermopower of this alloy for magnetic
fields up to 20kG and temperatures between 1.35 and 4.3K. As the
field is increased, the thermopower becomes rapidly less negative,
with larger percentage decreases occurring at lower temperatures.
This behavior is just what would be expected on the basis of the
hypothesis advanced above. Using the specific heat as a guide, we
would expect a field of 5kG to shift the thermopower peak extremum
up to about 4K, and this is just about what it does. We conclude
that the behavior of the magneto-thermopower of the 0.31% Co alloy
is both qualitatively and quantitatively in agreement with expectation at temperatures up to about 4.3K. Measurements on an apparatus
suitable for higher temperatures will be needed to ascertain whether
this agreement persists above 4.3K.

In order to test the hypothesis further in the accessible temperature range, we then turned to the more concentrated alloys.
The results obtained for these alloys and for the pure Pt control
are illustrated in Fig. 4. If the behavior of the Pt-0.31% sample
were due to the reasons hypothesized, then we would expect little
or no change in the thermopowers of either the pure Pt control or
the most concentrated alloy, Pt-3.74% Co, upon application of magnetic fields up to 20kG; for pure Pt because it contains no Co, and
for the Pt-3.74% Co alloy because its characteristic temperature
($\approx$66K) is so high that no vestige of the negative thermopower peak

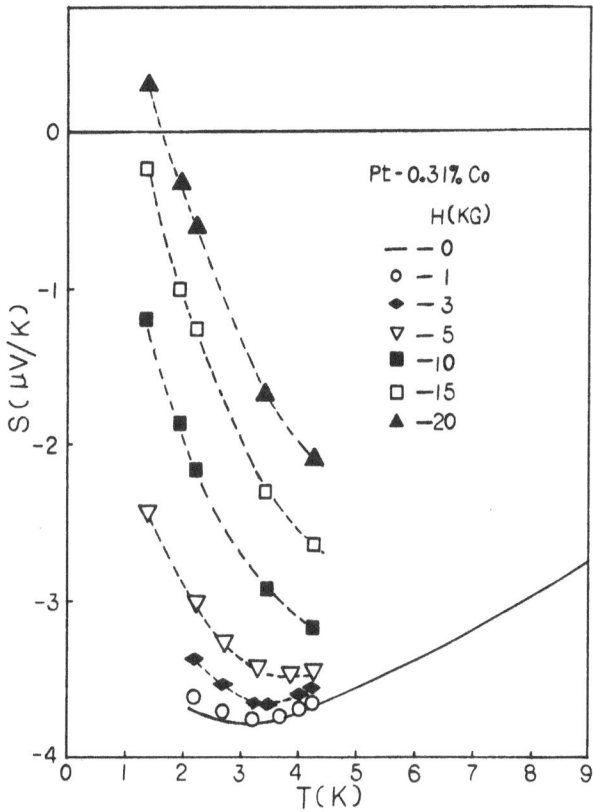

Fig. 3. The variation with magnetic field of the thermopower of the Pt-0.31% Co alloy.

remains in the temperature range of present investigation. Indeed, as shown in Fig. 4, the changes in thermopower with field are much smaller for these two samples than for any of the three where the negative thermopower peak does overlap the experimental temperature range. On the other hand, both the Pt-0.56% and the Pt-1.65% Co alloys would be expected to change with magnetic field in a manner similar to that induced by increasing the impurity concentration, and they both do! We conclude that all of our samples behave qualitatively as expected, and that the 0.31% alloy behaves quantitativly as expected in the accessible temperature range. To make a more definitive test, we need to extend our measurements to both higher temperatures and higher magnetic fields, and we are presently preparing to do both. We are also preparing more dilute alloys for study, in order to move the thermopower peak extrema further down in temperature and to see what happens to the magnitude of the

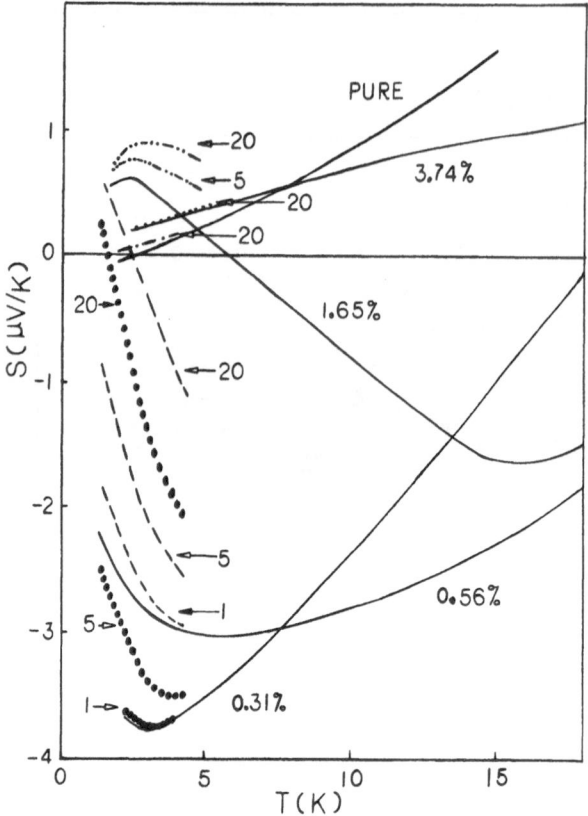

Fig. 4. The variation with magnetic field of the thermopowers of the pure Pt sample and of the four P̲t̲Co alloys. For the Pt-0.31% Co and Pt-0.56% Co alloys, data are shown for fields of zero, 1kG, 5kG, and 20kG. For the Pt-1.65% Co alloy, data are shown for zero, 5kG, and 20kG. For pure Pt and the Pt-3.74% Co alloy, data are shown only for zero and 20kG.

negative peak as the Co concentration becomes more dilute. We hope to be able to report the results of such measurements in the near future.

REFERENCES

1. J. Crangle and W.R. Scott, J. Appl. Phys. 3̲6̲, 921 (1965).
2. B. Tissier and R. Tournier, Sol. State Comm. 6̲, 895 (1972).
3. K.V. Rao, O. Rapp, Ch. Johannesson, J.I. Budnick, T.J. Burch, and V. Cannella, Conference on Magnetism and Magnetic Materials, A.I.P. Conf. Proc., 2̲9̲, 346 (1976).

4. J.C.G. Wheeler, J. Phys. C. (Sol. St. Phys), 2, 135 (1969).
5. P. Costa- Ribeiro, M. Saint-Paul, D. Thoulouze, and R. Tournier, Proc. 13th Int. Conf. Low Temp. Phys. (Timmerhaus, O'Sullivan, and Hammel, Eds., Plenum Press, N.Y.), Vol. 2, 520 (1974).
6. L. Shen, D.S. Schreiber, and A.J. Arko, J. Appl. Phys. 40, 1478 (1969).
7. O. Laborde, Thesis: De l'effet Kondo a l'etat verre de spins, Etude de la resistivite a tres basses temperatures des alliages dilues: AuFe, CuMn, et PtCo. L'Universite Scientifique et Medicale et L'Institute National Polytechnique de Grenoble, 1977 (Unpublished).
8. J.R. Kuhn, C.L. Foiles and J. Bass, Phys. Lett, 63A, 401 (1977).
9. B.M. Boerstoel and C.M. Van Baarle, J. Appl. Phys. 41, 1079 (1970).

DISCUSSION

R.B. Roberts: Are your thermopowers extracted from the old scale?

J. Bass: The thermopowers were all extracted using the new scale, as I recall. Below 4K we use a superconducting arm.

R.B. Roberts: I am concerned about the glitch in the 3% – 4% alloy.

C.C. Foiles: Whether one uses the old scale or the new scale, the important thing, I should think, is that there is a glitch at all. If it were a systematic error due to the scale it should show up in all four samples. It does not. It shows up in only one alloy.

R.B. Roberts: I would point out that there is a "glitch" in that scale, at exactly that temperature, of order 0.02 µV/K which is about 10% of the absolute magnitude of the thermopower you are looking at in that region. You won't see it in the other alloys because their thermopower is so much bigger and you don't see it as a large percentage of your result. But you do see it as a large percentage in that concentrated alloy because the thermopower is small then.

C.L. Foiles: We'll certainly look into that.

THERMOELECTRIC POWER AND RESISTIVITY OF Yb DILUTED IN $Ag_{1-x}Au_x$

F. F. Bekker

Universiteit van Amsterdam

Valckenierstraat 65
Amsterdam, The Netherlands

Ytterbium diluted in $Ag_{1-x}Au_x$ has attracted a lot of interest in the last 10 years. As soon as it was known that Yb is in a divalent non magnetic state in Ag[1,3] and in a trivalent magnetic one in Au[2,3] it became clear that for some value of x there should be a configuration cross-over in the $Ag_{1-x}Au_x$ host, with as a consequence a Kondo effect due to the s-f interaction.[4] This configuration cross-over however must also have strong influence on the screening cloud around the Yb impurity; this in turn means that not only the spin dependent scattering is affected but also the potential scattering.[5,6]

Ytterbium in Au was studied most extensively: the crystal field ground state was obtained from EPR measurements: Γ_7;[7] from the hyperfine splitting it was deduced that the overall interaction between localized f electrons and conduction electrons is antiferromagnetic.[8] Other local properties as well as susceptibility measurements indicate that the Kondo temperature must be smaller than 1 millikelvin.[9]

In this paper we want to study as a function of x the change of the spin dependent and potential scattering with the emphasis on the latter. Some of the measurements underlying this paper have been published,[5] others have been reported on the Low Temperature Conference (1975).[6] Here we will summarize our main experimental results and compare them with existing theories on the transport properties.

The resistivity experiments were fitted with

$$\rho(T) = a_o + a_1 \log T + a_2 T^3 + a_3 T^5 \tag{1}$$

the thermoelectric force: $E = \int_{T_b}^{T} S dT$ data with the integrated expression for the thermopower:

$$S = A_1 \frac{T}{T_o + T} + A_2 T + BT^3 \tag{2}$$

The first term in (2) is a phenomenological description of the Kondo effect in the diffusion thermopower first introduced by Guénault.[10] The T_o values for iron in silver and gold as given in the literature are small in comparison with our lowest experimental temperature ($T_b \simeq 1.4$ K); thus taking $T_o = 0$ has little influence on the values of A_1, A_2 and B. Therefore the first term in (2) is essentially a temperature independent contribution to the diffusion thermopower in our experimental temperature range: 1.5 < T < 15 K.

In Fig. 1 typical experimental resistivity results are shown; the results[1] for AgYb 2[a], which is the sample with the highest resistivity ($0.17 \times 10^{-6} \Omega$cm) from a series of three AgYb samples shows the weakest minimum. Therefore no scaling between the increase in resistivity and the depth of the minimum is observed. From the thermopower (in fact from the A_1 values) of the same samples, we deduced, using the Nordheim-Gorter rule, iron contents for these samples, which do scale with the slight minima in the resistivity. The minimum in AgYb 2[a], as well as the temperature independent part of the diffusion thermopower (A_1) can be explained with 1.0 ± 0.3 ppm Fe.

For the Ag.8Au.2Yb as well as for the AuYb samples the depths of the minima do scale with the impurity resistivity, so for these samples there is a Kondo effect due to Yb in the resistivity. With Fig. 2, which illustrates only the diffusion thermopower due to Yb in the three different hosts, (for the full analysis see Refs. 5 and 6), we want to stress that the Kondo effect in the resistivity in Ag.8Au.2Yb is accompanied by a temperature independent thermopower. After correction for the large host resistivity by means of the Nordheim-Gorter rule, $S \simeq 1$ μV/K. On the other hand in the AuYb case there is no trace of a Kondo effect in the thermopower, contrary to what one would expect from the resistivity minima for these alloys.

---

[1]corrected for the large host resistivity.

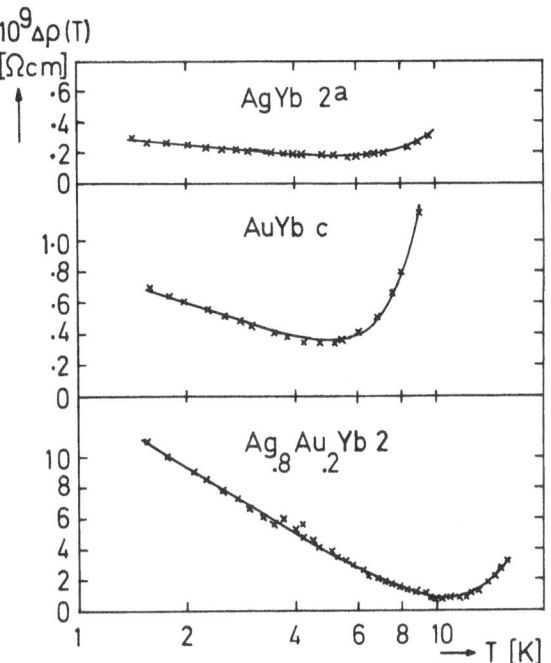

Fig. 1. Resistivity curves for Yb in a Ag, Au and Ag$_{.8}$Au$_{.2}$ host. The logarithmic slope in the upper curve is attributed to $\simeq$ 1 ppm Fe. Note the different scale for the Ag$_{.8}$Au$_{.2}$ host.

TABLE I

Parameters of eq.(1) and eq.(2)

Samples	Yb conc.	$10^9 \, a_1/C$	$10^6 \, A_1$	$10^6 \, a_o/C$	$10^9 \, A_2$
	C at %	Ωcm/at%	V/K	Ωcm/%	V/K^2
AgYb2[a]	0.10$\pm$0.03	no slope due to Yb	–	2.0$\pm$0.5	+3$\pm$3
Ag$_{.8}$Au$_{.2}$Yb2	0.16$\pm$0.03	–30$\pm$6	–1.0$\pm$0.2[1]	2.5$\pm$0.5	–
AuYbc	0.18$\pm$0.01	–1.8$\pm$0.1	–	6.0$\pm$0.2	+10.5$\pm$1.0

[1]Corrected for the large host resistivity.

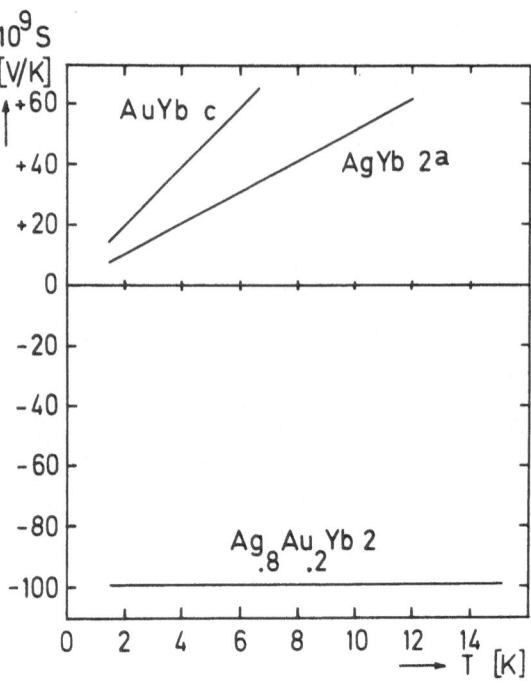

Fig. 2. Thermopower curves for Yb in Ag, Au and $Ag_{.8}Au_{.2}$. Only the part of the diffusion thermopower attributed to Yb is plotted.

The main quantitative results are shown in Table I. The results in the last two columns of this table each represent a series of samples with different Yb concentrations in the Ag; $Ag_{.8}Au_{.2}$ or Au hosts. From the slopes of the logarithmic part in the resistivity (a_1 from eq. (1)) it is clear that the Kondo effect in the $Ag_{.8}Au_{.2}$ host is more than an order of magnitude larger than in the pure Au host.

As a next step we will try to extract sets of phase shifts (η_0, η_1, η_2) which describe the impurity resistivity and linear contribution to the thermopower for Yb in Ag and Au. Such an analysis was first made for the non-magnetic impurities Sc, Y and Lu (all with one d electron in the free atomic state) in Ag.[11] The method has proved to be quite useful, for Pd impurities in Ag we derived phase shifts which are in excellent agreement with optical measurements.[12] For Pd in silver a virtual bound 4d state just below the Fermi level causes large negative A_2 values; for Sc, Y and Lu virtual bound d states just above E_F cause large impurity resistivities together with large positive A_2 values. Taking over these general ideas for the Yb alloys, we suggest that the increase in ρ and A_2 from Ag to Au host must be due to a virtual bound 5d state just above E_F, closer to E_F in Au because of the increasing strength of the potential: $Z = 1$ for Yb in Ag and $Z = 2$ in Au. Here Z is the difference in valency between impurity and host. The Friedel sum rule requires:

$$Z = \frac{2}{\pi} \eta_0 + \frac{6}{\pi} \eta_1 + \frac{10}{\pi} \eta_2 \tag{3}$$

We take the phase shifts $\eta_\ell = 0$ for $\ell \geq 3$. With the expressions for the impurity resistivity

$$\rho = \frac{12\pi^3 h}{e^2 \Omega_o k_F^4} \, C \sum_{\ell=0}^{2} (\ell+1) \sin^2(\eta_\ell - \eta_{\ell+1}) \tag{4}$$

and for the coefficient of the second term in eq.(2)

$$A_2 = \frac{\pi^2 k_B^2}{3|e|} \left\{ -\frac{2}{E} + \sum_{\ell=0}^{2} (\ell+1) \sin^2(\eta_\ell - \eta_{\ell+1}) - \ell \sin^2(\eta_{\ell-1} - \eta_\ell) \right.$$

$$\left. \times \frac{\partial \eta_\ell}{\partial E} \middle/ \sum_{\ell=0}^{2} (\ell+1) \sin^2(\eta_\ell - \eta_{\ell+1}) \right\}_{E=E_F} , \tag{5}$$

we have three equations with 6 variables. Here C is the impurity concentration in atomic parts, Ω_o is the atomic volume, k_F the Fermi wave vector and k_B the Boltzmann constant. The other symbols have their usual meaning. Relating the derivatives $\left(\frac{\partial \eta_\ell}{\partial E}\right)_{E_F}$ to the phase shifts $\eta_\ell(E_F)$ by:

$$\mathrm{tg}\eta_0 = C_o E^{1/2} \qquad \text{(low energy approx.)} \tag{6a}$$

$$\mathrm{tg}\eta_1 = C_1 E^{3/2} \tag{6b}$$

$$\mathrm{tg}\eta_2 = \frac{\Delta}{E_d - E} , \qquad \text{(virtual bound state approx.)} \tag{7}$$

the number of independent variables is reduced to four: $Z_o = \frac{2}{\pi} \eta_0$; $Z_1 = \frac{6}{\pi} \eta_1$; $Z_2 = \frac{10}{\pi} \eta_2$ and Δ. Here we assumed that the low energy approximation[13] holds up to the Fermi level for the $\ell = 0$ and $\ell = 1$ phase shifts. For the $\ell = 2$ phase shift we used the virtual bound state approximation[14] with halfwidth Δ and E_d: the centre of the state. Because of the lack of more experimental information on these systems we can only guess at the virtual bound state half width Δ.

For the 4d impurity Pd in Ag, $\Delta = 0.25$ eV is a rather well established value now; this is why we suppose that for the 5d impurity
Yb in Ag or Au the half width probably lies in the range $\Delta = 0.3$–0.7
eV.

If we assume that for Yb in Ag and Au the same Δ holds, it follows (See Table II) that the increase of Z from 1 to 2 is distributed
among the various components of the screening cloud as $\Delta Z_0 = 0.31$
± 0.11; $\Delta Z_1 = 0.09 \pm 0.01$, $\Delta Z_2 = 0.60 \pm 0.12$. Thus the increase of
the impurity potential corresponding to $\Delta Z = Z(Au) - Z(Ag) = 1$ mainly
effects the Z_0 and Z_2 values. The distance of the center of the
virtual bound 5d state to the Fermi energy $(E_d - E_f)$ is roughly two
times smaller for Yb in Au than for Yb in Ag.

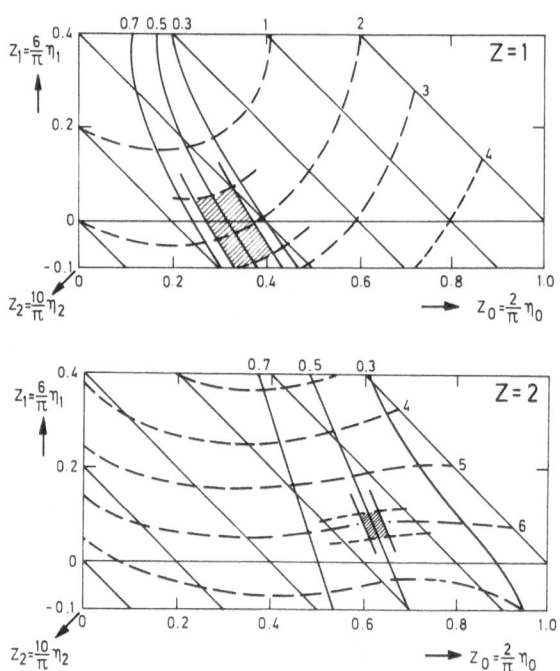

Fig. 3.   Computed curves of constant resistivity ρ (dashed curves)
and of constant diffusion thermopower A_2 (drawn) as function of the
phase shifts η_0 and for $Z = +1$ (fig. 3a) and $Z = +2$ (fig. 3b). The
values for the resistivity are given on the curves; the thermopower
curves shown are $A_2 = +3$ nV/K^2 (fig. 3a) and $A_2 = 10.5$ nV/K^2 (fig.3b)
in both figures for three values of the virtual bound state half
width Δ (0.3, 0.5 and 0.7 eV). The solution is determined by the
crossing of the ρ and A_2 curves. The shaded regions follow from the
uncertainties in ρ and A_2 values; they contain the solutions (Z_0, Z_1,
Z_2) for the $\Delta = 0.5$ eV. Note that the 45° lines are curves of constant Z_2.

TABLE II

Partial components Z_0, Z_1 and Z_2 to the screening and the distance between the center of virtual bound 5d state and Fermi energy, $E_d - E_F$, for the $Z = 1$ (AgYb) and $Z = 2$ (AuYb) cases.

		$\Delta = 0.3$ eV	$\Delta = 0.5$ eV	$\Delta = 0.7$ eV
	Z_0	0.39	0.33	0.27
	Z_1	0.01	−0.02	−0.04
$Z = 1$	Z_2	0.60	0.69	0.77
	$E_d - E_F$	1.57	2.27	2.84
	Z_0	0.84	0.62	0.49
	Z_1	0.08	0.08	0.06
$Z = 2$	Z_2	1.08	1.30	1.45
	$E_d - E_F$	0.85	1.16	1.43

We assumed in our analysis for AuYb that the whole impurity resistivity of 6.0 ± 0.2 $\mu\Omega$cm/at % is due to potential scattering. The fact that non magnetic Lu in Au introduces quite the same resistivity makes this assumption very plausible; straight evidence comes from magnetoresistance measurements in AuYb:[15] from these results it can be shown, that the spin dependent part of the resistivity is less than 1% of the total resistivity.

Using the de Gennes variant of the Kondo formula for the resistivity in second Born approximation one extracts a value for the exchange coupling $J_{sf} \approx 0.5$ eV; which leads to a spin dependent resistivity in first Born approximation also of about 1% of the total resistivity. For Yb in $Ag_{.8}Au_{.2}$ the spin dependent part of the resistivity estimated in the same way, results in about 20% of the 2.5 $\mu\Omega$cm/ at % Yb. We suggest that the absence of Kondo behaviour in the thermopower of AuYb is directly connected with the fact that the spin dependent part of the resistivity is such a minor part of the total impurity resistivity.

REFERENCES

1. J. Bijvoet, A.J. van Dam and F.van Beek, Solid State Comm. 4, 455 (1966).
2. P.E. Rider, K.A. Gschneidner and O.D. McMasters, Trans. Met. Soc. AIME 233, 1488 (1965).
3. V. Allali, P. Donze, D. Gainon and J. Sierro, J. Appl. Phys. 41, 1154 (1970).
4. J. Boes, A.J. van Dam and J. Bijvoet, Phys. Letters 28A, 101 (1968).
5. F.F. Bekker and C.A.J. van Duren, Physica 77, 609 (1974).
6. F.F. Bekker and C.A.J. van Duren, Proc. 14th Int.Conf. Low Temp. Phys. 1975, Vol.3, p. 406.
7. L.L. Hirst, G. Williams, D. Griffiths and B.R. Coles, J. Appl. Phys. 39, 844 (1968).
8. L.J. Tao, D. Davidov, R. Orbach and E.P. Chock, Phys. Rev. B4, 5 (1971).
9. A. Benoit, J. Flouquet and J. Sanchez, Phys. Rev. B9, 1092 (1974). G. Frassati, J.M. Mignot, A. Thoulouze and R. Tournier, Proc. 14th Int. Conf. Low Temp. Phys. 1975, Vol. 3, p. 402.
10. A.M. Guénault, Phil. Mag. 15, 17 (1969).
11. F.F. Bekker and T.P. Hoogkamer, Physica 84B, 67 (1976).
12. F.F. Bekker and N. Zuiderbaan, Physica 85B, 113 (1977).
13. L.S. Rodberg and R.M. Thaler, Intr. Quant. Theory Scatt., (Academic Press, New York-London, 1967).
14. J. Friedel, Nuovo Cim. Suppl. 7, 287 (1958). P.W. Anderson, Phys. Rev. 124, 41 (1961).
15. A. Friederich and A. Fert, Phys. Rev. Letters 33, 1214 (1974).

DISCUSSION

B.R. Coles: I think your neglect of the $l = 3$ phase shift for the silver and gold is essentially correct. However, for the silver-gold host, the alloy host, you then have an essentially intermediate valency situation. Now the proper treatment of that is unclear at the moment, but I am fairly certain that in such situations it effectively leads to a resonant scattering in the $l = 3$ phase shift at the Fermi surface. Therefore, a similar analysis would not be valid for this alloy.

F.F. Bekker: Yes, I agree. The question is, of course, if you can do this in a phase-shift analysis. The difficulty with that is that you can't handle the f electrons in a spin $-1/2$ way; you have a strong spin-orbit coupling, as pointed out by Lester Hirst. I think there is yet another complication: scattering is going on in the $l = 2$ and $l = 3$ channels, and there is interference between these two. That's the main reason why I was talking only about ytterbium in pure silver and gold.

INTERCONFIGURATIONAL FLUCTUATIONS IN THE DILUTE MAGNETIC ALLOY

AND THERMOPOWER MEASUREMENTS

Samuel P. Bowen

Department of Physics
Virginia Polytechnic Institute and State University
Blacksburg, Virginia 24061 USA

ABSTRACT

A new theoretical treatment of the dilute alloy problem which begins by treating the impurity correlation problem exactly is briefly discussed. The central result of the theory is the inter-configurational excitation energies of the impurity acquiring a temperature dependent shift through their coupling to the conduction electron gas. This shift induces a characteristic behavior in the low temperature S/T values which is confirmed for AuFe and CuFe systems. The need for new low temperature thermopower measurements for families of alloys to test this new theory is emphasized.

For many years our theoretical understanding of transition metal (TM) and rare earth (RE) impurities in simple metals has been dominated by the one electron approach of Hartree Fock (HF)[1] and by the virtual bound state (VBS) ideas of Friedel.[2] For atoms or arrays of atoms in which there is a dominant ground state configuration the validity of these ideas has been proven over and over. However, for atoms or systems with closely spaced configurations, strong many electron correlation, or fluctuations between two or more many-electron configurations, the one electron picture loses its physical coherence. The TM and RE atoms are precisely those for which the HF approximation is stretched beyond its applicability. Schrieffer and others[3] have many times observed that the VBS approach for these impurities is not sensible. Recently, Hirst[4] has presented the argument that an appropriate starting point for the description of an RE or TM impurity in a metal is the atomic or ionic picture in which the initial states of the

system are N-electron states on the impurity and a Fermi sea of
conduction electrons for the host. In this treatment the correlation in the impurity site is treated in zeroth order in that the
N-electron states $|\Psi_N(\alpha)>$ of the d or f shell each have different
energies $E_N(\alpha)$ (here we let α represent the quantum numbers of the
N-electron state).

The consequences for the traditional VBS theory of this ionic
approach are profound.

First, the atomic approach requires that we broaden our
understanding of what the VBS is approximating. In HF, the VBS
is the average energy with which it is possible to add (or remove)
an extra electron at the impurity site. Underlying the conception
of the VBS is the assumption that the impurity can be represented
well by only one N-electron state. When the many-electron states
of the impurity are taken as starting states, there are many
energies at which extra electrons can join or leave the impurity
site. These energies, the interconfigurational excitation energies
(ICEE), are the energy differences between, for example, all N and
N + 1 electron states having irreducible representations α and β,

$$\varepsilon_N(\alpha,\beta) = E_{N+1}(\beta) - E_N(\alpha).$$

The probability of an electron using one of these energies depends
on the impurity ground state configurations and can be shown[5] to be
proportional to the sum of the occupation probabilities

$P_N(\alpha)$ and $P_{N+L}(\beta)$ for the N and N + 1 electron states. The
usual definition of the VBS energy is just the average of all these
possible ICEE's,

$$\varepsilon_s - \mu = \frac{1}{2} \sum_{\alpha,\beta,N} (P_N(\alpha) + P_{N+1}(\beta)) (E_{N+1}(\beta) - E_N(\alpha) - \mu).$$

Here the Fermi energy of the host has been subtracted out from each
term. The VBS parameters are an extremely coarse measure of the
many electron dynamics of the impurities. As will be discussed
below and elsewhere many properties including the thermopower of
these dilute alloys can be understood if one of the ICEE's were
near the host Fermi energy, i.e. if

$$E_{N+1}(\beta') - E_N(\alpha') - \mu \approx 0 .$$

In such a circumstance, the VBS energy would not indicate such
a close lying level and would be dominated by ICEE's further from

the Fermi energy. In this case, the fine detail of the many-electron picture would be completely clouded by the VBS and HF approximations.

A second consequence of adopting an ionic many-electron description for the impurity is the radical alteration of the Friedel sum rule which must follow. In the usual derivation of the Friedel sum rule, the impurity potential is assumed to be non-fluctuating and the condition sought on the conduction electrons that the impurity potential be neutralized only by the conduction electrons. In the "ionic" picture the impurity states of lowest energy at the outset give a neutral impurity site and the types of interconfigurational transitions for the impurity are, for example, $d^N s^2 \rightarrow d^{N+1} s^1$ etc. This means that the conduction electrons are not as essentially involved in the electrostatic screening of the impurity site. Such a situation would obviate the essential condition for the Friedel sum rule.

The only approximate condition expressing the same physics as the Friedel sum rule that one has in the "ionic" approach is that the chemical potential of the two systems "impurity" and host be equal, or that

$$E_{N+1}(\beta') - E_N(\alpha') = \mu = \varepsilon_F(T=0). \tag{1}$$

This condition stretches the applicability of statistical mechanics in that one of our systems is quite small, and we should expect that at finite temperatures fluctuations about this equality should be important.

A third consequence of taking seriously the "ionic" picture is that the lifetime broadening Γ of the ICEE's is basically undetermined apriori and that either sophisticated many-electron calculations[6] or aposteriori fits to experiment will be required to determine the ICEE widths Γ.

A serious study of these ideas has been undertaken elsewhere[7] with promising results. In this paper the results of that study as they pertain to the thermoelectric power will be discussed. That theoretical study examined the S-shell Anderson model of an impurity under the assumption that the coupling V_{kd} is small. The ICEE's for this model impurity are $\varepsilon_{ds} - \mu$ and $\varepsilon_{ds} + U - \mu$ where we have subtracted off the host chemical potential. The thrust of our approach is to assume that one of the ICEE energies is small and to treat the interactions using a form of self consistent degenerate perturbation theory[8].

The results of this study are somewhat surprising. The ICEE
energies of the coupled impurity-host system acquire a temperature
dependent shift, which causes the ICEE to move relative to the
Fermi energy as the temperature changes. If we assume our Friedel-
like condition (1) holds at T = 0, the fluctuating behavior of the
ICEE is proportional to the shift function Λ plotted in Fig. 1.
If $e_1^{(r)}$ is the real part of the rth ICEE energy, it will have a
temperature variation proportional to the shift Λ (T) in Fig. 1.
Note that for a range of intermediate temperature Λ (T) is almost
linear in log (T). At lower temperatures its behavior is T^2. For
ICEE's far from the Fermi energy, $\left|e_1^{(r)}\right| > \left|e_2^{(r)}\right|$, ($e_2^{(r)}$ is the
imaginary part of an ICEE pole), the shift Λ is relatively unim-
portant. However for those ICEE's which satisfy (1) at T = 0, the
shift Λ dominates the system's behavior. In a lifetime approximation for the resistivity the impurity contribution is proportional
to a sum of Lorentzians $e_2^2/(e_1^2 + e_2^2)$ for each ICEE and for our
impurity satisfying (1) at T = 0, ρ is dominated by a Lorentzian
that approaches the Fermi energy as T → 0

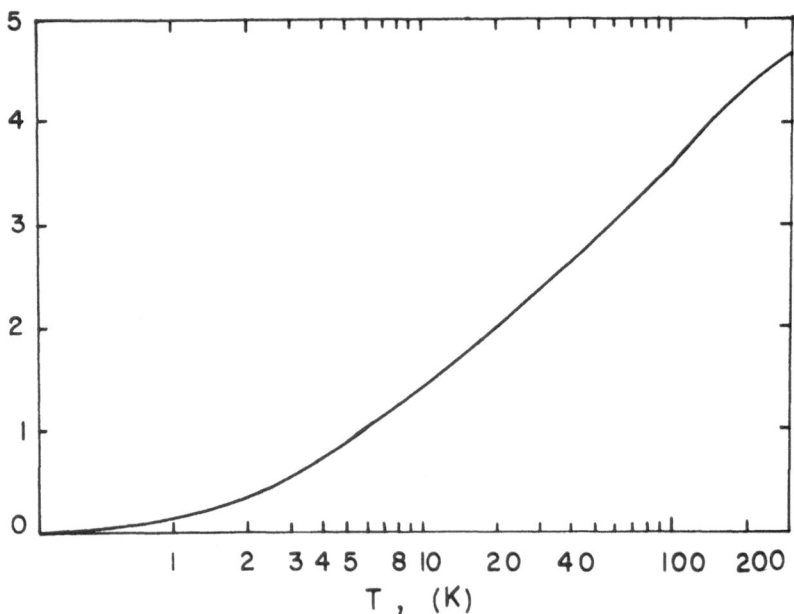

Fig. 1. The shift   of the impurity interconfigurational excitation energy ICEE for the coupled host-impurity system determined
from resistivity analysis for $\underline{Cu}Fe$ is plotted against temperature.

$$\rho = A + B/(1 + \alpha^2 \Lambda^2) \tag{2}$$

here A, B, and α are approximately constant in temperature. Such a parametrization has given rise to an excellent rendering of the impurity resistivity for CuAuFe alloys[9]. If one adopts the simplest approach to the theory of the thermoelectric power (TEP), it can be shown[10] that at low temperatures where the impurity contribution dominates

$$S/T = A' + B'/(1 + \alpha^2 \Lambda^2)^2 . \tag{3}$$

In anticipation of what follows it is well to point out the relationship between (3) and the Garland coefficient G which can be directly measured at low T. The simplicity of eqn (3) suggests that TEP studies of dilute alloys could be a very powerful too. It is not possible to demonstrate here because of space limitations, but the sign of A can be shown to be related to structure of the ICEE's which are not close to the Fermi energy. For the S = 1/2 Anderson model, if the dominant impurity fluctuations are between 1 and 2 electron A' <0 and B' >0, while if the dominant fluctuations are between 0 and 1, A' <0 and B' >0. Similar conditions can be expected for realistic many-electron atoms.

In order to briefly illustrate these ideas experimental TEP results for AuFe and for CuFe will be examined. Clearly if $\alpha \Lambda << 1$, and if Λ is proportional to log (T) there should be a temperature range in which S/T is linear in log (T). For AuFe the ICEE shift is proportional to log (T) down to .1K and so for the available range of temperatures $\Lambda \propto$ log (T). In Figure 2 we show the AuFe data for S/T of MacDonald et. al.[11] as a function of temperature. The data show the derived data, while the triangles are S/T multiplied by the resistivity to separate out any extra temperature dependence due to the impurity resistivity. For this sample (.006 at.% Fe) there is no significant resistivity effect. The solid line is a fit to equation (3) for T > - 5K with A = - 22, B = 24.5, α^1 = .4, and Λ = log (T) + .54.

For the CuFe system, the shift Λ (T) seems to change from log (T) to T^2 at temperatures around 4K and so such a simple fit is not possible. Figure 3, shows the data points derived from Christenson's[12] measurements on a (.05 at %) alloy. The fit to to the points reflects the combination of eqn (3) with the exact values of Λ which gave such a good rendering of the low temperature

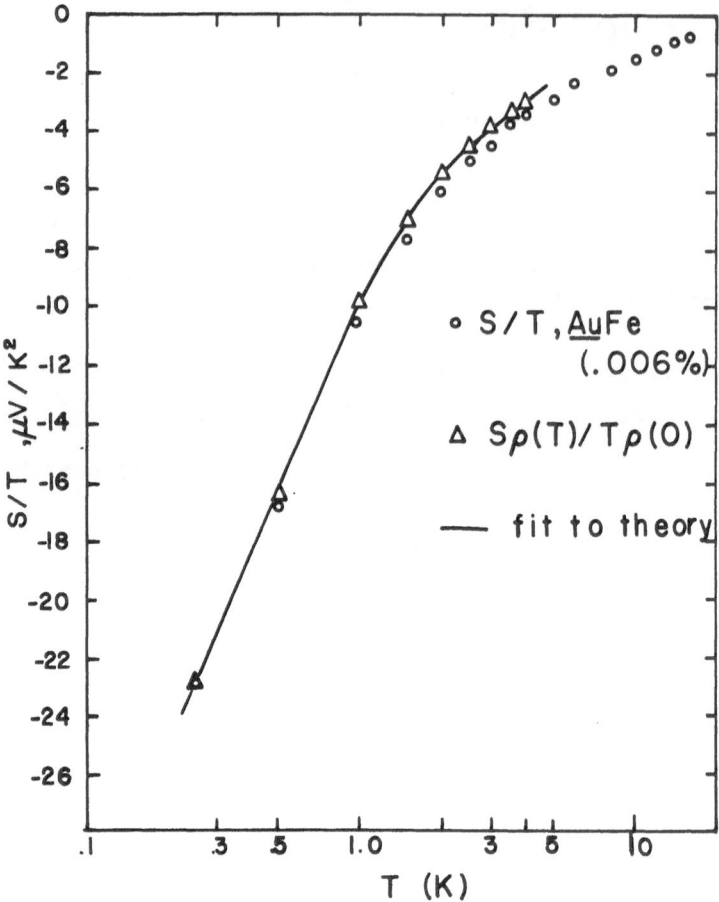

Fig. 2. The AuFe thermopower divided by temperature, S/T, and the same data renormalized by the system resistivity is plotted versus temperature. Theory is the solid line (see text).

resistivity for this system. The parameters are $A' = -.257$, $B' = 1.15$ and $\alpha = .13$ and the values of $\Lambda(T)$ from Fig. 1. The fact that the behavior was fixed by previous fits to resistivity data makes the closeness of the fit more compelling.

The quality and quantity of the experimental data on thermopower, or better, on the Garland coefficient, is not yet good enough to use such measurements as a good test of this ionic based theory. Expecially, the low temperature values of G or S/T do not exist today and could sensitively test these ideas.

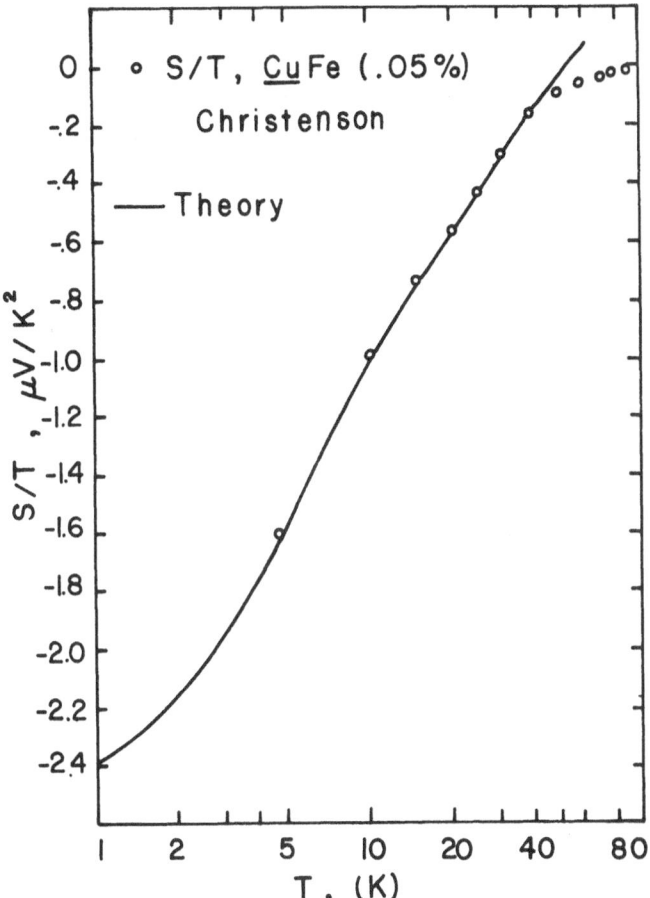

Fig. 3. C̲u̲Fe thermopower and theoretical fit to experiment plotted against temperature.

Finally, a comment about the sizes of the ICEE lifetime broadenings which are consistent with experiment is in order. The widths Λ for these two systems are quite small (A̲u̲Fe:.03K) and (C̲u̲Fe:2.8K) and represent a radical departure from the VBS perspective on similar VBS widths. Once one has recovered from the shock of such small widths for the many electron ICEE's, many other experimental results can be easily reconciled with this approach.[13]

Finally, a word about the largeness of the TEP peaks for the low T values of S/T. More detailed considerations elsewhere show that

$$A' = A_o \; (\tfrac{\eta}{\Gamma}) / (1+\alpha^2\eta^2)^2$$

where η is the ratio of the ICEE energies to Γ for the ICEE's which are not near to ε_F and Γ is the lifetime broadening of the ICEE. The largeness of A is then essentially due to small Γ and large η. For the S-shell model $A_o \approx 50\mu V/K^2$.

The "ionic" approach to the TM or RE impurity is radically different from the VBS or the traditional Kondo approach. The "ionic" approach is conceptually simple, treating impurity correlation at the start, and yet it appears capable of detailed tractable comparisons for several experiments. The existence of more high quality measurements of thermoelectric and other effects will enable the testing of this new perspective.

REFERENCES

1. P.W. Anderson, Phys. Rev. 124, 41 (1961).
2. J. Friedel, II Nuovo Cimento Suppl. 7, 287 (1958).
3. J.R. Schrieffer, J. Appl. Phys. 38, 1143 (1967).
4. L.L. Hirst, Kondens. Materie 11, 255 (1970).
5. S.P. Bowen, Physical Review., in press, J. Hubbard, Proc. Roy. Soc. 277A, 237 (1963).
6. See reference 4.
7. E. Feenberg, Phys. Rev. 74, 206 (1948). S.P. Bowen, J. Math. Phys. 16, 620 (1975).
8. E. Feenberg, Phys. Rev. 74, 206 (1948).
9. S.P. Bowen, Phys. Rev., in press, J. W. Loram, T. E. Whall, P.J. Ford, Phys. Rev. B2, 857 (1970).
10. F.J. Blatt, P.A. Schroeder, C.L. Foiles, D. Grieg, Thermo-electric Power of Metals, Plenum Press, N. Y., 1976.
11. D.K.C. MacDonald, W.B. Pearson, I.M. Templeton. Proc. Roy. Soc. (London) 266A, 161 (1962).
12. E.L. Christenson, J. Appl. Phys. 34, 1485 (1963).
13. S.P. Bowen, C.D. Williams, unpublished.

DISCUSSION

K.H. Fischer: How would you explain the data in aluminum-iron where, perhaps, we agree that in the Anderson model you get what Friedel calls a virtual bound state, which is, depending on the parameters, fairly broad. The usual approach is now to take into account the effect of the interaction U in some kind of perturbation theory, and this is the conventional way to explain the resistivity or the thermopower in aluminum-iron. The question is, whether or not you would agree with this approach, or whether you would suggest something completely different.

S.P. Bower: In our calculation we start out treating U exactly; we start out taking into account electron-electron interaction on the impurities exactly. We can recover the aluminum-iron situation by a slightly larger γ for the many-electron excitation energies. In other words, we can get the resistivity, the $(1-T^2)$ behavior very easily with a γ that's a little larger. I don't see any difficulty with those kind of systems, with systems we might call spin fluctuation or close to non-magnetic systems. Ours is a very different kind of picture, and the reason I am excited about it is that it starts out with a many-electron picture from the beginning.

LOW-TEMPERATURE THERMOPOWER AND OTHER TRANSPORT

PROPERTIES OF ALUMINIUM CONTAINING DILUTE POINT DEFECTS[+]

K. Böning

Physik-Department, Techn. Universität München, D-8046
Garching, Germany

ABSTRACT

We compare new experimental results of the low-temperature
thermopower and of the low-field Hall effect (and magnetoresistance)
of aluminium containing dilute point defects. The defects were
either nonmagnetic impurities (Ge, Mg, Zn, Ga, or Cu) or Frenkel
defects (FD, i.e. self-interstitials and vacancies) introduced by
reactor irradiation at 4.6 K. Some of the results can be compared
with 4 OPW calculations, which were performed using the realistic
Al Fermi surface (FS) and tabulated pseudo-potentials and without
adjusting any parameter.

The low-field Hall coefficient R_0 at 4.6 K was always found
to be consistent, even quantitatively, with the generalized Tsuji-
formula, see below. The low-temperature thermopower S was shown to
agree with the law $S = AT+BT^3$ below about 6-8 K (measurement at
1.3 K<T<13 K, superconducting reference).

The diffusion thermopower coefficient A was essentially independent of the concentration of isolated FD or impurities, but
changed drastically during FD agglomeration. This behaviour of A
parallels that of R_0 and demonstrates that the Al bandstructure
always remained sufficiently unchanged by the defects. The different values of A or R_0 as observed for different defect types are

---

[+]This work was supported by the German Bundesministerium für
Forschung und Technologie within the project "Nukleare
Festkörperforschung."

determined only by the different anisotropy (i.e. $\vec{k}$-dependence) of the relaxation time τ_k. This is all consistent with the Mott-formula for A. Both R_o and $-A$ depended in about the same way on the defect type, but the "wrong" sign of A shows that A is determined by the energy dependence of $v_k \tau_k$ (velocity v_k) which overcompensates that of the FS area elements dS. Our 4 OPW calculations gave virtually quantitative agreement for R_o but not for A (many body effects?).

The phonon drag thermopower coefficient B behaved totally different for impurities and for FD. For impurities B was again independent of the defect concentration and determined only by the anisotropy of τ_k, and both R_o and B depended in about the same way on the defect type. This behaviour of B is in qualitative and even semi-quantitative agreement with the Bailyn-formula, and there is no evidence of "phony phonon drag". In the FD case, however, B was approaching zero as a function of defect concentration and was independent of the anisotropy of τ_k. This anomalous behaviour obviously has to do with the exceptionally strong phonon scattering on the FD (resonance vibration modes).

I.   INTRODUCTION

In most real metals the Fermi surface (FS) touches Brillouin zone boundaries and, as a consequence, it deviates from simple spherical shape and the elctron wavefunctions vary with wavevector $\vec{k}$ instead of simply being plane waves. In a situation as it is today, where many fundamental questions concerning thermopower and other transport properties are still open even for the "simple" alkali metals [1-3], one might question that transport measurements on "complicated" metals could help for a better understanding of these phenomena. However, these "complicated" metals represent very sensitive systems where many transport coefficients (TC) respond drastically to any change of, e.g., the host bandstructure or the characteristics of electron scattering[4]. The reason is that dominant contributions to these TC just come from those FS regions where the local curvature is high, and this is just where the wavefunctions deviate most strongly from plane waves and the transport relaxation time τ_k assumes extreme values [4,5]. In this paper we want to show that careful transport measurements on such a "complicated" metal could, indeed, contribute appreciably to our fundamental understanding of these phenomena.

We always consider low temperatures where the rate of electron scattering on phonons is negligible with respect to that on dilute point defects ("dirty limit"). The point defects are either substitutional, nonmagnetic impurities or Frenkel defects (FD), i.e. self-interstitials and vacancies introduced by reactor irradiation at 4.6 K. This low-temperature irradiation technique allows new types of investigations, since the concentration of virtually

identical defects can be increased during **irradiation** ("isolated"
FD) and the configuration of these defects can be changed during
subsequent isochronal annealing (agglomerated FD). This is all
possible in the same sample and for the same measuring temperatures
around 4 K, so the relative accuracy is very high [6,7].

The metal we have chosen is aluminium, for many reasons. Al
is a model candidate for pseudopotential theory, its FS is well
known [8,9] and differs from a sphere only in the vicinity of intersecting Brillouin zone boundaries where the local mean curvature
$1/\kappa$ is higher by one or two orders of magnitude (both signs!)[4].
Finally, Al is a good candidate for reactor irradiation experiments.

The problems we want to study are of the following nature[7]:
Is a given TC "normalized", i.e. independent of defect concentration?
Is it determined by a possible change of the Al bandstructure or by
the anisotropy ($\vec{k}$-dependence) of the relaxation time τ_k? What can
we learn about this anisotropy of τ_k for various types of point
defects in Al (and about possible configuration changes of the
defects)? Are there other influences on the TC we can observe?
Finally, we will investigate the influence of various types of
point defects on a given set of TC, and compare those TC where we
have a rather good knowledge about what is going on (low-field
Hall effect, magnetoresistance) with those where the problems are
much larger (diffusion and phonon drag thermopower). This comparison with theoretical models is essentially qualitative, but in many
**cases quantitative comparison by means of accompanying 4 OPW computer
calculations is also possible.**

In Section II we mainly discuss the low-temperature irradiation
technique. In Section III we report on experimental and theoretical
work on the low-field Hall effect and magnetoresistance, and in
Section IV about similar work on the low-temperature thermopower.
Conclusions will be given at the end of each major section, when
necessary.

II. POINT DEFECTS

We have studied two kinds of point defects in polycrystalline
Al at low temperatures (dilute defect concentrations c<<1):

a) Substitutional impurities (nonmagnetic) which have been
alloyed into Al by melting (Ge, Mg, Zn, Ga, or Cu).

b) Frenkel defects (FD), i.e. self-interstitials (I) and vacancies (V) which have been introduced into pure Al samples by reactor
irradiation at 4.6 K. The principle is outlined in Fig. 1. A fast
reactor neutron may shoot a lattice atom from its site, which

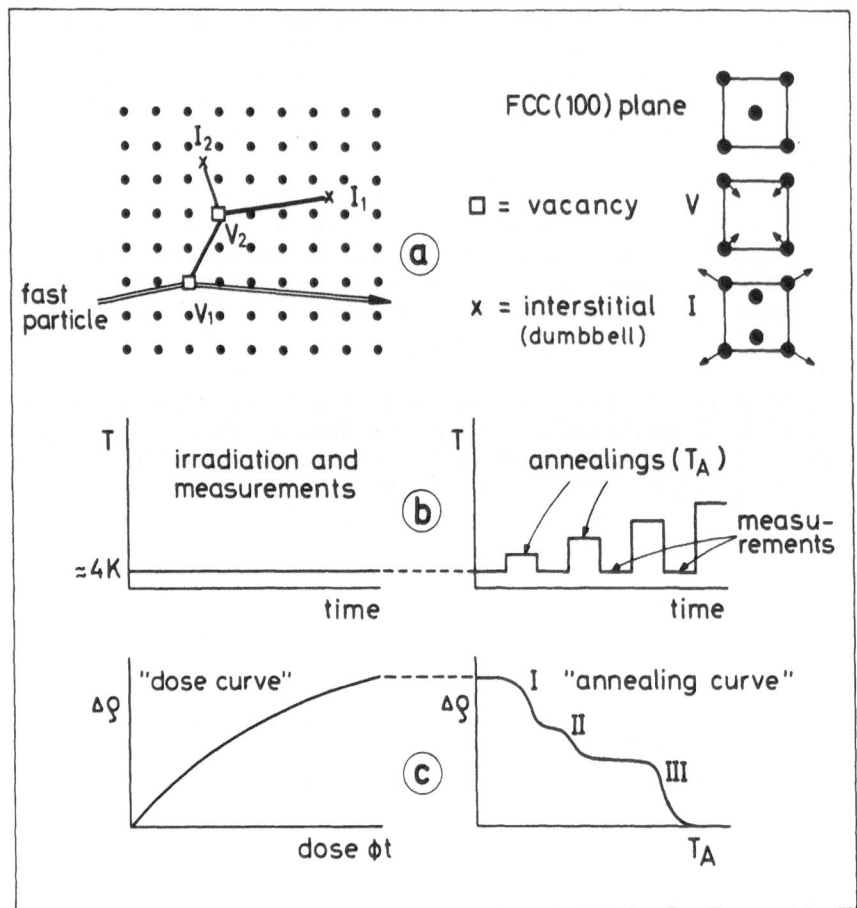

Fig. 1: Schematic illustration of a reactor irradiation experiment on Al. (a): production and configuration of the Frenkel defects. (b): histogram of the experimental procedure. (c): typical result for the change of residual resistivity $\Delta\rho$.

usually induces further displacements (cascade!). So an equal number of I and V are produced which are "frozen in" at 4 K[10]. The single I has [100] dumbbell configuration and its nearest neighbours are shifted by about 10% of the lattice parameter (Fig. 1a)[11,12]. Due to the cascade effect single I as well as agglomerates of up to about 3I have been observed in reactor irradiated Al[13,14], but this very low degree of agglomeration is insignificant for electronic transport properties[6]. A schematic view of our experimental set-up will be given in Section IV.1.

In a typical experiment (Fig. 1b) the sample temperature T is

always about 4 K during irradiation and during all interruptions
for the measurements. During subsequent isochronal annealing there
are constant holding times at the temperatures T_A but the measurements are done again at about 4 K.

Usually the change $\Delta\rho$ of residual electrical resistivity is
used as a measure of the Frenkel Defect concentration c (in Al
$\Delta\rho/c = 0.40$ nΩcm/ppm)[7,11]. During irradiation ("dose curve") $\Delta\rho$
increases nonlinearly with neutron dose ϕt since the probability
increases that a new I recombines spontaneously with an existing
V, and vice versa (Fig. 1c). During annealing ("annealing curve")
$\Delta\rho$ disappears in three stages in Al as a function of T_A[15,16]. In
stage 1 (T_A<50 K) close I-V-pairs begin to recombine at about 18 K,
but then also single I get mobile and can avoid recombination with
A only by clustering[15]. In stage 2 (50 K<T_A<170 K) smaller I
clusters migrate which leads to the growth of larger I clusters as
(111) loops. In stage 3 (170 K<T_A<250 K) the V become mobile and
"eat up" the I loops until full FD recombination is obtained
($\Delta\rho=0$). A somewhat different annealing model is outlined in Ref.[16].

III.   HALL EFFECT AND MAGNETORESISTANCE

We are interested in the "low-field case" which is defined by
sufficiently small values of the effective magnetic field H/ρ,
where ρ is the electrical resistivity[17,18]. Further, the low-field
magnetoresistance coefficients (transverse and longitudinal) have
always been observed[5,19,20] to behave very similar to the low-field
Hall coefficient R_0 but are much more complicated theoretically[21,22],
whence we will concentrate on R_0 in this paper.

III.1.   LOW-FIELD HALL COEFFICIENT R_0

We assume a cubic metal as Al and an anisotropic transport
relaxation time τ_k also having cubic symmetry[18,21]. On the other
hand, the exact solution of the Boltzmann equation would be a
relaxation time τ_k having less than cubic symmetry (due to the
electric field $\vec{E}$) and depending on $\vec{H}$[23]; the reduced symmetry of τ_k
could be easily allowed for which is not done here only for simplic-
ity[21,22], and the $\vec{H}$-dependence of τ_k is actually neglected here but
for low fields H/ρ this is plausible[21] and experimentally supported
by R_0 being virtually H-dependent[22]. So, for a cubic and H-
independent τ_k the low-field Hall coefficient R_0 is given by the
generalized[18] Tsuji-formula[17]

$$R_0 = \frac{e^3}{12\pi^3\hbar^2\sigma^2} \int (1/\kappa) \, v_k^2 \tau_k^2 ds \qquad (1)$$

tures T_O the temperatures of the susceptibility maximum, of the maximum in the Schottky anomaly, and of either the abrupt change in
slope or the derivative maximum in the electrical resistivity. Thes
temperatures are indicated in Fig. 1 as open symbols for a range of
Co concentrations. We see that above about 1 at.% Co, all three pro
erties give a single curve for T_O. Below 1% Co, however, the characteristic temperatures diverge somewhat for the different properties.

In a recent letter[8] we reported that measurements from 4.2K to
300K of the thermopower of PtCo alloys containing from 0.31 to 3.74
at.% Co revealed pronounced negative peaks which decreased in magnitude and moved rapidly to higher temperatures with increasing Co
concentration. The extrema of these peaks were present in this
temperature range only for the three most concentrated alloys, containing 0.56, 1.65, and 3.74% Co. To complete the picture of the
thermoelectric behavior of these alloys, we have now extended our
thermopower measurements to 1.35K, thereby observing the extremum of
the peak for the 0.31% alloy, and we have also studied the magneto-
thermopowers of these alloys from 1.35K to 4.5K. This paper contains the results of these measurements.

Fig. 2 shows the thermopowers of the four alloys in zero magnetic field for the temperature range 1.35 to 90K, along with that

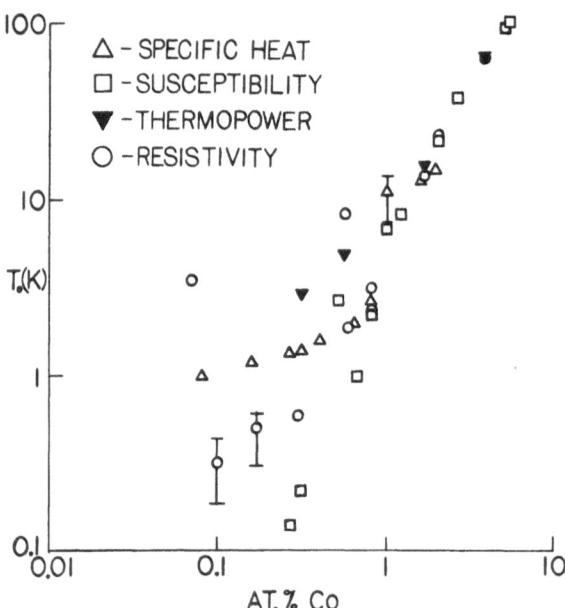

Fig. 1. The variation with Co concentration of the characteristic
temperatures T_O determined from measurements of electrical resistivity, magnetic susceptibility, electronic heat capacity, and ther
power.

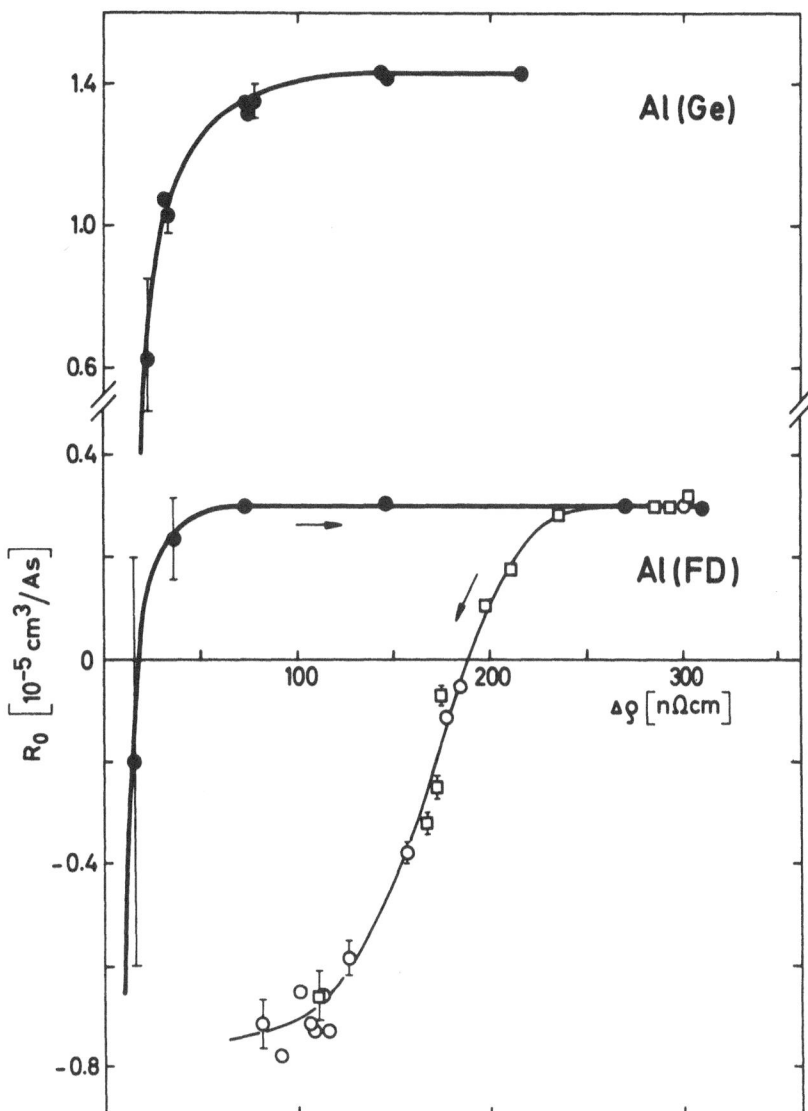

Fig. 2: Low-field Hall coefficient R_O versus increase of residual
resistivity $\Delta\rho$ due to point defects in Al, which are either substitutional Ge impurities[4] ($c_{max} \approx 3000$ ppm) or Frenkel defects FD
($c_{max} \approx 800$ ppm), at 4.6 K.   The irradiation results Al(FD) consist
of a "dose curve" (full symbols, $\Delta\rho$ increasing[4] and an "annealing
curve" (open symbols, $\Delta\rho$ decreasing)[20]; the latter is composed of
two runs which have been scaled to the same $\Delta\rho_{max}$. From experimental
reasons the R_O values have always been determined at log
$\{H/\rho \ [kOe/\Omega cm]\} = 7.6$ which in the case of Al(Ge) is not quite low
enough for the "true" low-field limit.[4]

of R_O (not shown in Fig. 2)[6]. In the Al(FD) annealing curve R_O decreases strongly which is also due to FD clustering. The interpretation of the Al(Ge) curve is equivalent to that of the Al(FD) dose curve.

In Fig. 2 both the Al(FD) dose curve and the Al(Ge) curve show that a definite "plateau region" of $\Delta\rho$ exists where the normalization of R_O is verified. This means that the Al bandstructure or FS remained virtually unchanged by the defects[4] and that (concentration-dependent) defect clustering was still negligible. With respect to Al(Ge) this result implies that the well-known changes of the Al FS due to charged impurities[24] are still too small in the concentration range we used to affect R_O significantly (these FS changes would be proportional to $\Delta\rho$). So the different values of R_O in Fig. 2 are a consequence only of the different anisotropy of τ_k due to the defects.[4].

This anisotropy of τ_k is very strong and always of the same kind, as can be seen by evaluating R_O for a three-group-model of the Al FS[4,5]. According to eq. (1) and the local values of $\overline{1/\kappa}$ we divide the Al FS into a large spherical part S_- which gives a small negative contribution per surface element dS and, on both sides of intersecting Brillouin zone boundaries, into small cylindrical parts S_{++} and S_{--} which give very high positive and negative contributions, respectively, per element dS. One can show from the geometry of the Al FS[4,5] that in the case $\tau_{++} \approx \tau_{--}$ ("isotropic scattering") both these contributions to R_O just cancel approximately and the remaining spherical part S_- then leads to the free-electron like result

$$R_o^{isotropic} = \frac{1}{ne} = -3.5 \cdot 10^{-5} \ cm^3/As \qquad (4)$$

where n is the electron density of Al. This strongly negative value of R_O is actually observed in pure Al in the case of high-temperature phonon scattering[17,20]. For all types of point defects, however, which have been investigated so far,* the experimental value of R_O at 4 K was always considerably more positive than that isotropic value. This means that all types of point defects must

---

*Further results, not contained in Fig. 2, are: $R_O[10^{-5} \ cm^3/As]$ = -0.4 and -1.0 for Al(Mg) and Al(Zn), resp.[4]; see also Figs. 5 and 7.

lead to a universal anisotropy of the form $\tau_{++} > \tau_{--}$ ("anisotropic scattering"[4,5]. Especially for FD or Ge impurities this anisotropy of τ_k must be very strong, since the positive contribution of S_{++} must overcompensate the negative contribution of both S_{--} and S_- in order to enable the change of sign of R_0.

III.3. 4 OPW CALCULATIONS

The physical origin of this universal anisotropy of τ_k was supposed to lie essentially in the $\vec{k}$-dependence of the electron wavefunctions. To confirm this hypothesis, we have calculated[4,21] 4 OPW wavefunctions $\Psi_k(\vec{r})$ for various $\vec{k}$-vectors on the Al FS using the well-known Al formfactors[8,9]. For better illustration these complex $\Psi_k(\vec{r})$ have been converted into charge densities $\rho_k(\vec{r}) = e|\Psi_k(\vec{r})|^2$ and plotted on (100) planes of the Al FCC lattice[4]. It was found, indeed, that for those $\vec{k}$-vectors which belong to the S_{--} region of the FS the electron charge was mainly localized on the lattice sites, whereas for $\vec{k}$-vectors belonging to S_{++} it was mainly localized between the lattice sites. So a scattering potential $\Delta V(\vec{r})$ which is centered at a lattice site will scatter a S_{--} electron much more than a S_{++} electron, and this behaviour generally explains our universal experimental result $\tau_{++} > \tau_{--}$ for the case of substitutional impurities and vacancies[4]. The smaller the real-space volume in which $\Delta V(\vec{r})$ is localized (i.e. the more $\Delta V(\vec{r})$ is s-like as was independently found by Sorbello[25], the higher will be the degree of this anisotropy $\tau_{++} > \tau_{--}$ and the more positive will be R_0.

We then extended these 4 OPW calculations to calculate R_0 quantitatively for substitutional impurities in Al[21,22]. So $\Psi_k(\vec{r})$, $\vec{v}_k$ and $\overline{1/\kappa}$ were evaluated for 48 x 109 (or more) $\vec{k}$-points on the al FS. The scattering probabilities $P_{kk'}$ were obtained in Born approximation using tabulated impurity formfactors, renormalized to the Al electron density, and applying appropriate strain field corrections[21]. The Boltzmann equation was solved by iteration (subject to the "low-field approximation" H = 0, see Section III.1) and the obtained anisotropic τ_k were inserted into eq. (1). The results[21] are shown in Table 1. Obviously the values of R_0 are of the right order of magnitude, they have the correct sign and also the sequence of R_0 when going from Ge to Mg and Zn impurities is obtained correctly. Keeping in mind that no parameter has been adjusted, we consider this to be a very encouraging result. The transverse and longitudinal magnetoresistance coefficients P_t and P_l have been calculated in this way, too, and the results are not bad, either[21]. We do not yet have agreement for the case of FD in Al, probably because the Born approximation may no longer be applicable for these strong scattering potentials.

Table 1:   Low-Field Hall Coefficient R_o [10^{-5} cm^3/As]

	Al(Ge)	Al(Mg)	Al(Zn)
4 OPW Calculation	+ 3.6	− 1.2	− 2.7
Experiment	+ 1.8	− 0.4	− 1.0

III.4.   CONCLUSIONS

We conclude this section by saying that the low-field Hall
coefficient R_o seems to be reasonably well understood for our
system of dilute point defects in Al.   While the Al bandstructure
and FS remain sufficiently unchanged by the defects, R_o is determined by the anisotropy of the relaxation time τ_k only.   Applying
a three group model to our experimental results we see that there
exists a universal anisotropy of the form $\tau_{++}>\tau_{--}$ for all types of
point defects.   Our 4 OPW calculations show that this τ_k anisotropy
is a direct consequence of the $\vec{k}$-dependence of the Al wavefunctions
$\Psi_k(\vec{r})$ and we even obtain virtually quantitative agreement for R_o
without having fitted any parameter.   The behaviour of the low-
field magnetoresistance is quite analogous.

IV.   THERMOPOWER

IV.1. EXPERIMENT

Fig. 3 shows our experimental set-up[26,27], i.e. the thermopower
measuring cryostat mounted on top of the liquid helium irradiation
facility of the research reactor Munich.   The sample holder (diameter 14 mm, length 130 mm) is fastened on the end of a 13 m long
stainless steel capillary, through which it can be evacuated or
filled with He exchange gas and through which most of the electrical
leads go to the measuring system.   The irradiation is at 4.6 K but
the Al specimen wire (diameter 2 mm, length 85 mm) actually is at
15 K in this experiment due to the indirect cooling conditions.
After irradiation the sample is pulled up (at 4.6 K) into the central insert of the measuring cryostat, which is then rotated by
180° and the sample pushed down into the measuring chamber.   There
it slides into two pressure contacts which provide a virtually

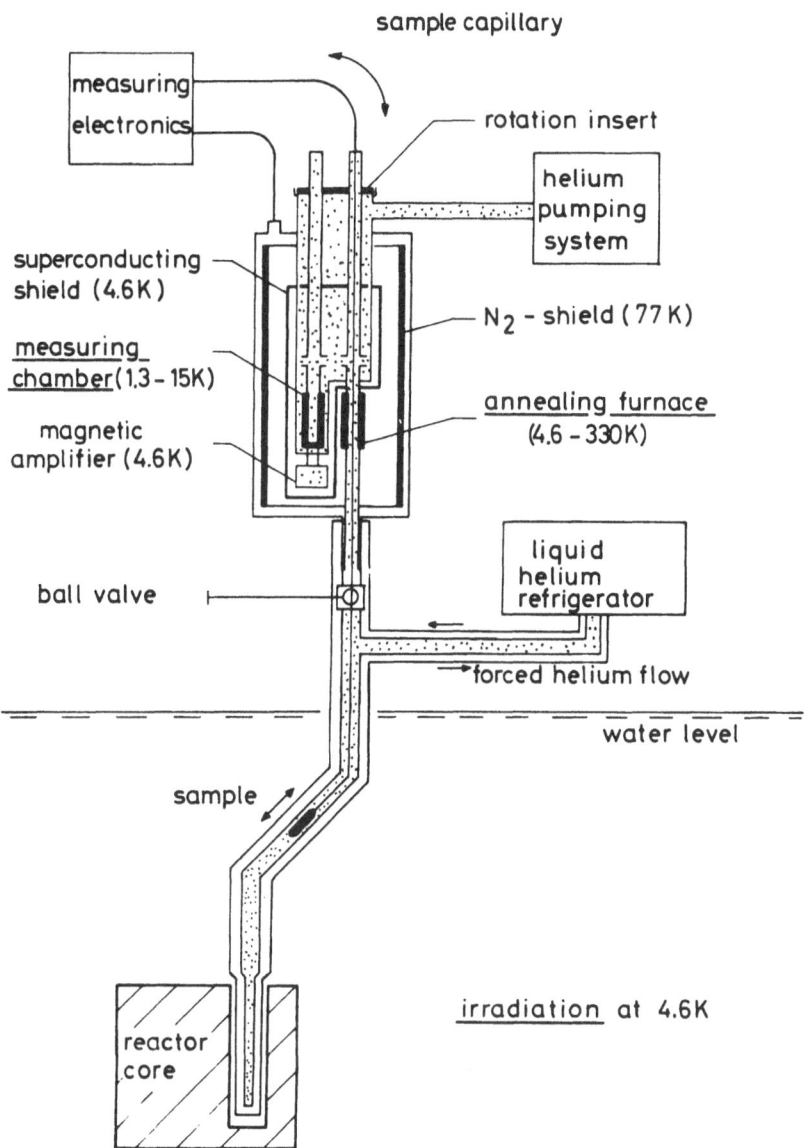

Fig. 3: Thermopower measuring cryostat and low-temperature reactor irradiation facility.

superconducting connection between the thermovoltage contacts of the specimen and a superconducting magnetic preamplifier. For the measurements the sample temperature T can be changed within 1.3 and 15 K by pumping on the He bath or by ohmic heating of the measuring chamber. After that the sample is either transferred back into the reactor core for further irradiation, or is annealed in the furnace[26,27].

Our measurements give directly the absolute thermopower S(T) of the specimen since S = 0 for the superconducting reference material. Further, we apply the differential technique where the temperature difference T along the specimen is always very small (typically 0.1 K). For more experimental details see Ref. 26, 27. Using this set-up we again performed both irradiation and annealing experiments on pure Al samples[26,27] and measurements without irradiation on dilute Al alloys[28,29].

Our first result was that the low-temperature law[30]

$$S(T) = AT + BT^3 \tag{5}$$

was fulfilled in all our measurements[26-29] up to about 6-8 K (for details see Section IV.4). So we can separately discuss the diffusion coefficient A and the phonon drag coefficient B, as will be done in the following two sections.

IV.2. DIFFUSION THERMOPOWER COEFFICIENT A

Theoretically, the diffusion coefficient A is described by the Mott-formula[30]

$$A = \frac{\pi^2 k_B^2}{3e\sigma} \left(\frac{d\sigma(\varepsilon)}{d\varepsilon} \right)_{\varepsilon=\varepsilon_F} \tag{6}$$

which is an exact consequence of the linearized Boltzmann equation[27,31] (irrespective of the assumption of a relaxation time τ_k having cubic symmetry in this paper, see Sec. 3.1.[23,27]). Here k_B is the Boltzmann constant and $\sigma(\varepsilon)$ is given in eq. (2) with the integral being taken over a surface of energy being close to the Fermi energy ε_F[31]. Inserting eq. (2) into (6) the energy derivative can be worked out to some extent[32] and the result is

$$A = \frac{ek_B^2}{36\pi\hbar^2} \left[2 \int \overline{(1/\kappa)} \tau_k dS + \int \frac{1}{v_k^2} \vec{v}_k \cdot \vec{\nabla} (v_k \tau_k) dS \right] . \tag{7}$$

All quantities are as in Section III.1. Both integrals go over
the Fermi surface (FS) now; the first considers the local change
of surface area (dS) with energy, and the second contains the energy
dependence (i.e. the $\vec{k}$-space gradient) of the remaining quantities
$v_k \tau_k$.

Qualitatively, the diffusion coefficient A should behave
strictly analogous to the low-field Hall coefficient R_o, as
follows from eqs. (6) or (7) and the arguments given in
Section III.1. That is, A should be "normalized", i.e. independent
of the concentration c of dilute identical defects, and A should
be strongly dependent on the anisotropy (i.e. $\vec{k}$-dependence) of τ_k.
The first term of eq. (7) is very similar to eq. (1) for R_o but
the energy dependence contained in the second term is a specific
feature of A.

Most of our experimental results of A are shown as a function
of point defect resistivity $\Delta\rho$ in Fig. 4, which closely parallels
Fig. 2 for R_o. The irradiation results again consist of "dose
curves" (full symbols, $\Delta\rho$ increasing) and "annealing curves" (open
symbols, $\Delta\rho$ decreasing) which have been obtained on five pure Al
samples irradiated to different neutron doses[26,27]. The behaviour
of A is qualitatively like that of R_o, i.e. A is essentially constant during irradiation but changes drastically during annealing,
even up to positive values. The interpretation of these Al(FD)
results also is qualitatively the same as that of R_o, see
Section III.2. However, A seems to be more sensitive to FD clustering because during irradiation A begins to change already at
lower values of $\Delta\rho$ (i.e. around 230 nΩcm) than was the case for R_o.
The unirradiated dilute Al(Ge) alloys[28,29] behave quite normally,
but for Al(Zn) the anomalous $\Delta\rho$-dependence of A is very probably
due to Zn atom clustering (see Figure Caption).

In Fig. 4 both the Al(FD) dose curve and the Al(Ge) curve
again exhibit a definite "plateau region" of $\Delta\rho$ where the normalization of A is verified. This means that the Al bandstructure or
FS can be considered as sufficiently unchanged by the defects also
in the case of A. The different values of A for different defect
types, the, can be only a consequence of the different anisotropy
of τ_k due to the defects. Some more arguments are given in the
analogous discussion of R_o in Section III.2

All this behaviour of A is qualitatively consistent with the
Mott formula, eq. (6) or (7). We will now investigate this point
somewhat more quantitatively. In Fig. 5 we compare the typical
values of A, for the different defect types, with those of R_o by
plotting both coefficients on linear scales. However, the sign of
A had to be inverted for this plot, i.e. the most negative values
of A appear on the top of the scale. More explanations are given

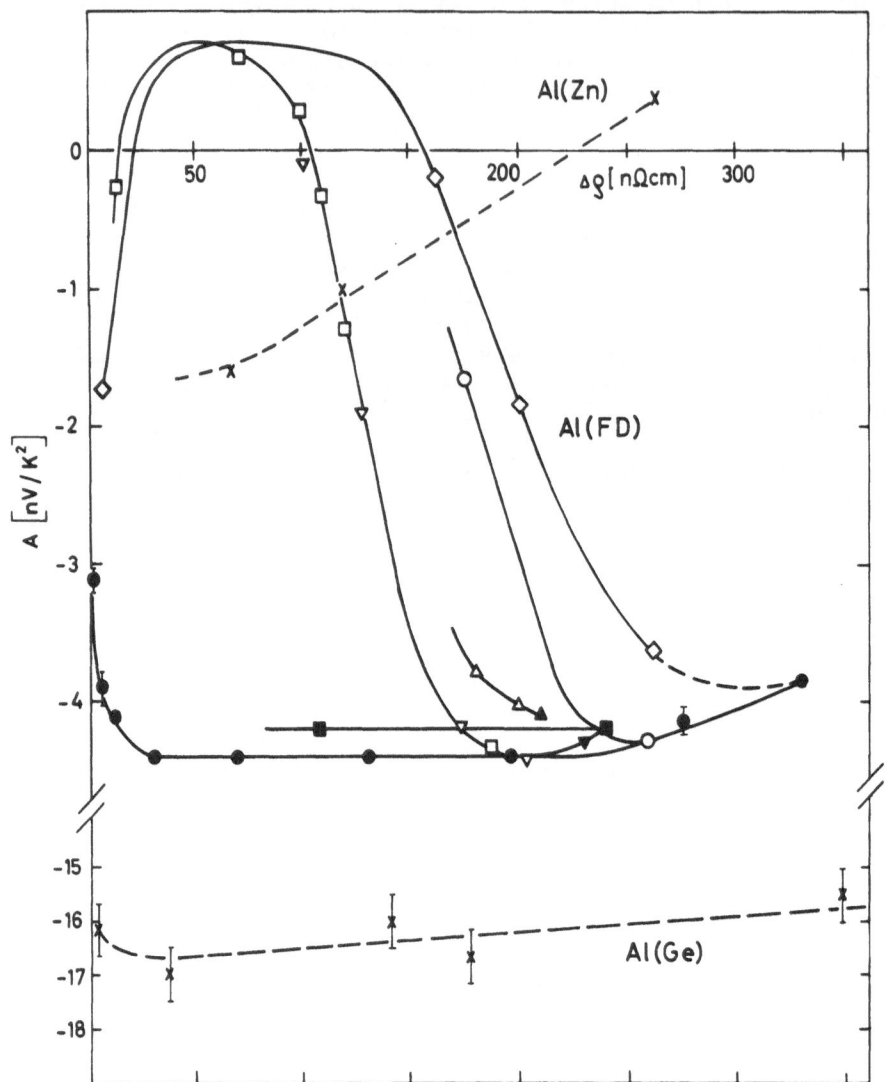

Fig. 4: Diffusion thermopower coefficient A versus increase of
residual resistivity $\Delta\rho$ due to point defects in Al. The irradiation
results Al(FD)[26,27] consist of "dose curves" (full symbols, $\Delta\rho$
increasing) and "annealing curves" (open symbols, $\Delta\rho$ decreasing)
of five pure Al samples irradiated to different values of $\Delta\rho_{max}$.
Two series of alloy measurements are also shown[28,29]; in the Ge
case there was $c_{max} \approx 5000$ ppm but in the Zn case there was
$c_{max} \approx 11000$ ppm because of the small specific resistivity of Zn
impurities.

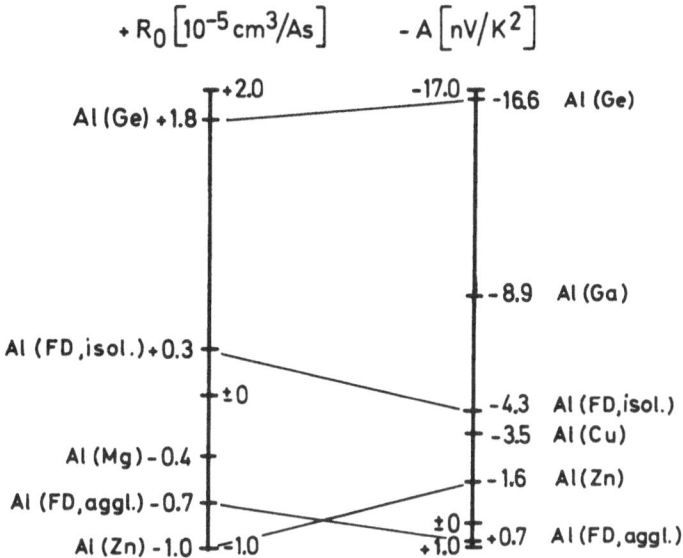

Fig. 5: Comparison of the low-field Hall coefficient R_0[4,20] and the diffusion thermopower coefficient A[26-29]. Both coefficients are taken from the "plateau regions" in Figs. 2 and 4 and are plotted on linear scales, but the scale of A is inverted. Al(FD, isol.) means the dose curve value and Al(FD, aggl.) a typical annealing curve value. R_0 is always the "true" low-field limit (compare caption of Fig. 2).

in the Figure Caption. We can see in Fig. 5 that a rough but definite correlation exists between the values of R_0 and -A; e.g. Al(Ge) is placed on the top and Al(Zn) in the bottom region of both scales. If in eq. (7) only the first term would contribute, then such a correlation with eq. (1) would be expected due to the pronounced similarity of both formulae; however, the sign of both R_0 and A would have to be the same, then! So we conclude that the second term in eq. (7) always behaves in an opposite way to the first term and that it systematically overcompensates the first term. It is astonishing, indeed, that in this complicated situation a correlation between R_0 and A still exists; this is not yet understood.

We have applied our 4 OPW calculations (Section III.3) to the calculation of A for substitutional impurities in Al, using Eq. (6). For that we evaluated $\sigma(\epsilon)$ on two different energy surfaces ϵ_F and $\epsilon_F + \Delta\epsilon$ with $\Delta\epsilon \simeq \epsilon_F \cdot 10^{-4}$ being small enough to ensure that $\sigma(\epsilon)$ was varying linearly with $\Delta\epsilon$. The preliminary results[22] are shown in Table 2; although the values of A are again of the right order of magnitude and have the correct sign, the sequence of A when going

Table 2:   Diffusion Thermopower Coefficient A [nV/K^2]

	Al(Ge)	Al(Zn)
4 OPW Calculation	- 7.4	-11.2
Experiment	-16.6	- 1.6

from Ge to Zn impurities is obtained incorrectly now. So this result is certainly unsatisfactory; it seems that the first term in eq. (7) still dominates in the calculations.

The theoretical values in Table 2 have been obtained neglecting any energy dependence of the pseudopotentials, but if we include the usually accepted form of energy dependence[33] the results change only insignificantly[22]. The reason for this quantitative discrepancy could perhaps be that the energy dependence of the pseudopotentials is much stronger than expected, or more probably that higher-order corrections[2] to electron-scattering contribute appreciably to the value of A (Nielsen-Taylor-Effect).

We conclude this section by saying that the diffusion thermopower coefficient A is determined only by the anisotropy of τ_k due to the dilute point defects, with the Al bandstructure always remaining virtually unchanged. The Mott-formula is always obeyed qualitatively, but the quantitative results of our 4OPW calculations are not correct. Both R_o and $-A$ depend in about the same say on the defect type, but the "wrong" sign of A shows that A is determined by the energy dependence of $v_k \tau_k$ which overcompensates that of dS.

IV.3.   PHONON DRAG THERMOPOWER COEFFICIENT B

Theoretically, the phonon drag coefficient B is described by the Bailyn-formula[30,34] which can be written for low temperatures as[27,34]

$$B = \frac{2e}{3T^3 \sigma} \sum_q \omega_q \frac{\partial N_q^o}{\partial T} \sum^{(q)} \alpha(\vec{q},\vec{k}) \, (1/\kappa_{\parallel}) v_k \tau_k \quad . \tag{8}$$

Here $\vec{q}$ is the phonon wavevector (a polarization index is left out for simplicity), ω_q is the phonon frequency and N_q^o the phonon equilibrium distribution function. The factor $\alpha(\vec{q},\vec{k})$ means the

relative probability that the phonon $\vec{q}$ is scattered sending an
electron from $\vec{k}$ to $\vec{k} + \vec{q}$, relative to all other interactions the
phonon $\vec{q}$ may enter. $\sum(q)$ considers all transitions on the FS
(reduced zone scheme) for a given $\vec{q}$. The other quantities are as
before, but $1/\kappa_\parallel$ is the local curvature of the FS normal section
defined by the vectors $\vec{v}_k$ and $\vec{q}$. Eq. (8) is valid for small values
of q when both states $\vec{k}$ and $\vec{k} + \vec{q}$ are lying on the same circle of
curvature radius $\kappa_\parallel$ on the FS and have the same values of $v_k\tau_k$. In
this case the second sum is virtually independent of q and the first
sum, at low T, gives rise to a cubic T dependence[35] and so to a
T-independent value of B.

If phonon scattering on point defects is negligible as is the
case for substitutional impurities and low T (Rayleigh-scattering)[30],
$\alpha(\vec{q},\vec{k})$ is independent of defect concentration. Then the phonon
drag coefficient B should qualitatively behave strictly analogous
to the low-field Hall coefficient R_O (Section III.1) and diffusion
coefficient A (Section IV.2): B should again be "normalized", i.e.
independent of the concentration of dilute identical defects, and
B should again be strongly dependent on the anisotropy of τ_k. The
second sum in eq. (8) has some similarity with eq. (1) for R_O but
the influence of the phonons as contained in the factor $\alpha(\vec{q},\vec{k})$ is
a specific feature of B.

Most of our experimental results of B are shown as a function
of point defect resistivity $\Delta\rho$ in Fig. 6; this figure corresponds
in all details to Fig. 4 where A was shown for the same experiments.
The behaviour of the two series of dilute Al alloys is as expected,
i.e. B is normalized and determined by the anisotropy of τ_k, and
the qualitative discussion is exactly as in the case of R_O (see
Section III.2) or A. The behaviour of B for the irradiated samples
Al(FD) is highly anomalous, however: for all five samples the
dose curves (full symbols) and annealing curves (open symbols) fall
upon each other in spite of the anisotropy of τ_k being strongly
different in both cases (compare Figs. 2 and 4); further, B is not
normalized but tends to approach zero with increasing FD concentration. We will postpone the discussion of this FD case until toward
the end of this section.

In the case of substitutional impurities the behaviour of B
was qualitatively consistent with the Bailyn-formula, eq. (8). For
a semi-quantitative test we again compare the typical values of B,
for the different defect types, with those of R_O by plotting both
coefficients on linear scales. This is done in Fig. 7. We can see
again that a rough but definite correlation exists between the
values of R_O and B: in both cases Al(Ge) is most positive, Al(Mg)
is intermediate and Al(Zn) most negative. Such a correlation is
expected from the similarity of eqs. (1) and (8) and also the sign
is correct, now. E.g., in the three-group model of Section III.2

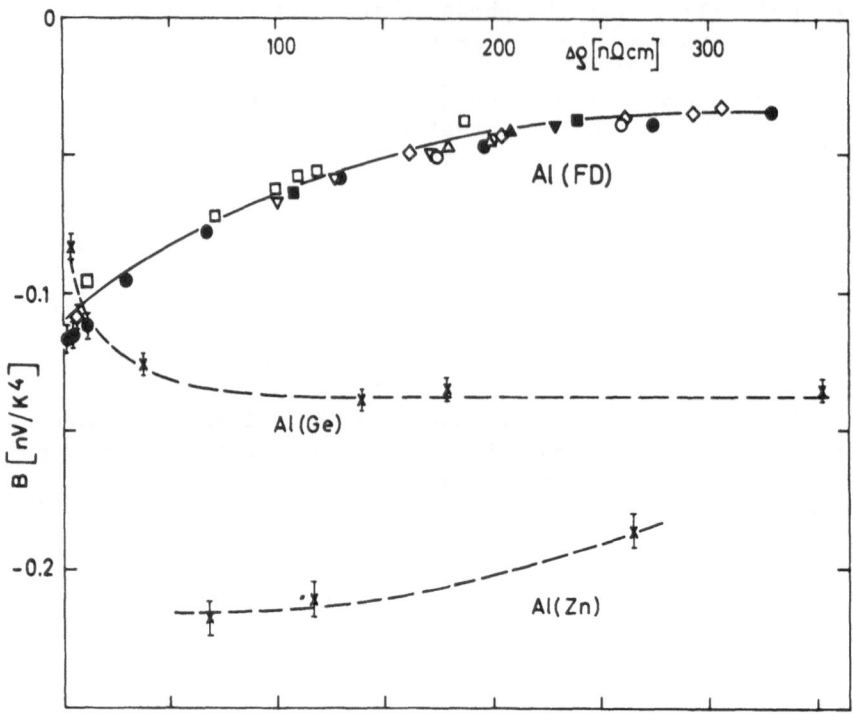

Fig. 6: Phonon drag thermopower coefficient B versus increase of
residual resistivity Δρ due to point defects in Al. This figure
corresponds in all details to Fig. 4 since it is based on the
same measurements; see Caption of Fig. 4.

a large value of τ_{++} gives strongly positive contributions to both
R_0 and B. Only the details of the weighting procedure of τ_k over
the FS are different in both cases, of course. As a result, we see
that there is not only qualitative but also semi-quantitative agreement between the experimental values of B and the Bailyn-formula.
Our 4 OPW calculations have not been extended to the case of B.

It was argued in the literature[2] that some part of the term
BT^3 in eq. (5) might perhaps not be related to phonon drag at all
but to higher-order scattering contributions to the diffusion term
AT. We can say, however, that at least the present free-electron
(FE) theory of this effect[2] cannot explain our results, since inserting the appropriate parameters into this theory the values of B
have the wrong sign and are about two orders of magnitude too
small.[26]. What we find is obviously phonon drag rather than "phony
phonon drag".

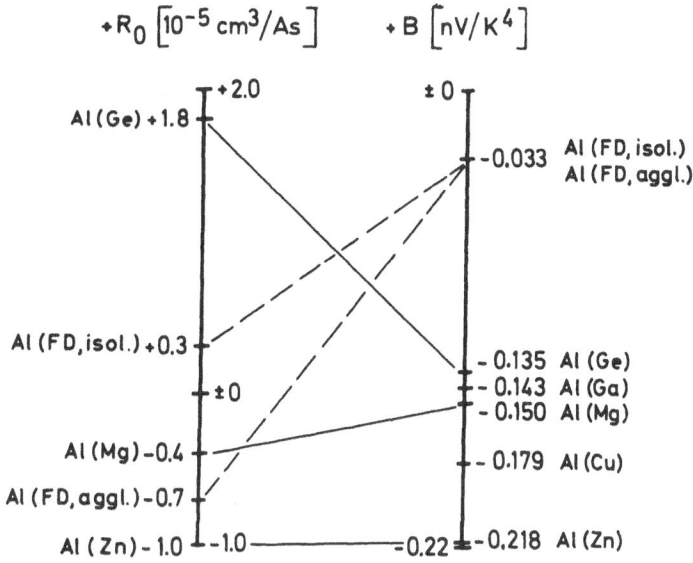

Fig. 7: Comparison of the low-field Hall coefficient R_0[4,20] and the phonon drag thermopower coefficient B[26-29]. For more explanation see Caption of Fig. 5. In the Al(FD) case the high-dose value of B has been taken (Fig. 6). For Al(Mg) the value of B is from Ref.[36], see Section IV.5.

Finally, we will discuss the anomalous behaviour of B in the irradiation experiments Al(FD) where B was found to depend strongly on the FD concentration and to be independent of the anisotropy of τ_k (Fig. 6). We further see from Fig. 7 that the value of $|B|$ for FD is extremely small compared to the case of substitutional impurities. It is very likely, now that this peculiar behaviour of B is a consequence of the extremely strong phonon scattering of the FD[27]. The single interstitials having dumbbell configuration (Fig. 1a) as well as multiple interstitials produce pronounced strain fields and are subject to various resonant vibration modes, which strongly couple with phonons of appropriate symmetry[37]. Inelastic neutron scattering measurements on reactor irradiated Al(FD) have confirmed this strong phonon-FD interaction[38]. In Fig. 6 the decrease of $|B|$ with FD concentration is definitely related to this strong phonon scattering. The corresponding decrease of $\alpha(\vec{q},\vec{k})$ in eq. (8) simultaneously reduces the influence of the anisotropy of τ_k on B. Nevertheless, the total disappearance of this influence is not well understood; probably it comes from averaging effects since the various scattering processes are expected to vary strongly with the phonon energy and with the directions of the phonon wave and polarization vectors.

As a summary of this section, the phonon drag thermopower coefficient B of Al was found to behave totally different for substitutional impurities and for FD. For impurities B was again determined by the anisotropy of τ_k due to the dilute point defects, with the Al bandstructure again remaining sufficiently unchanged. Both R_0 and B depend in about the same way on the type of impurity, and all this behaviour of B is qualitatively and semi-quantitatively consistent with the Bailyn-formula. "Phony phonon drag" (in the present FE form) cannot explain our results. In the case of FD, however, B was found to approach zero and to be independent of the anisotropy of τ_k, which obviously is a consequence of the extremely strong phonon scattering on the FD.

IV.4. VALIDITY RANGE OF THE LOW-TEMPERATURE LAW

The low-temperature law of thermopower $S = AT + BT^3$, see eq. (5), will break down when either A or B becomes temperature-dependent. This was between about 6 and 8 K in our experiments (Section IV.1). What is the physical origin of this breakdown?

We can safely rule out the possibility that the anisotropy of τ_k could change in the temperature range of our measurements (1.3 - 13 K), since in practically all cases electron scattering on the phonons was negligible with respect to that on the point defects. This argument is confirmed by the fact that experimentally[26-29] the breakdown temperature T_B of this low-temperature law was essentially independent of the defect concentration and only dependent of the defect type (i.e. of the anistropy of τ_k).

So we conclude that the diffusion coefficient A was always independent of T and that the origin of the breakdown lies in the phonon drag coefficient B. The mechanism is probably that B begins to change with T when the thermally excited phonon wavevectors q become large enough to connect electron states $\vec{k}$ and $\vec{k} + \vec{q}$ on the FS which no longer belong to the same circle of curvature radius $\kappa_\parallel$ and have different values of $v_k\tau_k$, see eq. (8). If this is true, then the breakdown temperature T_B should be the lower (for different defect types) the larger is τ_k on those FS regions where $|1/\kappa_\parallel|$ is extreme, since then just those FS regions contribute most to B where the above condition is violated first. In the case of Al our 4 OPW calculations have shown[22] that the magnitude of the local mean curvature $1/\kappa$ was, on the average, about 1.5 times larger on the S_{--} regions than on the S_{++} regions (Section III.2). So we would expect that the more negative is B the lower should be T_B. Roughly, this behaviour has been actually observed in our

experiments[26-29]: the approximate values of T_B are 5.5 K for Al(Zn)
and Al(Cu), 7 K for Al(Ge), 8 K for Al(Ge and Al(FD), and this
sequence just corresponds to that of B in Fig. 7. Further, for
$T>T_B$ the deviation of the real thermopower S from the extrapolated
low-temperature law was always found to be positive, which is also
consistent with this mechanism. So it seems that the Bailyn-theory
could also explain qualitatively some of the details of the breakdown of the low-temperature T^3 law.

IV.5. COMPARISON WITH THE LITERATURE

Finally, we will very briefly compare our results with literature data on Al containing nonmagnetic defects; a more extensive
discussion will be given in Refs.[27,29].

Absolute thermopower measurements (Section IV.1) on two
dilute Al(Mg) alloys have been performed by deVroomen et al.[36].
The two samples obviously suffered from magnetic trace impurities
and so the values of A were strongly different (+ 1.7 and -9 nV/K^2),
but B was the same (Fig. 7). The low-temperature law, eq. (5), was
fulfilled up to nearly 8 K and then the deviations from this law
were positive, which is all consistent with our results. Averback
et al.[39] investigated a series of very dilute Al alloys and found
the low-temperature law to be fulfilled up to the highest measuring
temperature of about 5.5 K. For their most concentrated Al(Cu)
sample they obtained A = - 5.4 nV/K^2 and B = - 0.175 nV/K^4. We see
that again B agrees with our result (Fig. 7), but the small difference in A (Fig. 5) could be either due to unknown impurities in
their sample, or due to Cu agglomeration effects in ours[28,29].

Reference measurements (ΔS) of Al containing defects versus
pure Al have been performed between 4.2 K and about 300 K[29,40-42].
An interpretation of the low-temperature data faces the problem,
which trace defects might have determined the thermopower of the
"pure" Al reference. Assuming reasonable values for this "pure"
Al, our absolute measurements quite nicely agree with those reference measurements and extend their data to temperatures below 4 K,
showing a new peak of ΔS in most cases. All this will be discussed
elsewhere[27,29].

APPENDIX

Since I have a few extra minutes, I will briefly compare the
results of our "absolute" measurements with those of "reference"
measurements taken from the literature (Section IV.5). This is
shown in Fig. 8, where the thermopower difference $\Delta S = S_{alloy} - S_{pure}$

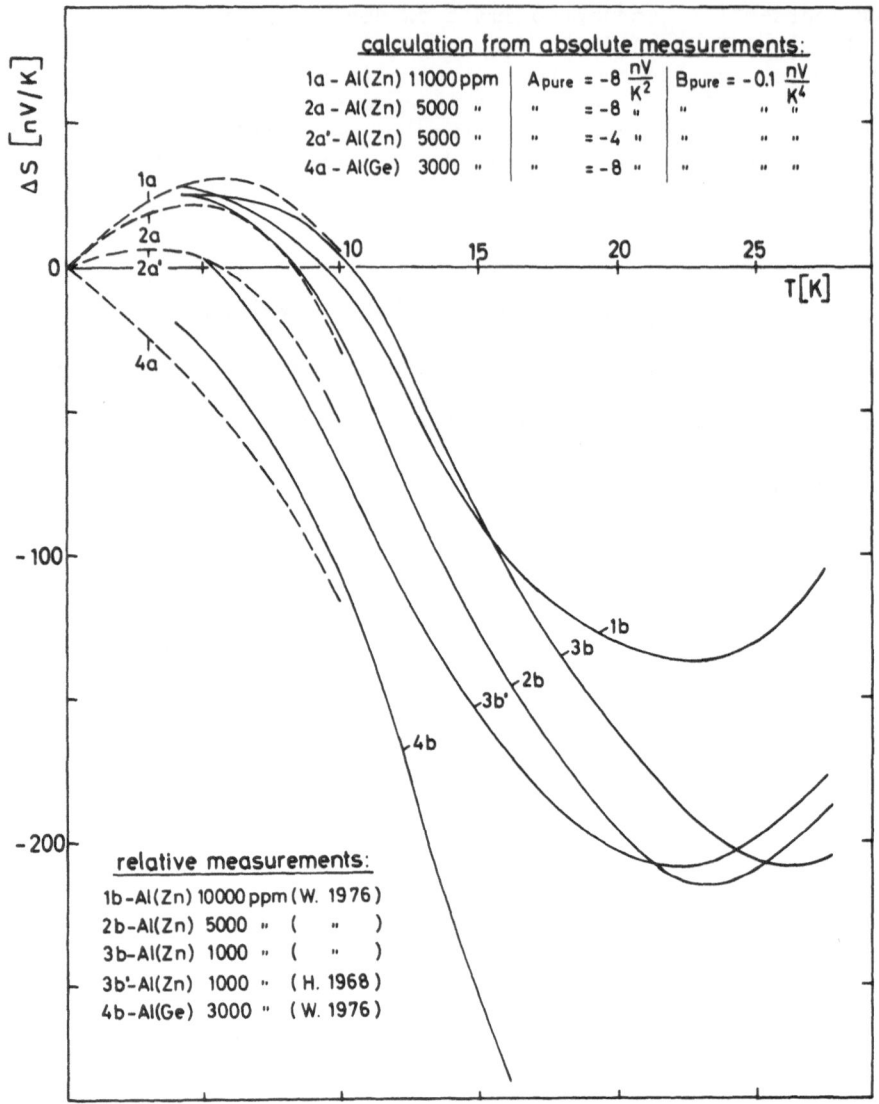

Fig. 3: Thermopower difference $S - S_{alloy} - S_{pure}$ between some dilute Al alloys and "pure" Al. The curves "b" are directly taken from the literature, and the curves "a" are calculated from our data. For explanations see text (appendix).

between some dilute Al alloys and "pure" Al is plotted versus T[26,29]. The full lines (curves "b") show typical experimental results of Huebener (labelled H. 1968)[40] and Wang (labelled

W. 1976[42]). All the authors[40-42] found essentially two peaks of ΔS in the temperature range 5 - 250 K. A positive peak at about 80 K (not shown in Fig. 8) was always attributed to a suppression of phonon drag due to phonon scattering at the impurities. The negative peak at about 25 K, however, was either interpreted as being due to a change of the phonon drag contribution in the alloys[40], or as being due to a change of the diffusion term when considering higher-order scattering contributions[43]. The peculiarities occurring at the lowest temperatures (5 - 10 K) have not been discussed.

The dashed lines in Fig. 8 (curves "a") have been derived from our absolute thermopower measurements (1.3 - 10 K). Here S_{alloy} was directly taken from experiment (compare Figs. 4 and 6), but S_{pure} was estimated assuming $S = AT + BT^3$ and using reasonable values of A and B for the "pure" Al reference material. These values A_{pure} and B_{pure} are taken from low-temperature data as compiled in Ref.[27] and are also listed in Fig. 8. Now we can argue as follows. The difference in the coefficients A is responsible for the initial slope of the S curves. From Fig 5 we see that for $A_{pure} \approx - 8$ nV/K^2 this difference is positive for Al(Zn), Al(Cu) and very probably also for Al(Mg), and negative only for Al(Ge). The difference in the coefficients B is responsible for the initial curvature of the ΔS curves. From Fig. 7 we see that for $B_{pure} \approx - 0.1$ nV/K^4 this difference is negative for all alloys. There is rather nice agreement of the corresponding curves "a" and "b" in Fig. 8 (apparently Huebener used a different Al reference material than Wang, compare curves 3b/3b' and 2a/2a'). For all alloys except Al(Ge) a new peak of ΔS shows up at low temperatures. We conclude that it is this peak at about 5 K which is determined by the differences in the diffusion coefficient A (Section IV.2), whereas the peak around 25 K originates from the differences in the phonon drag coefficient B (Section IV.3).

ACKNOWLEDGEMENT

I am grateful to W. Gläser and H. Vonach for their permanent interest and support, and especially to G. Sieber, K. Pfändner, and P. Rosner for the very enjoyable cooperation in all this work.

REFERENCES

1. Int. EPS Study Conf. on "Transport properties of normal metals and alloys below Θ_D" in Cavtat near Dubrovnik, May 9-12, 1977.
2. P.E. Nielsen and P.L. Taylor, Phys. Rev. B 10, 4061 (1974).
3. C.R. Leavens and M.J. Laubitz, J. Phys. F 6, 1851 (1976).
4. K. Böning, K. Pfändner, P. Rosner, and M. Schlüter, J. Phys. F 5, 1176 (1975).
5. W. Kesternich, H. Ullmaier, and W. Schilling, J. Phys. F 6, 1867 (1976).
6. K. Böning, W. Mauer, K. Pfändner, and P. Rosner, Rad. Effects 29, 177 (1976).
7. K. Böning, paper presented at the Conf., see Ref. 1; to be published.
8. N.W. Ashcroft, Phil. **Mag.** 8, 2055 (1963).
9. J.R. Anderson and S.S. Lane, Phys. Rev. B 2, 298 (1970).
10. M.T. Robinson, USERDA report CONF - 751006 (1975), p. 1.
11. H.G. Haubold, in Ref. 10, p. 263.
12. A. Seeger, E. Mann, and R.v.Jan, J. Phys. Chem. Sol. 23, 639 (1962).
13. B.v. Guérard and J. Peisl, in Ref. 10, p. 287 and to be published.
14. W. Mansel, H. Meyer, and G. Vogl, to appear in Rad. Effects.
15. W. Schilling, P. Ehrhardt, and K. Sonnenberg, in Ref. 10, p. 470.
16. A. Seeger, in Ref. 10, p. 493.
17. C.M. Hurd, The Hall Effect in Metals and Alloys, Plenum Press, New York, London (1972).
18. K. Böning, Phys. kondens, Materie 11, 177 (1970).
19. K. Böning, H.J. Fenzl, E. Olympios, J.M. Welter, and H. Wenzl, phys. stat. sol. 34, 395 (1969).
20. K. Böning, H.J. Fenzl, J.M. Welter, and H. Wenzl, phys. stat. sol. 40, 609 (1970).
21. K. Pfändner, K. Böning, and W. Brenig, solid state comm. 23, 31 (1977).
22. K. Pfändner, Ph.D. thesis in preparation at the Techn. Universität München; K. Pfändner, K. Böning, and W. Brenig, to be published.
23. R.S. Sorbello, J. Phys. F 4, 503 (1974).
24. J.P.G. Shepherd and W.L. Gordon, Phys. Rev. 169, 541 (1968).
25. R.S. Sorbello, J. Phys. F 4, 1665 (1974).
26. G. Sieber, Ph.D. thesis, Techn. Universität München (1976).
27. G. Sieber, G. Wehr, and K. Böning, J. Phys. F7, 2503 (1977).
28. G. Schmitt, Diplomarbeit (master's thesis), Technische Universität München (1976).
29. G. Sieber, G. Schmitt, K. Böning, S.Y. Wang, and R.R. Bourassa, to be published.
30. R.D. Barnard, Thermoelectricity in Metals and Alloys, Taylor and Francis LTD, London (1972).

31. P.L. Taylor, A Quantum Approach to the Solid State, Prentice
     Hall Inc., New Jersey (1970).
32. J.P. Jan, Can. J. Phys. 46, 1371 (1968).
33. R.W. Shaw, Phys. Rev. 174, 769 (1968).
34. M. Bailyn, Phys. Rev. 157, 480 (1967).
35. A.M. Guénault, J. Phys. F 1, 373 (1971).
36. A.R. DeVroomen, C.van Barle, and A.J. Cuelenaere, Physica 26,
     19 (1960).
37. H.R. Schober, V.K. Tewary, and P.H. Dederichs, Z. Physik B 21,
     255 (1975).
38. K. Böning, G.S. Bauer, H.J. Fenzl, R. Scherm, and W. Kaiser,
     Phys. Rev. Lett. 38, 852 (1977).
39. R.S. Averback, C.H. Stephan, and J. Bass, J. Low Temp. Phys.
     12, 319 (1973).
40. R.P. Huebener, Phys. Rev. 171, 634 (1968).
41. T. Rybka and R.R. Bourassa, Phys. Rev. B 8, 4449 (1973).
42. S.Y. Wang, Master's Thesis, University of Oklahoma (1976).
43. A.W. Dudenhoeffer and R.R. Bourassa, Phys. Rev. B5, 1651 (1972).

DISCUSSION

P.B. Allen: Isn't it true that in your neutron irradiation experiments the damage is rather inhomogeneous, that cascade processes
lead to defects that cluster strongly?

K. Böning: We do have clustering during irradiation at 4 K, but
the clustering is not very pronounced in our case. Now in copper,
by comparison, the atoms have very large ion cores with the space
between them rather small. So if you shoot a Cu atom through a
Cu lattice you produce a very high local density of displacements.
Measurements show clusters in copper of about 25 interstitials. In
aluminum, however, the situation is rather different. Aluminum has
much smaller ion cores and much more space between the atoms. In
aluminum the results are that there are small clusters of between
one and three interstitials, and there are a lot of these tiny sub-
clusters dispersed in a large cascade.

R.R. Bourrassa: I have two comments. First, concerning Fig. 8, I
simply want to point out that if the value for B (pure metal) were
to be -0.2 instead of -0.1 then the curvature of your data on this
curve would be up instead of down, which would change the conclusion that you are drawing that the low temperature peak is consistent
with phonon-drag. Second, as regards the A coefficient, if the
breakdown of the Jan equation is correct in that the first term is
mainly the anisotropy of the Fermi surface and the second term is
the $\tau(v)$ part, that would be consistent with our results, presented
yesterday, that the anisotropy term is positive and the other term
is negative.

C. Uher: I take it your thermometers are mounted on the sample.
What is the effect of irradiation on thermometer calibration?

K. Böning: That's a really important point. There was a previous
publication, about 15 years ago, saying that there should be only
a very small effect on calibration of carbon resistors thermometers.
But we found a very large effect. So what we had to do is the
following. After irradiation we had to keep the sample at 4K in
the irradiation tube, out of the reactor core, for about one day
to allow the decay of the short-living radioactivity. After that
we moved the sample into the cryostat where we had to do a new
calibration before every measurement. The resistors were Allen-
Bradley carbon resistors, but it seems that a new type has been
developed during the last few years that is different from the old
type available 15 years ago.

A.M. Guénault: May I ask, with reference to Fig. 8, where the
cross-over temperatures occur. I refer to the sort of thing I was
talking about on the first day for the pure noble metals, where
you go from the dirty regime at low temperatures to that where
phonon scattering comes in at higher temperatures. I think that is
yet another example of how you can get a very complicated temperature dependence from this sort of thing.

K. Böning: I would say two things in response. First, for the
dilute alloys which we have been using this cross-over temperature
would be very high. But for the pure reference material, this
cross-over occurs, of course, at much lower temperatures. If the
coefficient A of the pure metal were strongly temperature dependent
in this region, that could create a problem. We have determined
for a typical pure aluminum sample at what temperature the electrical resistivity due to phonon scattering is about one-third of the
residual resistivity. The calculation was for a sample with a
resistance ratio R_{300K}/R_{4K} of 1000; these S measurements were
usually done with reference material not much better than that. We
found that only at about 22K the phonon resistivity is about one-
third of the residual resistivity. So the residual impurities
really dominate the low-temperature behavior of the "pure" metal.

QUASI-ONE-DIMENSIONAL CONDUCTORS

P. M. Chaikin*

Department of Physics
University of California
Los Angeles, California 90024, USA

ABSTRACT

The highly anisotropic systems which are referred to as quasi one-dimensional conductors are characterized by rather small bandwidths (< 1 eV) and simple tight binding bandstructures. The thermoelectric power is therefore larger than in typical metals and more easily analyzed to obtain information on anisotropic bandstructure and cation-anion charge transfer. The metal-insulator transitions present in many of these systems are also easily probed by thermopower. Perhaps most interestingly these narrow band conductors show sizeable effects relating to the on-site Coulomb interaction. The correlated electrons retain a spin degree of freedom which is observable in the thermopower as an "entropy per carrier" contribution of $\frac{k}{e} \ln 2$ or $- 60$ μV/K. All highly conducting TCNQ salts of the form $R^+(TCNQ)_2^-$ saturate to this value at room temperature.

INTRODUCTION

Over the past several years there has been considerable interest in "quasi one-dimensional" solid state systems, crystals which are characterized by having highly anisotropic physical properties.[1] The great interest which centers on these compounds

---

*Research supported by National Science Foundation under Grant No. DMR76-83421.

results not only from an attempt to understand the cause of the
anisotropy but from general physical arguments which show that
one-dimensional systems are intrinsically unstable against a number of different low temperature distortions (due to a divergence
of the generalized response functions) to insulating magnetic and
non-magnetic ground states. Moreover, statistical mechanics shows
that one-dimensional systems cannot undergo any phase transitions
above absolute zero, so that there are large temperature regions
which are dominated by fluctuation effects. This apparent contradiction, that one-dimensional systems are unstable but cannot
distort, coupled with the fact that many one-dimensional problems
are exactly soluble, explains why these compounds are fascinating.

The materials which are used in these studies fall into several categories. Transition metal mixed valence salts such as
$K_2Pt(CN)_4Br_{0.3} \cdot 3H_2O$ (KCP) are tetrahedrally coordinated with Pt-Pt d
electron overlap along chains which are separated by the cyanide
groups and by water molecules.[2] Poly-Sulfur Nitride $(SN)_x$ is a
metallic polymer which, however, has a great deal of coupling between chains and is therefore not as one-dimensional as the other
compounds.[3] Mercury chain salts such as $Hg_{2.86}AsF_6$ (of which we
shall hear more later in this conference) has Hg chains arranged
as lines in planes and is thus one-dimensional in two directions.[4]
The last category, with which the author has had most experience,
and which will be discussed in the remainder of this talk will be
the charge transfer salts of Tetracyanoquinodimethane (TCNQ).

The TCNQ molecule is planar and a good electron acceptor due
to the cyanide groups attached at either end of the benzene like
ring. It is thus possible to form charge transfer salts with a
variety of cations. As examples, the Alkali TCNQ's such as
NaTCNQ are stable and have been known and investigated for many
years.[5] Most of these compounds with metal cations are insulating
or semi-conducting, albeit for interesting reasons. However, when
the cation is an organic molecule such as Tetrathiofulvalene (TTF)
the compounds formed can have relatively high conductivities
(100-1000 $(\Omega\text{-cm})^{-1}$) and metallic temperature dependence to their
conductivity.

In these highly conducting salts the planar TCNQ molecules
form segregated stacks with a separation of about 3.5A between
molecules along the stack, and about 10A between stacks.[6] The π
electrons have wavefunctions which stick out of the plane and
overlap with the molecules above and below. This leads to the
formation of bands with small (.01-.5 eV) bandwidth along the

stacks and considerably smaller bandwidths in the perpendicular
direction where the overlap is less. This produces the quasi one-
dimensionality which is observed. The anisotropic bandstructure
is responsible for the metal-insulator transitions and the smallness of the bandwidth implies that many of the interactions (coulomb,
electron-phonon) which have little effect in most metals can be of
fundamental importance in these salts.

The TCNQ salts present a fertile ground for thermoelectric
power studies. The small bandwidths give rise to thermopowers
which are an order of magnitude or more larger than what is observed in typical metals. The low thermal conductivities and
needle shapes of the crystals make them very easy for thermopower
measurements. The metal-insulator transitions are easily detected
by the characteristically different temperature dependences and
sharp changes in slope. The bandstructure consisting of anisotropic
tight binding bands makes the thermopower relatively easy to
interpret. The measurements are also of substantial importance in
that other transport measurements such has Hall effect are very
difficult to perform due to anisotropy and crystal imperfection.
The TCNQ salts are therefore one of the few systems where thermopower measurements produce more information about the material
studied than they do about mechanisms of thermoelectricity.

The most famous member of the TCNQ family is TTF-TCNQ. At
the time of its preparation it was the most highly conducting
organic charge transfer salt. What aroused physicists interest,
however, was the temperature dependence of its conductivity which
rose from a room temperature value of about 500 $(\Omega\text{-cm})^{-1}$ to a maximum of 15-several hundred times that value at about 60K.[7] The
ratio of room temperature to maximum conductivities is still
controversial due to problems related to the anisotropy and
perfection of crystals prepared by different groups.[8] (It is
interesting to note that the thermopower measurements are reproducible from group to group and largely unaffected by crystal
perfection. This is a result of thermopower being a zero-
current measurement.) The original conductivity data showed a
sharp metal insulator transition near 53K and more detailed work
showed at least two transitions. The character of these transitions has been well established, through several beautiful
diffuse x-ray and neutron diffraction studies, as one-dimensional
(1-d) Peierls' transitions.

The thermoelectric power of TTF-TCNQ is shown in Fig. 1.[9] We
focus our attention first on the high temperature region well
above the transitions. For a single 1-d tight binding band the

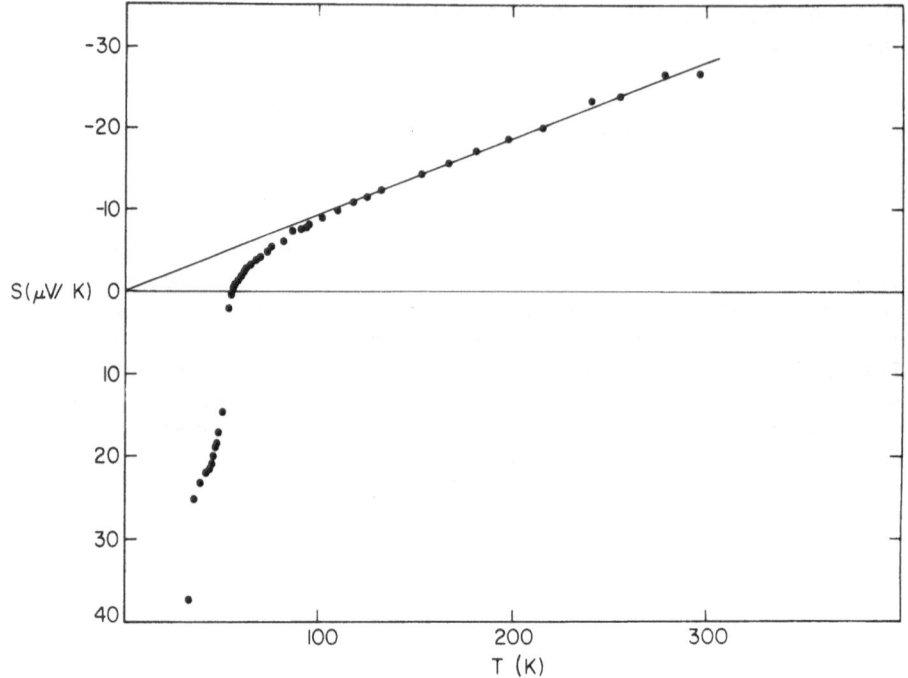

Fig. 1. Absolute thermopower of one single crystal of TTF-TCNQ.

thermopower may be developed from the Boltzman equation as:

$$S = \frac{-\pi^2 k_B^2 T}{3 |e|} \left\{ \frac{\cos(\pi\rho/2)}{2 |t| \sin^2(\pi\rho/2)} + \frac{\tau'(\varepsilon)}{\tau(\varepsilon)} \right\} \Bigg|_{\varepsilon=\varepsilon_F} \tag{I}$$

where t is the transfer integral, (bandwidth = 4t), ρ is the electron density per site, τ is the scattering time, τ' is its time derivative and T is the temperature. As usual, two terms are present, the first depending on the band structure and the second on the scattering time. Unfortunately, little is known about the dominant scattering processes in these systems. However, the mean free path is quite small (0.3-1 lattice constants at room tempera-ture)[8] and it is conceivable that the energy dependence of the scattering is sufficiently smeared out (at 300K) that the second term does not significantly contribute. As yet we have no way of treating it and no indication that it plays a major role.

If the scattering term is not temperature dependent and the
bandstructure is smooth on the scale of kT, we would expect the
classic linear temperature dependence to the thermopower. As can
be seen from Fig. 1, over the range 100-300K TTF-TCNQ is a textbook example of a metal (and thus strikingly different from the
more common metals). The magnitude of the thermopower is more
than an order of magnitude more than seen in typical metals
as anticipated. Taking the value of $\rho=.59$ obtained from the x-ray
studies[10] we can then use equation I to find the bandwidth, which
works out at .45eV. This determination was one of the earliest
made and agrees quantitatively with more recent measurements from
optical properties.[11]

In the temperature region of the phase transitions there is
considerable structure in the thermopower. Very close to the first
transition at 53K the sign of the thermopower changes in a sharp
manner. Between 39 and 50K the value is fairly constant at
$+20\mu V/K$. Below 39K the thermopower takes on the characteristic temperature dependence of a semiconductor (1/T). As might be expected, samples from different labs have vastly different thermopowers in the semiconducting state where the chemical potential
is extremely sensitive to impurities.[9] The fact that the thermopower appears to be intrinsic down to 39K shows that above this
temperature the material is not semiconducting. This result was
the earliest indication that more than one transition was taking
place (although we did not realize this consequence until further
evidence was found). A more detailed explanation of the thermopower in the transition region was not available until alloying
experiments were performed. These will be discussed later.

The anisotropy of TTF-TCNQ was investigated both by cutting
crystals in such a way as to make the a axis the long dimension,
and by growing crystals with the a axis along the needle axis.
Translation along the a axis takes one from a TCNQ molecule to a
TTF molecule. The results of this study are shown in Fig. 2.[12] The
striking characteristic of this figure is the extreme anisotropy
of the thermopower along the two axes, with opposite signs over
almost all of the temperature range. This is one of the most
anisotropic thermopowers yet found. The two directions do not
cross zero at the same value indicating that the zero value near
the transition is not the manifestation of a collective state
(at least not three dimensionally). It is also worth noting that
the thermopower along the a axis does not have the simple textbook
metallic behavior observed along the b axis.

Since the bands are probably of a tight binding nature, we
attempted to fit both the temperature dependence and the anisotropy

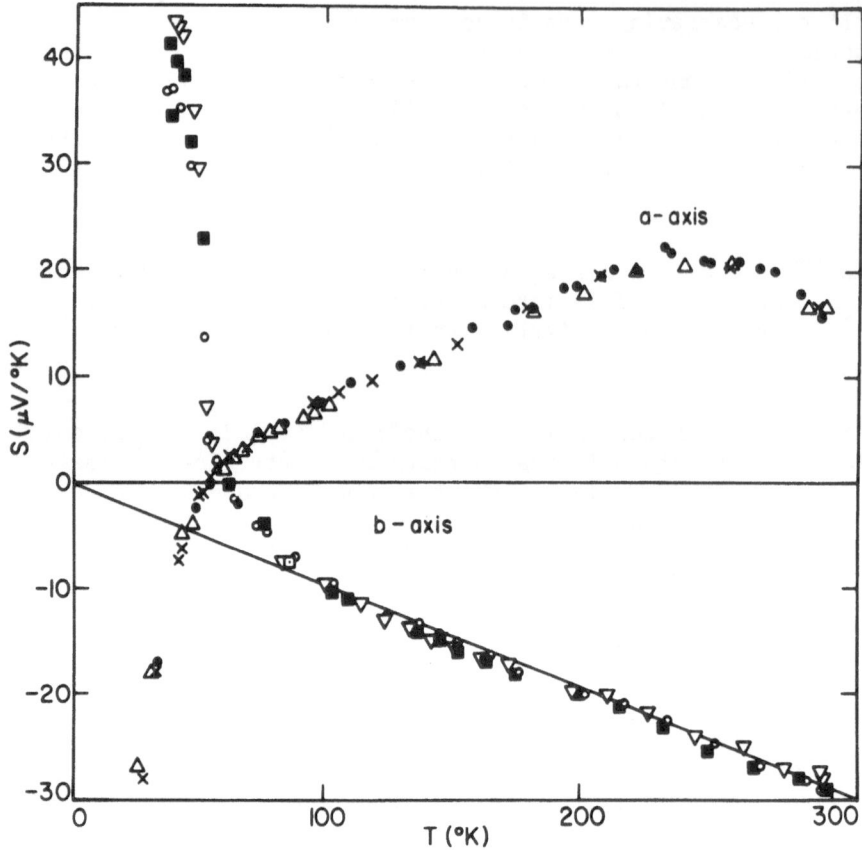

Fig. 2. Absolute thermopower for TTF-TCNQ along the conducting
chains (b̲ axis) and in a direction from TTF to TCNQ molecules
(a̲ axis).

of the thermopower to a bandstructure. We found that it was necessary to include transport in the b̲ direction by broad bands associated with both the TTF and the TCNQ chains and to include a
smaller overlap between the two, in the a̲ direction. Again only
the band term in the thermopower was included, scattering effects
being neglected. After trying several configurations, it was
found that only inverted bands with interband hybridization even
qualitatively fit the data (see reference 13). We were thus able
to propose a bandstructure for TTF-TCNQ which not only reproduced
the thermopower, but explained much of the temperature dependence
of the susceptibility and other properties and was subsequently

confirmed qualitatively by Huckel calculations.[14] The idea of inverted hybridized bands has had significant implications for many of

the other TCNQ salts investigated more recently.[15]

We have found that both the cation and the anion bands can contribute to the thermopower in these samples and we should like to explore the consequences of this two band model. From the Boltzmann equation the thermopower will have the form:

$$S = \frac{\sigma_1 S_1 + \sigma_2 S_2}{\sigma_1 + \sigma_2} \tag{II}$$

where σ_1 is the conductivity associated with each band and $S_{1,2}$ is given by equation I with separate densities, transfer integrals and scattering times for the two bands. For the cases of interest, the densities are related by $\rho_1 + \rho_2 = 2$, as the band fillings are determined by interband charge transfer. Note that according to equation I a band that is less than half filled ($\rho < 1$) will have a negative (electron-like) thermopower, while a more than half filled band will have the opposite sign, neglecting the $\tau'(e)$ term. As the charge transfer in these salts is always less than one electron per molecule the TCNQ chain will contribute a negative thermopower, and the cation chain positive. We can then write the band structure contribution for two one dimensional, tight binding, charge transfer bands as:

$$S = -\frac{\pi^2 k^2 T}{6|e|} \left\{ \frac{(\tau_1 - \tau_2)\ \cot\pi\rho/2}{(|t_1|\tau_1 + |t_2|\tau_2)\ \sin\pi\rho/2} \right\} \tag{III}$$

where the subscript 1 refers to the TCNQ band. It is interesting that independent of ρ and t we obtain zero if the scattering times are equal.

TSeF–TCNQ AND ALLOYS WITH TTF–TCNQ

To test these ideas we have investigated an isostructural compound and organic alloys. The thermopower of TSeF–TCNQ is shown in Fig. 3.[16] This compound differs from TTF–TCNQ in the substitution of Selenium for Sulfur in the cation. The thermopower is small, positive and certainly not linear in temperature dependence. We believe this to be the result of the similarity of the scattering times on the two chains as discussed above. Since the bandwidths are comparable on the two chains the conductivity is shared by the two chains as contrasted to TTF–TCNQ where the anion chain dominates (evidenced by the large negative thermopower). Similar conclusions have been drawn from a variety of other experiments as well as the alloy measurements described below. The

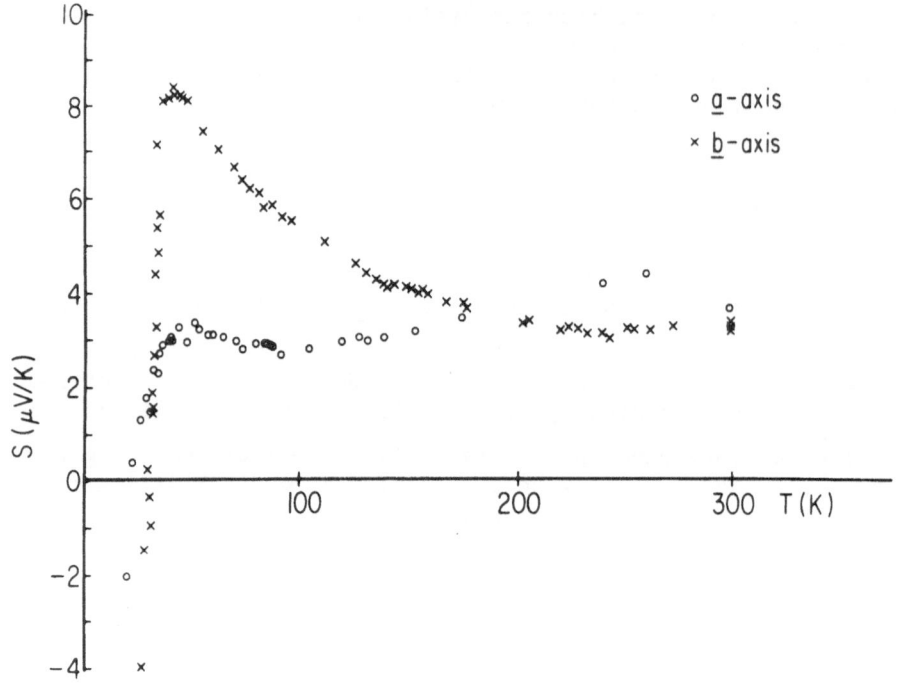

Fig. 3. Thermopower and its anisotropy in TSeF-TCNQ.

anisotropy seen in Fig. 3 is considerably less than one observes in TTF-TCNQ. We attribute this behavior to increased interchain coupling, as suggested by other workers,[18] but it also may be due to a pathological combination of bandstructure and scattering times. Fig. 3 indicates the presence of a single transition in TSeF-TCNQ in agreement with other experiments.

Up to this point we have ignored the energy dependence of the scattering time and also any phonon drag effect. In TTF-TCNQ in the region between 100-53K and in TSeF-TCNQ over the entire temperature range there are substantial deviations from linearity in the temperature dependence. In both cases we find an increasing positive contribution of comparable size as temperature is lowered. We suggest that this is the result of an increase in the energy dependence of the scattering on the TCNQ chain with $\tau'(e)$ being negative as in the case of scattering from a softening phonon. The soft phonon associated with the Peierls transition and accompanying Kohn anomaly and located at $2k_f$ has been observed in neutron scattering experiments.[19] It is however unusual that the conductivity in these salts is seen to be increasing in the region where the thermopower is becoming more positive.[7] One would also expect this phonon to contribute to a phonon drag term.[20] However, the scattering at $2k_f$ is across a Fermi surface composed of parallel sheets, and is therefore not an Umklapp process. Given the bandstructure one would thus predict a negative phonon drag contribution rather than positive as is observed. The lack of phonon drag effects may be related to the small mean free paths and imperfections in these crystals.

The effects of making alloys of the form $TTF_{1-x}TSeF_x-TCNQ$ are shown in Fig. 4.[21] Since the alloys are formed from isostructural components we expect the TCNQ chain to remain uniform and the cation chain to contain randomly arranged TTF and TSeF ions. For the transport properties the anion chain should be largely unaffected, while the cation chain should show the effects of disorder. We will discuss the thermopower behavior at high temperatures first. As argued above, the sign of the thermopower is negative for the TCNQ chain and positive for the cation chain. The primary effect of disorder at high temperatures is to decrease the mean free path and scattering time. According to equation II the thermopower of the cation chain then has a reduced contribution to the total thermopower. As alloys are made from either side of the phase diagram the thermopower should go toward more negative values.

In Fig. 5 we have plotted the room temperature thermopower as a function of doping. We see that the positive thermopower contribution of the cation chain is indeed reduced from the pure values for TTF-TCNQ or TSeF-TCNQ and that the samples reach a maximum negative value at the 50-50 alloy concentration. In the region of the phase transitions we also see an interesting effect upon alloying. Whereas pure TTF-TCNQ has a thermopower which is positive in the region from 53-39K, slightly doping the cation chain with TSeF produces a characteristic which is never positive. We attribute this behavior to the additional scattering on the cation chain which limits its conductivity so that it does not dominate the conductivity even below the semiconducting transition of the TCNQ chain. The following scenario then emerges for TTF-TCNQ: at high temperatures the conductivity is dominated by the TCNQ chain yielding a negative thermopower. At 53K there is a gap which opens on this chain and reduces its conductivity, between 53K and 39K the TTF chain is still conducting and the thermopower reflects this by its positive value, below 39K both chains are insulating. This picture has been confirmed by many other experiments on these systems.[22]

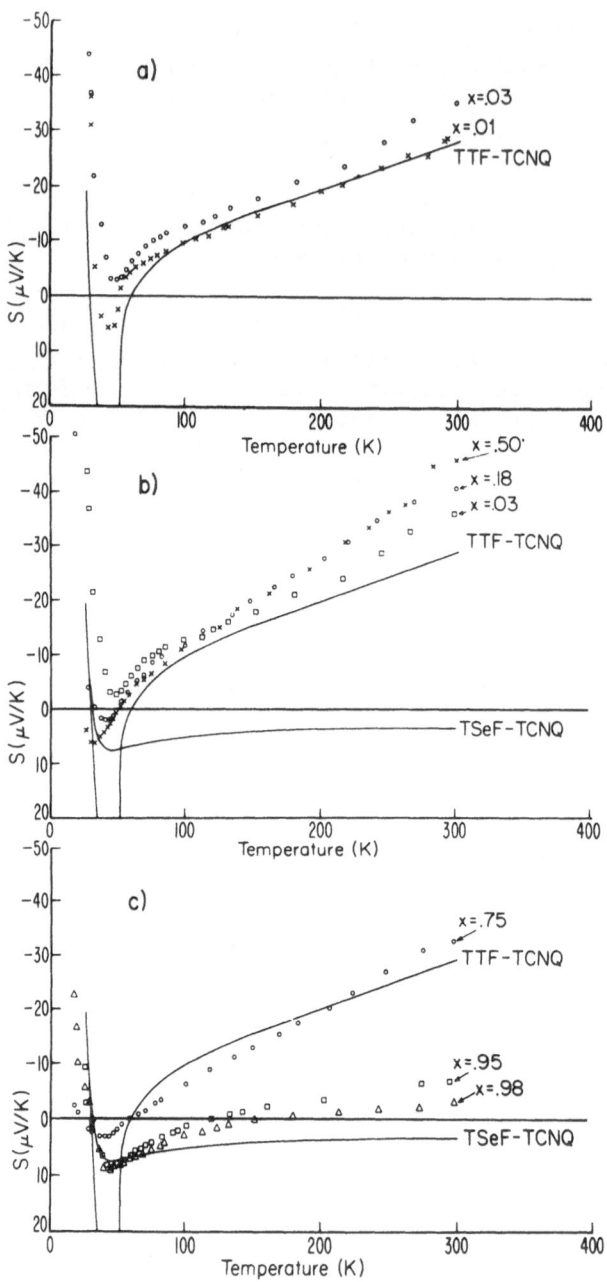

Fig. 4. Thermopower of a series of organic alloys of the form $TTF_{1-x}TSeF_x$-TCNQ.

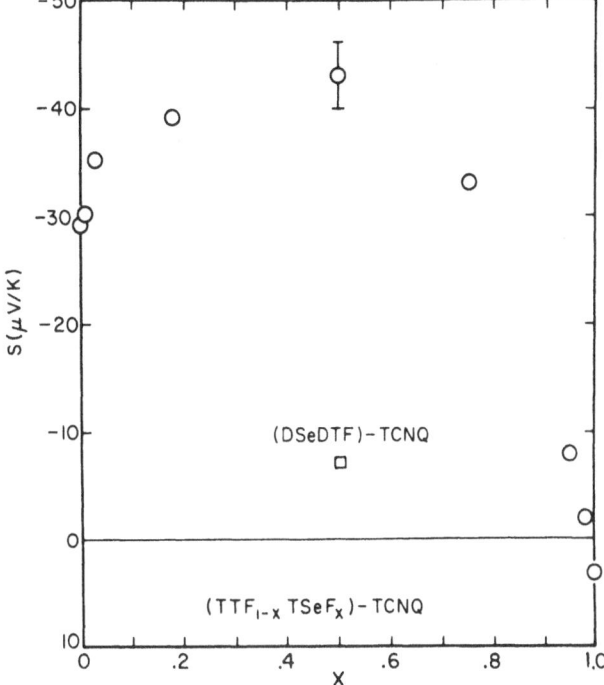

Fig. 5. Room temperature thermopower of $TTF_{1-x}TSeF_x$-TCNQ.

Focusing on the low temperature properties, we note that with increased doping from either side of the phase boundary the thermo-electric power behaves less as an ideal semiconductor (which would have $S \propto 1/T$ and a large value). We believe this to be the result of the disorder introduced by the alloying. In one-dimensional systems the effect of disorder is to introduce localized states into the semiconducting gap. While these states will not consider-ably affect the electrical conductivity, they provide a low energy path for the electrons. The thermopower thus remains low while the resistivity is rapidly increasing with decreasing temperature.

THERMOPOWER IN CORRELATED SYSTEMS

Of the phenomena which we have investigated using thermopower on quasi-one-dimensional systems the results we have obtained for 1:2 TCNQ are the most beautiful and perhaps have the most conse-quence for the science of thermoelectricity. For these salts we have used some simple ideas to quantitatively explain the observed thermopower and obtain information about coulomb correlations. The

essence of the argument is that the thermopower measures entropy
per carrier (only in certain limits) and the entropy associated
with a free spin is just k(ln2). The spin contribution to the
thermopower is then k/e(ln2) or –60μV/K. All highly conducting
TCNQ salts with 1:2 stoichiometry saturate to this value at room
temperature.

At the beginning of this talk it was noted that the small
bandwidths associated with the organic charge transfer salts imply
that care must be taken with the treatment of coulomb correlation
effects which are to a large extent absent in conventional metals.
A model which·is often used to describe these effects is the
Hubbard model. The Hamiltonian includes a term for electron transfer from one site to another and an onsite interaction. The first
term leads to the formation of tight binding bands while the
second correlates the electrons apart. The Hamiltonian is written
as:

$$H = -t \sum_{i,\sigma} (c^{\dagger}_{i,\sigma} c_{i+1,\sigma} + c^{\dagger}_{i+1,\sigma} c_{i,\sigma})$$

$$+ U \sum_{i,\sigma} n_{i\sigma} n_{i-\sigma} \tag{IV}$$

where t is the transfer integral and U is the onsite coulomb interaction.

The Boltzmann equation cannot handle the transport properties
within the context of the Hubbard model. For interacting systems
it is necessary to perform calculations using the Kubo formalism.
The thermopower is then given by:

$$S = -\frac{S^{(2)}/S^{(1)} + \mu/e}{T} \tag{V}$$

where $S^{(2)}$ and $S^{(1)}$ are the transport coefficients involving
respectively the heat flux-velocity (Q,v) and velocity-velocity
correlation functions.

$$S^{(2)} = \frac{1}{2} \beta e \int_0^{\infty} \left\{ \text{Tr} \left[\exp\beta \left(\mu \sum_i n_i - H \right) \right] \right.$$
$$\left. \times [Qv(\tau) + v(\tau)Q] \right\} d\tau \tag{VIa}$$

$$S^{(1)} = \frac{1}{2} \beta e^2 \int_0^\infty \left\{ \mathrm{Tr} \left[\exp\beta \left(\mu \sum_i n_i - H \right) \right] \right.$$

$$\left. \times \left[vv(\tau) + v(\tau)v \right] \right\} d\tau \tag{VIb}$$

In general these equations are quite difficult to evaluate, but we will content ourselves with looking at some exact results obtained from the high temperature limit.[23] Suppose that the Hamiltonian which describes the system may be broken into two terms, those which have eigenvalues much greater or much less than the temperature.

$$H = H_1 + H_2 \qquad \qquad \beta H_1 \rightarrow 0$$

$$\beta H_2 \rightarrow \infty \tag{VII}$$

In this limit all terms in H_2 have vanishing contributions to the traces in equations VI due to the exponential factor. The only states which contribute are those with less energy than kT and the only temperature dependence is through the exponent $e^{-\beta\mu\sum_i n_i}$. However, from the definition of the chemical potential:

$$\frac{\mu}{T} = - \left(\frac{\partial\sigma}{\partial N} \right)_{E,V} \quad , \quad \sigma = k_B \ln g \tag{VIII}$$

where σ is the entropy and g is the degeneracy. For our high temperature limit the degeneracy involves counting the states in H_1 and becomes independent of temperature. We then have:

$$\frac{\mu}{T} = - k_B \left(\frac{\partial \ln g}{\partial N} \right) \rightarrow \beta\mu = \text{constant} \tag{IX}$$

The exponents and transport coefficients become independent of temperature at high T so that the thermopower becomes:

$$S(T \to \infty) \to - \frac{S^{(2)}/S^{(1)}}{T} + \frac{\mu}{eT}$$

(X)

$$\to \frac{\mu}{eT} = - \frac{k_B}{|e|} \frac{\partial \ln g}{\partial N}$$

In this limit the thermopower is a measure of the change in entropy upon addition of a carrier or simply "entropy per carrier". We can now easily calculate the thermopower for a variety of interactions by calculating the degeneracy. In Fig. 6a we illustrate a configuration for spinless Fermions. With N Fermions and N_A sites the degeneracy is $g = (N_A!)/(N!)(N_A-N)!$ and the thermopower is:

$$S = - \frac{k_B}{e} \ln[(1-\rho)/\rho] \qquad \text{spinless Fermions} \qquad \text{(XI)}$$

where ρ is the density, N/N_A. If the system contains electrons with spin (Fig. 6b) but a large onsite coulomb interation (U in equation IV) the configurational degeneracy is as above but there is an additional factor of 2^N for the spin degeneracy. The thermopower becomes

$$S = - \frac{k_B}{e} \ln[2(1-\rho)/\rho] \qquad U \gg kT \qquad \text{(XII)}$$

Finally if there is no onsite coulomb repulsion (Fig. 6c) we must sum over configurations with different numbers of spin up and spin down electrons. The result is:

Fig. 6. Some possible configurations of particles on sites for a) Spinless Fermions, b) Electrons in a system with large onsite repulsion ($U=\infty$), c) Non-interacting electrons.

$$S = - \frac{k_B}{e} \ln[(2-\rho)/\rho] \qquad kT \gg U \qquad\qquad (XIII)$$

We now focus our attention on the 1:2 TCNQ salts. With one electron per two sites $\rho = 0.5$ and we would expect the thermopower to saturate to a value of $k/e \ln 3 = -95\mu V/K$ if $kT \gg U$, $k/e \ln 2 = -60\mu V/K$ if $U \gg kT$ and $k/e \ln 1 = 0$ if $V \gg kT$ and the spins are quenched (by magnetic ordering). As can be seen from Fig. 7 all of the highly conducting 1:2 salts of which we know tend toward $-60\mu V/K$ at room temperature.[24-27] Note that this is true despite the varied low temperature behaviors which result from the weaker interactions present in the different salts.

In addition to the correlated thermopower in the high temperature limit, there are several other limits which have been calculated. For $t=0$ we know S for any ρ, as a function of $U/k_B T$, for $t=0$, $U=\infty$, any ρ, as a function of $V/k_B T$ (nearest neighbor inter-

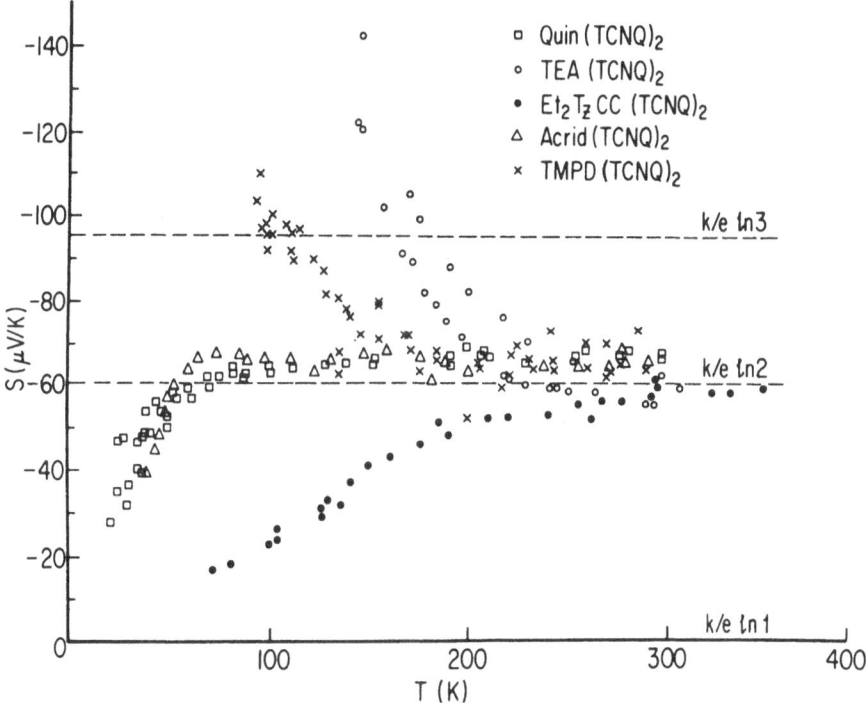

Fig. 7. Thermopower of highly conducting salts of the form $(Donor)^+(TCNQ)^-_2$.

action is V), for U=∞, any ρ as a function of bandwidth and temperature.[29] The interesting result is that if ρ=0.5 and U=∞ the thermopower is independent of bandwidth, nearest neighbor interaction and temperature, and equal to $-60\mu V/K$. We therefore believe that we have strong evidence that sizeable coulomb correlations exist in these salts. It is also interesting to note that the small deviations from $-60\mu V/K$ may be related by equation XII to partial charge transfer between cation and anion.

In conclusion, we have shown that the thermopower of the quasi one-dimensional organic conductors show a variety of different behavior relating to the strengths of different electronic interactions and the transitions they produce. These materials are perhaps unique systems for such studies but many of the results on correlated hopping may find application in other conducting systems.

REFERENCES

1. For several review papers in this field see Low Dimensional Cooperative Phenomena, edited by J.H. Keller (Plenum Press, New York, 1974).
2. H.R. Zeller, Adv. Solid State Phys. 13, 31 (1973).
3. R.L. Greene, G.B. Street and L.J. Sutter, Phys. Rev. Lett. 34, 577 (1975).
4. B.D. Cutforth, W.R. Datars, R.J. Gillespie and A. van Schyndel, Advances in Chemistry Series 150 (1976).
5. J. Kommandeur, in Low Dimensional Cooperative Phenomena, edited by H.J. Keller (Plenum Press, New York, 1974).
6. T.E. Phillips, T.J. Kistenmacher, J.P. Ferraris and D.O. Cowan, Chem. Comm. 14, 471 (1973).
7. L.B. Coleman, M.J. Cohen, D.J. Sandman, F.G. Yamagishi, A.F. Garito and A.J. Heeger, Solid State Commun. 12, 1125 (1973).
8. G.A. Thomas et al., Phys. Rev. B13, 5105 (1976); Marshall J. Cohen et al., Phys. Rev. B13, 5111 (1976).
9. P.M. Chaikin, J.F. Kwak, T.E. Jones, A.F. Garito and A.J. Heeger, Phys. Rev. Lett. 31, 601 (1973).
10. F. Donoyer, R. Comes, A.F. Garito and A.J. Heeger, Phys. Rev. Lett. 35, 445 (1975).
11. A.A. Bright, A.F. Garito and A.J. Heeger, Phys. Rev. B10, 1328 (1974); P.M. Grant, R.L. Greene, G.C. Wrighton and G. Castro, Phys. Rev. Lett. 31, 1311 (1973).
12. J.F. Kwak, P.M. Chaikin, A.A. Russel, A.F. Garito and A.J. Heeger, Solid State Commun. 16, 729 (1975).
13. U. Bernstein, P.M. Chaikin, and P. Pincus, Phys. Rev. Lett. 34, 271 (1975).

14. A.J. Berlinsky, J.F. Carolan and L. Weiler, Solid State Commun. 15, 795 (1974).
15. D.Jerome and M. Weger, Nato Summer School on One Dimensional Metals, Bolzano 1976, edited by H.J. Keller (Plenum Press, New York, 1977).
16. P.M. Chaikin, R.L. Greene, S. Etemad and E.M. Engler, Phys. Rev. B13, 1627 (1976).
17. T.D. Schultz and S. Etemad, Phys. Rev. B13, 4928 (1976); S. Etemad, Phys. Rev. B13, 2254 (1976).
18. Y. Tomkiewicz, to be published.
19. R. Comes, S.M. Shapiro, G. Shirane, A.F. Garito and A.J. Heeger, Phys. Rev. Lett. 32, 1518 (1975).
20. See for example Thermoelectric Power of Metals by F.J. Blatt, P.A. Schroeder, C.L. Foiles and D. Greig (Plenum Press, New York, 1976).
21. P.M. Chaikin, J.F. Kwak, R.L. Greene, S. Etemad and E. Engler, Solid State Comm. 19, 1201 (1976).
22. Y. Tomkiewicz, A.R. Taranko and J.B. Torrance, Phys. Rev. Lett. 36, 751 (1976).
23. P.M. Chaikin and G. Beni, Phys. Rev. B13, 647 (1976).
24. J.F. Kwak, G. Beni, P.M. Chaikin, Phys. Rev. B13, 641 (1976).
25. R. Somoano, V. Hadok, S.P. Yen, A. Rembaum and R. Deck, J. Chem. Phys. 62, 1061 (1975) Fig. 4 of this paper has the sign of the ordinate axis inverted (R. Somoano, private communication).
26. J.P. Farges and A. Brau, Phys. Status Solidi(b) 64, 269 (1974).
27. L.I. Buravov, D.N. Fedutin and I.F. Schegolev, Sov. Phys. JETP 32, 612 (1971).
28. G.Beni, Phys. Rev. B10, 2186 (1974).
29. J.F. Kwak and G. Beni, Phys. Rev. B13, 652 (1976).

DISCUSSION

B.R. Coles: We were assured in an earlier paper that the one thing that is constant like the Northern Star is the high temperature value of 60 μV/K. The last slide of yours showed 30 μV/K.

P.M. Chaikin: That's only for $\rho = \frac{1}{2}$ and for smaller bands. It turns out in this compound, in TTF-TCNQ, where the thermopower appears to be linear for no good reason, ρ is not equal to one-half, and the band widths are much greater than in the compounds I showed you. In that case the band width is larger than the coulomb interaction, and so, in a Hubbard model, you would not expect the spins to be free.

PHASE TRANSITIONS IN THE THERMOPOWER OF TTF-TCNQ SINGLE CRYSTALS*

C.K. Chiang, A.F. Garito and A.J. Heeger

Department of Physics and Laboratory for Research on
the Structure of Matter
University of Pennsylvania
Philadelphia, Pennsylvania 19104 USA

ABSTRACT

We report the b-axis thermoelectric power of single crystal
TTF-TCNQ in the temperature range from 30 K to 65 K. From the
thermoelectric power measurement we can observe clearly two of
the three phase transitions found in the structure studies. In
the vicinity of the principal metal-insulator phase transition,
the data were analyzed to compare with the results of electrical
resistivity and specific heat.

Considerable experimental effort has been devoted to the study
of the 1 D organic conductor tetrathiafulvalene-tetracyanoquin-
odimethane (TTF-TCNQ)[1]. Current interests include recent x-ray
and neutron scattering results demonstrating that the change from
metallic to insulating behavior as revealed by transport measurements corresponds to three structural phase transitions at 53, 49
and 38 K, where the 38 K transition is first-order[2-4]. These three
phase transitions have been observed in further high precision
measurements such as electrical resistivity[5]. Since the thermopower is one of the most sensitive electronic transport properties
of a metal, further characterization of the phase transition region
may be obtained by high precision measurements of the thermopower
of TTF-TCNQ. In this paper, we present initial high resolution
thermopower data measured along the b-axis of TTF-TCNQ single
crystals from 30-65 K.

*Work supported by the National Science Foundation DMR-21667.

The TTF-TCNQ single crystals used in these measurements were prepared in our laboratory as previously described[6]. The thermopower was measured along the crystallographic b-axis by a differential method in which the sample was mounted between a pair of calibrated pure silver reference leads. Silver paint was used to make electrical and thermal contacts, and the silver leads were thermally anchored to copper blocks where a temperature gradient ΔT could be maintained. The apparatus had been carefully designed to minimize stray voltage. In each run the sample temperature was allowed to drift slowly through the phase transitions at a rate of 0.1 - .2K/min. The temperature gradient ΔT and thermal voltage ΔV were recorded continuously. The thermal gradient was maintained at approximately 0.5 - 1 K. The thermopower was calculated from the ratio of ΔV and ΔT. The data were checked against the results of static measurements by Chaikin, et al.[7], agreeing with an uncertainty of less than 15%.

Figure 1 shows the temperature dependence of the b-axis thermopower (S) of single crystal TTF-TCNQ over the phase transition region (30-65 K). The principal metal-insulator phase transition appears as a smooth second-order transition located at 53 K. The shape of the transition is step-like, and the increase of thermopower is about 40 $\mu V/K$. The second major feature of the data in the phase transition region is a distinct break in the thermopower near 38 K. This discontinuity in slope corresponds to the first-order phase transition at 38 K where a clear hysteresis effect about 1.5 K in width is observed. Moreover, the hysteresis-like behavior continues into the intermediate temperature range between the 38 and 54 K transitions. Similar temperature cycling effects in this range were observed by Ellenson, et al. in elastic neutron scattering measurements of the low temperature satellite peaks[8]. Figure 1 is a typical thermal cycle in which the sample was cooled down to 4.2 K and then warmed up through the phase transition region. The third phase transition (49 K) cannot be clearly defined in these data.

In order to further characterize and examine the phase transition region, we have taken the derivative of the thermopower dS/dT, and the results are shown in Fig. 2. In the dS/dT curve, the principal metal-insulator transition appears as a sharp peak centered at 53 K and the 38 K phase transition is a distinct step. Between 38 and 48 K, dS/dT remains constant, giving no indication of the 49 K transition.

The principal metal-insulator transition at 53 K has been extensively studied by specific heat (C_e) and b-axis electrical resistivity (ρ) measurements. Through simple band theory, the

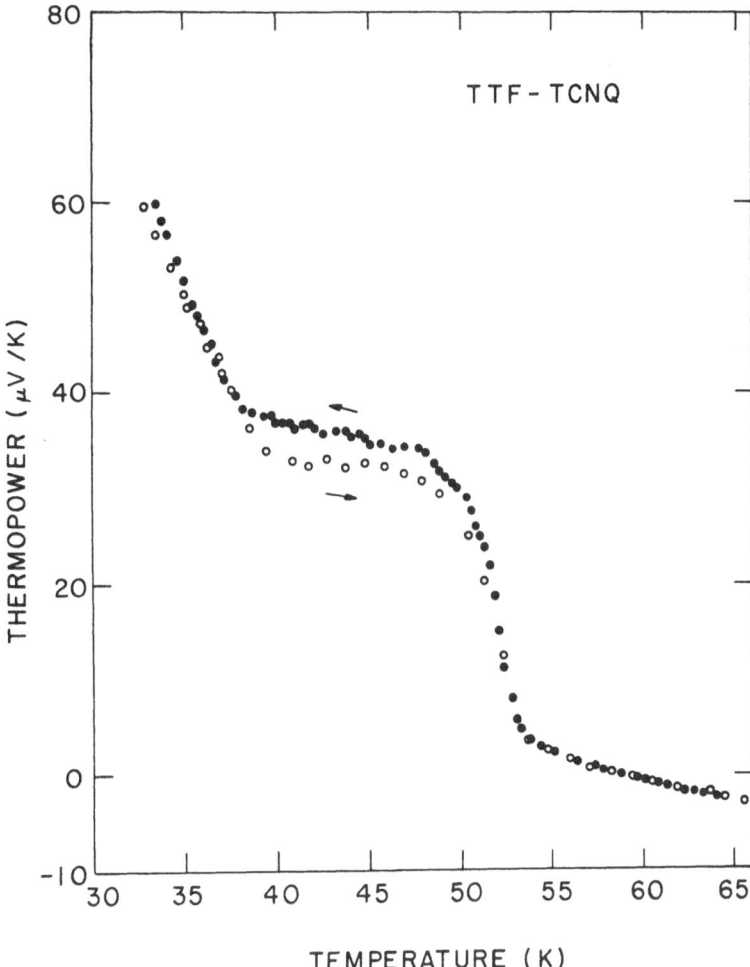

Fig. 1. Temperature dependence of the b̲-axis thermopower (S) of TTF-TCNQ single crystals in the phase transition region.

thermopower results can be compared with each of these quantities by the approximate relations[10,11]

$$\frac{dS}{dT} \sim \frac{C_e}{eT} \tag{1}$$

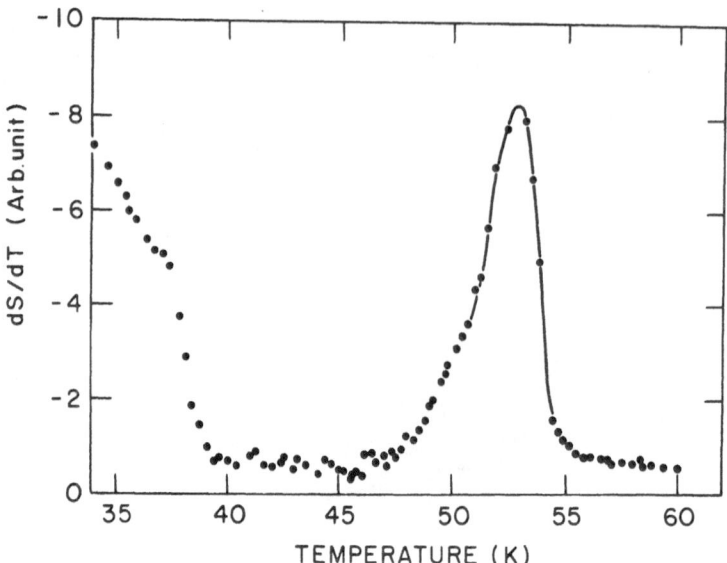

Fig. 2. Derivative of the thermopower (dS/dT) versus temperature.

and

$$\frac{dS}{dT} \sim \frac{\pi^2 k_B T}{3e\Delta E_F} \frac{d\ln\rho}{dT}; \quad T \sim T_c \tag{2}$$

where C_e is the electronic specific heat; e, electronic charge;
ΔE_F, the shift in the Fermi energy; k_B, Boltzmann's constant.
Figures 3 and 4 show the comparison of dS/dT at the principal
metal-insulator transition with d $\ln\rho$/dT, and $\frac{C_e}{T}$, respectively.
The resistivity data were measured in our laboratory using samples
from the same batch and are in good agreement with Horn et al.[5]
The specific heat anomaly was reported by Craven, et al[6]. The
qualitative similarity in the general behavior of the three
quantities is striking, particularly between dS/dT and d $\ln\rho$/dT.
This observation then suggests that the fluctuations at the phase
transition may also be conveniently studied using dS/dT.

For temperatures less than Debye temperature, the thermoelectric power includes an additional phonon drag contribution[10,11].

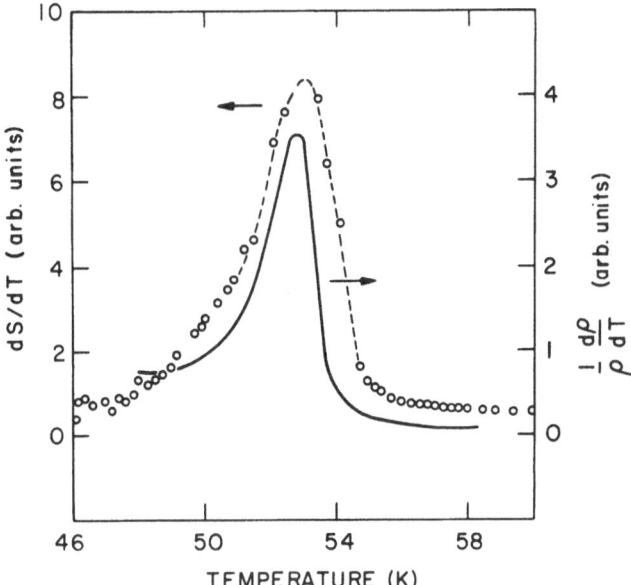

Fig. 3. Comparison of the thermopower derivative and the logarithmic derivative of the b-axis electrical resistivity
[(1/ρ) dρ/dT] versus temperature.

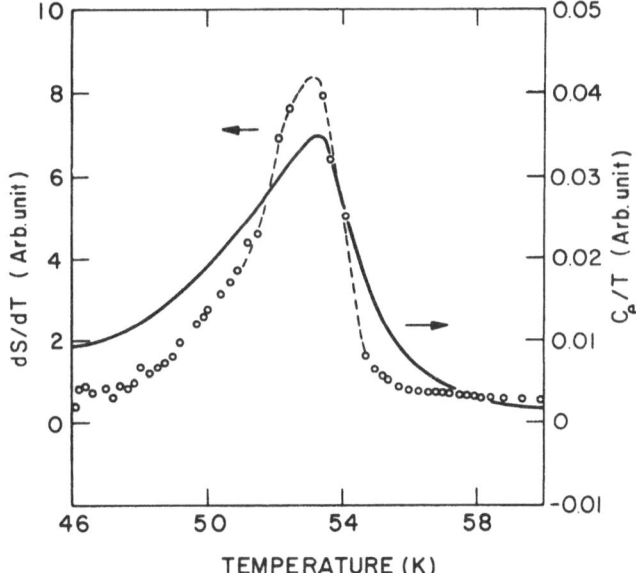

Fig. 4. Comparison of the thermopower derivative and the electronic
specific heat (C_e/T) of ref. 9, versus temperature.

The Debye temperature of TTF-TCNQ is 113 K, and thus in the phase transition region, the phonon drag thermopower should be considered especially since an anomaly has been observed in the phonon spectrum at $q = 2k_F$[12]. However, a suitable prescription from microscopic theory for including this term in a Peierls distorted system remains lacking.

In summary, high resolution thermopower measurements of TTF-TCNQ single crystals in the phase transition region clearly reveal the principal metal-insulator transition at 53 K and the first-order 38 K transition which is accompanied by a relatively large hysteresis 1.5 K in width. However, the 49 K transition is not clearly defined by the thermopower results. The derivative of the thermopower allows a comparison with the specific heat and the logarithmic derivative of the electrical resistivity at the principal metal-insulator transition.

REFERENCES

1. For a review see, for example Low Dimensional Cooperative Phenomena, ed., H.J. Keller, (Plenum, New York 1975); Chemistry and Physics of One-Dimensional Metals, ed. by H.J. Keller (Plenum, New York 1977).
2. F. Denoyer, R. Comès, A.F. Garito and A.J. Heeger, Phys. Rev. Lett. 35, 445 (1975).
3. R. Comès, S.M. Shapiro, G. Shirane, A.F. Garito and A.J. Heeger, Phys. Rev. Lett. 35, 1518 (1975).
4. S. Kagoshima, H. Anzai, K. Kajimura and T. Ishigoro, J. Phys. Soc. Japan 39, 1143 (1975).
5. P.M. Horn and D. Rimai, Phys. Rev. Lett. 36, 809 (1976).
6. See, for example A.R. McGhie, F. Yamagishi, P. Nigrey and A.F. Garito, Ann. New York Acad. Sci. (In press); P. Nigrey, J. Cry . Growth (In press).
7. P.M. Chaikin, J.F. Kwak, T.E. Jones, A.F. Garito and A.J. Heeger, Phys. Rev. Lett. 31, 601 (1973).
8. W.D. Ellenson, R. Comès, S.M. Shapiro, G. Shirane, A.F. Garito and A.J. Heeger, Solid State Commun. 20, 53 (1976); W.D. Ellenson, R. Comès, S.M. Shapiro, G. Shirane and A.F. Garito, Phys. Rev. (In press).
9. R.A. Craven, M.B. Salamon, G. DePasquali, R.M. Herman, G. Stucky and A. Schultz, Phys. Rev. Lett. 32, 769 (1974).
10. See, for example, J.M. Ziman, Electrons and Phonons (Oxford 1960).
11. R.D. Barnard, Thermoelectricity in Metals and Alloys, (Wiley, New York 1972).
12. G. Shirane, S.M. Shapiro, R. Comès, A.F. Garito and A.J. Heeger, Phys. Rev. B 14, 2325 (1976).

DISCUSSION

P.B. Allen: Could I ask a rude question. How high is the conductivity at the 60K phase transition? I mean, the giant conductivity
peak, is that up by a factor of twenty, or what?

C.K. Chiang: That peak depends on how pure your sample is. The
better the sample, the higher the conductivity. Larry Coleman's
data shows that there is a 3% chance that you might get a factor
of 100; but that's very rare. If you pick any sample from our
laboratory, I can guarantee 20 to 25 times.

THERMOELECTRIC POWER OF $Hg_{2.86}AsF_6$

G. A. Scholz and W. R. Datars
Department of Physics

D. Chartier and R. Gillespie
Department of Chemistry

McMaster University
Hamilton, Ontario, Canada

ABSTRACT

Linear chain mercury compounds were investigated by measuring the thermoelectric power of $Hg_{2.86}AsF_6$. The temperature dependence of the thermopower above 100K along the $\hat{a}$ and $\hat{b}$ directions exhibits a negative slope which indicates conduction by electrons along the mercury chains and a Fermi energy of 4.6 eV. The slope of the temperature dependence of the thermopower along the $\hat{c}$ axis exhibits a positive slope indicating hole conduction along the $\hat{a}$ direction. A transition in the region 180–200 K is observed along the $\hat{c}$ direction but not along the $\hat{a}$ direction. A large positive thermopower peak at 13 $\pm$ 3 K along both directions is considered to be due to phonon drag. Models are presented to explain the electrical conductivity.

1. INTRODUCTION

Compounds containing linear chains of mercury cations have been of interest since the discovery of $Hg_{2.86}AsF_6$ and $Hg_{2.91}SbF_6$.[1,2] They are good metallic compounds with a conductivity along the mercury chains of approximately $10^4(ohm\ cm)^{-1}$ at room temperature. The conductivity increases by a factor of 10^3 as the temperature is varied from room temperature to 4.2 K.[3] Furthermore, electrical and optical measurements have shown that they are anisotropic metallic conductors with an anisotropy in the conductivity of the order of 100 at all temperatures from room temperatures down to 4.2 K.[4-6]

These results have been confirmed in more recent experiments inves-
tigating the electrical and optical properties of $Hg_{2.86}AsF_6$.[7,8]

The structure of $Hg_{2.86}AsF_6$ may be viewed in two parts. The
$(AsF_6)^-$ octahedra shown in Fig. 1 are situated in a tetragonal
lattice with a = 7.54 Å and c = 12.32 Å. There are four $(AsF_6)^-$
ions per unit cell. This framework of ions has linear, non-
intersecting, mutually perpendicular channels in which the second
part made up of infinite mercury chains is situated. The shortest
distance between two chains in consecutive planes occurs where the
chains cross and is 3.85 Å. The mercury atoms in the chains have
a nominal formal charge of +0.35 and are separated on average by
2.64 Å. However, the intrachain mercury-mercury distance is com-
mensurate with the tetragonal lattice.

We have studied the thermoelectric power of $Hg_{2.86}AsF_6$ to obtain
further information on the nature of the electrical conduction in
this compound. The Seebeck coefficient is a direct measurement of
the intrinsic properties of a metal even when anisotropy and crystal
imperfections result in some irreproducibility in electrical conductivity measurements.[9] The Seebeck coefficient, which is independent of geometry, is a zero-current measurement so that breaks in a
current path, unless accompanied by large breaks in the heat flow
paths, do not produce appreciable effects. Furthermore any decom-
position resulting in the formation of elemental mercury is detected,

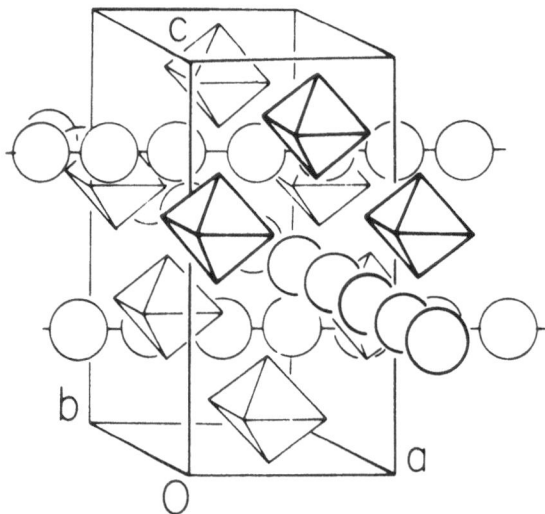

Fig. 1. An isometric view of $Hg_{2.86}AsF_6$ showing the chains of
mercury atoms (circles) running through the lattice compound of
AsF_6 ions (octahedra).

particularly at the melting point of mercury, as an additional component in the Seebeck coefficient. Thus, with this method, it is possible to make sure that measurements are made on samples that have not decomposed.

The sample was mounted in the sample holder inside a dry box and was held firmly in place by two copper rods, one of which was spring-loaded. For the experiments, small heaters on the copper rods heated the sample directly and a larger heater was also provided to heat the whole sample holder. G.E.7031 varnish was used to provide good thermal contact between the heater wires and the copper and to electrically isolate copper-constantan thermocouples attached to the copper rods. The voltage probes were soldered directly on to the copper rods as close as possible to the thermocouples. Electrical leads were thermally anchored wherever required to minimize unwanted heat flow and thermal gradients. A copper can surrounding the sample holder assembly provided an isothermal environment for the sample and also isolation from the atmosphere.

The thermopower was measured by using a slowly alternating temperature gradient. The thermoelectric voltage was measured while current was supplied to the heater on one of the copper rods in contact with the sample until the temperature difference across the sample was 1/2°C and then the current switched to the heater on the other copper rod to produce a temperature difference of about 1/2°C in the opposite direction. Finally, the first heater was turned on again until the temperature gradient disappeared. This produced a hysteresis curve of ΔV vs. ΔT, in which the hysteresis depends on the rate of heating but the slope is independent of the heating rate. The temperature of the sample does not vary by more than 1°C, so that, providing the thermopower does not vary rapidly with temperature, the hysteresis curve has a linear slope and forms a closed loop. The measurements thus represent an average thermopower over approximately 1°C for both signs and varying magnitude of the temperature gradient.

The accuracy of the method for small samples and with mechanical contacts was evaluated in experiments with lead and nickel pellets of the same size as the $Hg_{2.86}AsF_6$ crystals.

The absolute thermopower of copper (S_{Cu}) was evaluated from the average value of S_{CuPb} by subtracting S_{Pb} determined by Christian et al.[10] In the higher temperature region, the data are in excellent agreement with the accepted values of the diffusion thermopower.[11] S_{Cu} was subtracted from the thermopower of $Hg_{2.86}AsF_6$ - Cu to obtain the absolute thermopower of $Hg_{2.86}AsF_6$.

The thermopower along the $\vec{c}$ axis ($S_{\vec{c}}$) was determined from

crystal platelets which could be inserted into the holder with the
$\vec{c}$-axis along the temperature gradient. Suitable crystals for the
measurement of the thermopower along the $\vec{a}$ and $\vec{b}$ axes ($S_{\vec{a},\vec{b}}$) were
not obtained. However, the net thermopower for a polycrystalline
sample can be written as[12]

$$S = \frac{\sigma_{\vec{c}}S_{\vec{c}} + \sigma_{\vec{a},\vec{b}} S_{\vec{a},\vec{b}}}{\sigma_{\vec{c}} + \sigma_{\vec{a},\vec{b}}}$$

Since $\sigma_{\vec{c}} \ll \sigma_{\vec{a},\vec{b}}$, the thermopower measured for a polycrystalline
sample is approximately $S_{\vec{a},\vec{b}}$. Four polycrystalline samples were
used to determine $S_{\vec{a},\vec{b}}$ in this way.

The temperature dependence of $S_{\vec{a},\vec{b}}$ is shown in Fig. 2 for the

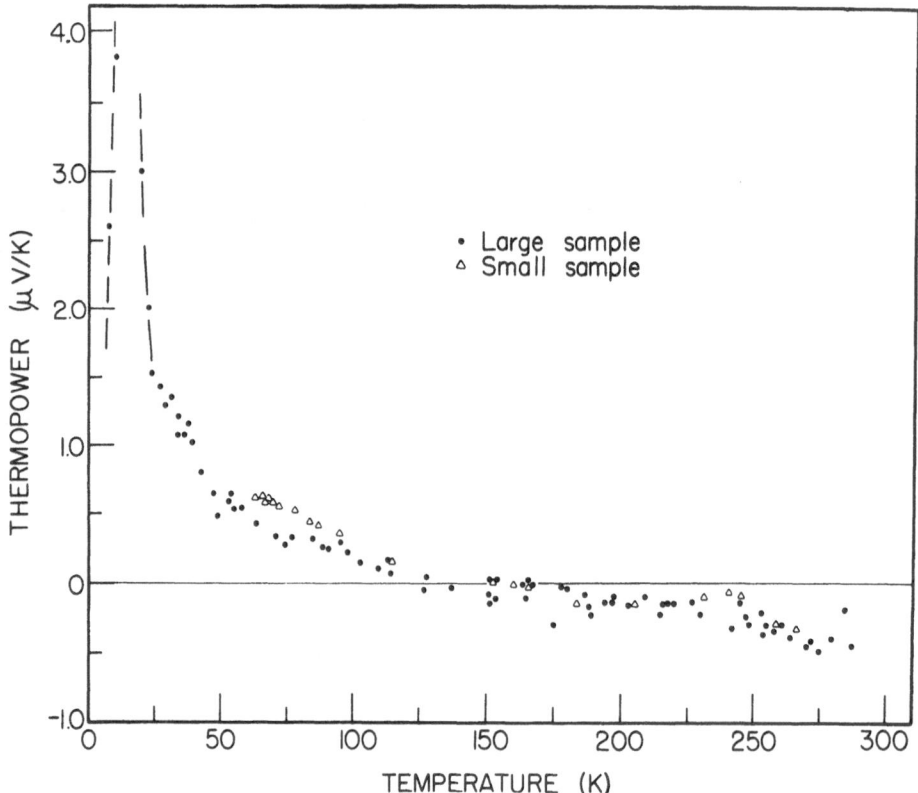

Fig. 2. Thermopower of $Hg_{2.86}AsF_6$ along the $\hat{a},\hat{b}$ direction as a
function of temperature.

largest and smallest samples, their length along the temperature
gradient differing by a factor of five. The identical results for
both samples demonstrates the validity of the measuring technique.
Above 140 $\pm$ 20 K, the thermopower is negative and varies linearly
with temperature with a negative slope. Only the general trend of
the temperature dependence is indicated below 25 K since in this
region the errors are large. However, it is evident that there is
a relatively large positive peak centred at 13 $\pm$ 3 K. Although the
thermopower of the Hg$_{2.86}$AsF$_6$ - Cu couple was measured accurately,
the error in S$_{\vec{a},\vec{b}}$ is larger because of the uncertainty in S$_{Cu}$.

Measurements on three platelets were made to determine S$_{\vec{c}}$.
Again, results from all crystals were similar and those for the
largest are displayed in Fig. 3. The thermopower is positive at all

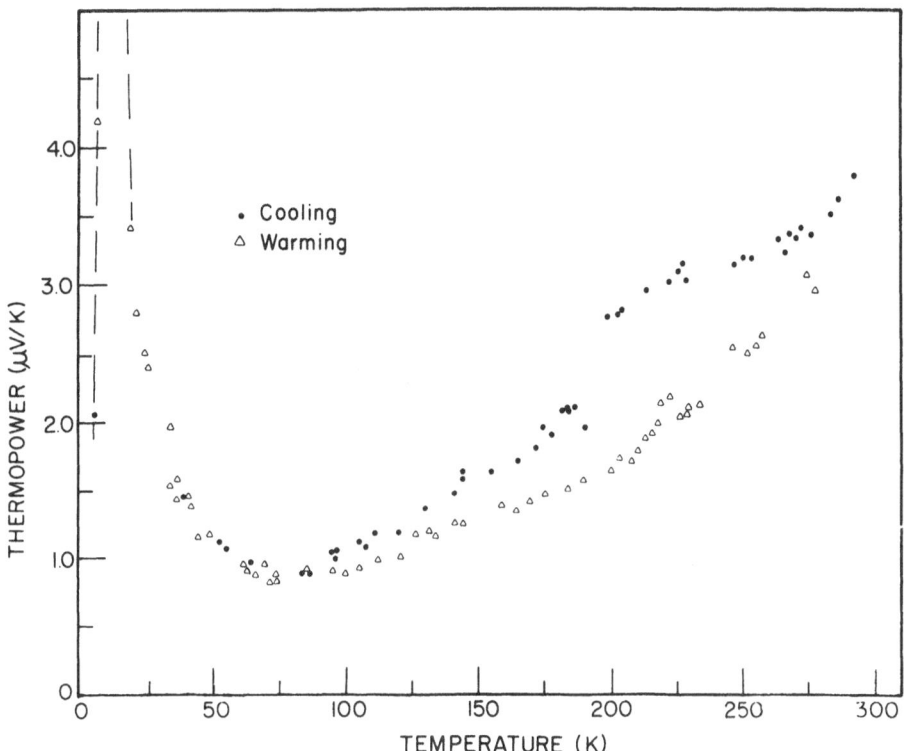

Fig. 3. Thermopower of Hg$_{2.86}$AsF$_6$ along the $\hat{c}$ direction as a function of temperature. The solid dots are the results obtained during
the first cool down of a sample. The results indicated by triangles
were obtained on warming the sample and subsequently cooling it.

temperatures with a minimum at $\sim$ 75 K. On cooling from room temperature, $S_{\vec{c}}$ varies linearly with temperature with a positive slope
which is interrupted by a transition to a smaller thermopower at
190 $\pm$ 10 K. There is a relatively large positive peak at 13 $\pm$ 3 K
similar to that in $S_{\overrightarrow{ab}}$. On warming, hysteresis was observed above
$\sim$ 75 K. When the samples were cycled again, the original cooling
curve was not retraced but instead the initial warming curve was
followed.

$S_{\vec{a},\vec{b}}$ of $Hg_{2.86}AsF_6$ appears to vary linearly with temperature
above $\sim$ 120 K and has a negative slope (Fig. 2). The thermopower
behaves like the diffusion term for a degenerate electron gas. The
diffusion thermopower (S_D) of a free-electron metal obeying Fermi-
Dirac statistics for which $T_F \gg T$ and in which a single relaxation
time exists and is inversely proportional to energy is

$$S_D = - \frac{(\pi k)^2}{6|e|} \frac{T}{E_F} \tag{1}$$

where E_F is the Fermi energy and T is the temperature.[9] A fit of
the linear portion of the thermopower data of Fig. 2 to Eq..(1)
yields a Fermi energy at 4.3 $\pm$ 0.5 e V and the negative slope indicates that conduction is by electrons. The linear temperature dependence is characteristic of a metal and does not depend on its
dimensionality. Thus, it cannot be used to determine the possible
one-dimensional nature of the compound.

It is of interest to compare E_F determined from the thermopower
results with that expected for the electron density in the channels
containing the mercury atoms. We assume the AsF_6^- octahedra to be
spherical with a radius of 2-3Å and that there are 1.65 conduction
electrons per mercury atom with a formal charge of +0.35 required
for charge neutrality. Then the Fermi energy is 4.1 eV. Alternatively, a one-dimensional free electron density of states may be
used to estimate E_F although the use of a one-dimensional density
of states may over-simplify the problem. In this model there are,
on average, 2.86 mercury atoms per unit cell distance of 7.54Å each
with a nominal valence of +1.65. This electron density corresponds
to a Fermi energy of 3.7 eV. The values of E_F for both models are
in reasonable agreement with the measured value of 4.3 eV and are
not sufficiently different to enable one to distinguish between
the two models.

The thermopower $S_{\vec{c}}$, shown in Fig. 3 increases with temperature
above 100K. Therefore, whereas electrons are the dominant carriers
along the $\vec{a},\vec{b}$ axes, holes are the majority carriers in the $\vec{c}$

direction. This clearly demonstrates the anisotropic nature of the conductivity. The conduction along the $\vec{c}$ axis must cross the regions where the mercury chains cross. Here, the mercury atoms are separated, on average, by only $\sim 3.1\text{Å}$ compared to much larger distances anywhere else in the $\vec{c}$ direction. Thus, we can visualize electrical conduction in the $\vec{c}$ direction as occurring at chain crossings with subsequent diffusion along the chains in the $\vec{a}$ and $\vec{b}$ directions to the next set of chain crossings. Thus, the conductivity along $\vec{c}$ does not have a one-dimensional path.

Between 180 K and 200 K there is evidence of a transition with a $\sim 70\%$ decrease in the value of $S_{\vec{c}}$ on cooling. This is not thought to be due to elemental mercury since it would then have to occur much closer to 234 K, the melting point of mercury. This transition is also observed in differential thermal analysis experiments[13] and in the $\vec{c}$-axis resistivity measurements[4] where the change in the resistivity is of similar magnitude to the change in $S_{\vec{c}}$. This transition may result from an ordering of mercury atoms from their somewhat random intrachain spacings so that, for example, their interchain distance is minimized wherever chains cross. This explanation seems reasonable since no anomalies are observed for $S_{\vec{a},\vec{b}}$ in this temperature range and therefore the transition must be related in some way to the interchain conduction process.

$S_{\vec{a},\vec{b}}$ becomes positive below 140 ± 20 K with a maximum at 13 ± 3 K and then decreases to zero at still lower temperatures. This peak is typical of phonon drag thermopower (S_g) enhancement. Generally, it has been found that peaks in S_g occur in the temperature range $0.1\ \theta_D < T < 0.2\ \theta_D$, where θ_D is the Debye temperature, i.e. the Debye temperature is presumably in the range 65-130 K. According to Ziman,[14] S_g should be proportional to T^{-1} for $T \gtrsim \theta_D$. Subtracting the diffusion component from the total thermopower, we obtain S_g which when plotted against T^{-1} is indeed linear in the range 100-25 K, as shown in Fig. 4. A plot of $S_{\vec{a},\vec{b}}$ versus T^{-1} is also quite linear below 100 K since S_D contributes very little at low temperatures.

The positive nature of S_g for electron conduction in the $\vec{a},\vec{b}$ directions can occur when electron scattering by Umklapp U-processes is more frequent than scattering by normal N-processes.[14] However, at low temperatures U-processes will "freeze out" if the Fermi surface is not close to the Brillouin zone boundaries but can exist down to the lowest temperatures if there is a multiple-connected surface or if the Fermi surface is close to the Brillouin zone boundary. Multiple-connected surfaces along the $\vec{a},\vec{b}$ directions causing the positive phonon decay thermopower can be visualized by constructing a two-dimensional reciprocal lattice using the inter

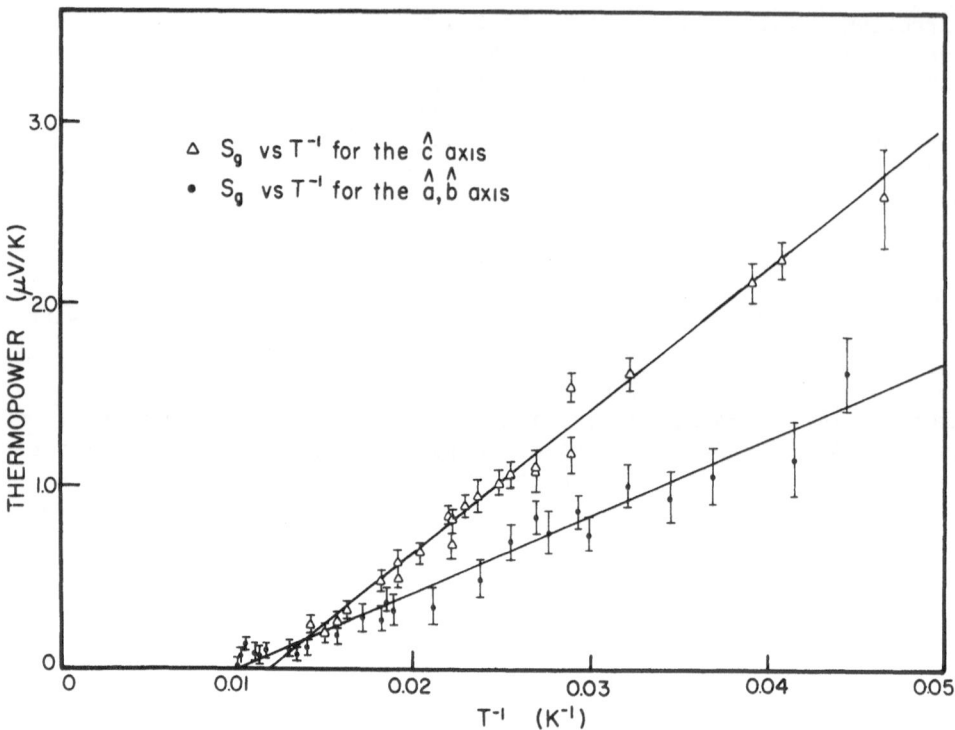

Fig. 4. Thermopower <u>versus</u> inverse temperature for the $\hat{c}$ axis and the $\hat{a},\hat{b}$ axes of $Hg_{2.86}AsF_6$.

and intrachain mercury distances of a particular plane for unit cell parameters. The Brillouin zones thus constructed do indeed generate connected Fermi surfaces when a free-electron Fermi sphere is used in a single orthogonalised-plane-wave construction.

Below $\sim$ 75 K, S_c is very similar to $S_{\vec{a},\vec{b}}$. N-processes can now be the cause of the positive phonon decay component since electrical conduction is hole like. However, since conduction in the $\vec{c}$ direction also involves diffusion along the chains in the $\vec{a}$ and $\vec{b}$ directions, we may also be observing U-processes occurring along the $\vec{a}$ and $\vec{b}$ directions.

The present results may help to explain the large temperature dependence of the electrical conductivity of the linear chain mercury compounds which is comparable to that of many pure metals. The large phonon drag peak indicates that phonon scattering is important and the low Debye temperature indicates a significant density of low-energy phonons. These properties give a strong temperature

dependence to the phonon resistivity and a large resistance ratio.[14]
A large phonon-drag component also acts to reduce the resistivity.

We conclude from the thermoelectropower experiment that there
is metallic electron dominated conduction along the $\vec{a}$ and $\vec{b}$ directions of $Hg_{2.86}AsF_6$. Holes are the majority carriers in the $\vec{c}$
direction. This amply confirms the anisotropic nature of the compound. The thermopower of $Hg_{2.86}AsF_6$ is described by $S = S_D + S_g$
where S_D is the diffusion component and S_g is the phonon drag component. The Fermi energy is evaluated to be 4.3 ± 0.5 eV from S_D.
S_g is the dominant below 100 K and appears to be caused largely
by Umklapp processes. The Debye temperature is estimated to be
between 65 and 130 K. There is a change in the thermopower be-
tween 180 K and 200 K along the $\vec{c}$ axis but not along the $\vec{a},\vec{b}$ axes.
This indicates possible ordering between neighbouring mercury
chains or a change in the crystal lattice along the $\vec{c}$ axis.

REFERENCES

1. I.D. Brown, B.D. Cutforth, C.G. Davies, R.J. Gillespie, P.R.
    Ireland and J. Vekris, Can. J. Chem. 52, 791 (1974).
2. B.D. Cutforth, Ph.D. Thesis, McMaster University (1976),
    unpublished.
3. B.D. Cutforth, W.R. Datars, R.J. Gillespie and A. van Schyndel,
    Adv. Chem. 150, 56 (1975).
4. B.D. Cutforth, W.R. Datars, A. van Schyndel and R.J. Gillespie,
    Solid State Commun. 21, 377 (1977).
5. E.S. Koteles, W.R. Datars, B.D. Cutforth and R.J. Gillespie,
    Solid State Commun. 20, 1129 (1976).
6. E. Batalla, W.R. Datars, B.D. Cutforth and R.J. Gillespie,
    (to be published).
7. C.K. Chiang, R. Spal, A. Denestein, A.J. Heeger, N.D. Miro and
    A.G. MacDiarmid, Solid State Commun. 21, 197 (1977).
8. D.L. Peebles, C.K. Chiang, M.J. Cohen, A.J. Heeger, N.D. Miro
    and A.G. MacDiarmid, Phys. Rev. 15, 4412 (1977).
9. R.D. Barnard, "Thermoelectricity in Metals and Alloys", (1972).
10. J.W. Christian, J.P. Jan, W.P. Pearson and I.M. Templeton,
    Proc. Roy. Soc. (London) A245, 213 (1958).
11. W.B. Pearson, Solid State Physics (U.S.S.R.) 3, 1411 (1961).
12. R.P. Hübner, Solid State Physics 27, 63 (1972).
13. A. van Schyndel, W.R. Datars, B.D. Cutforth and R.J. Gillespie,
    (to be published) (1976).
14. I.M. Ziman, "Electrons and Phonons", Clarendon Press, Oxford,
    (1960).

DISCUSSION

W.B. Muir: I was wondering when you showed that phase change with
the hysteresis loop, was that repeatable around the cycle as you
went up and down, or did it only do this once?

W.R. Datars: It only did it the first time it was cooled down.
After that, the lower curve was followed, that is, the first warm-
up curve was followed. However, this pattern was reproducible on
all samples that we measured. With each sample the first cool-down
was different. This may seem strange to you and, in fact, is. However, different measurements on other properties indicate that
something unusual happens on the first cool down.

K. Boning: I would say that what you have described as a hysteresis
could also be evidence of some defects annealing out, and not of a
hysteresis.

W.R. Datars: Yes, perhaps, I think this is a possibility. In this
case it is not a true hysteresis but an effect that really occurs
only during the first cool-down of the material.

THERMOELECTRIC POWER OF LINEAR MERCURY CHAIN COMPOUND[†]

C.K. Chiang, R. Spal[*], A. Denenstein and A.J. Heeger
Department of Physics and Laboratory for Research on
the Structure of Matter,

and N. D. Miro and A. G. MacDiarmid
Department of Chemistry and Laboratory for Research on
the Structure of Matter

University of Pennsylvania
Philadelphia, Pennsylvania 19104   USA

ABSTRACT

We have studied the thermoelectric power of the incommensurate linear chain compound $Hg_{3-\delta}AsF_6$. At high temperatures, the electron diffusion thermopower in this anisotropic metal is linear in temperature. At low temperatures, phonon drag thermopower contributions can be identified.

INTRODUCTION

Recently a new class of compounds, containing infinite chains of mercury ions, with highly anisotropic electrical conductivity and anisotropic optical properties, have been reported[1-4]. The $Hg_{3-\delta}AsF_6$ system remains metallic to low temperatures and exhibits anisotropic superconductivity[2] and an anisotropic Meissner effect

---

†This work supported by the National Science Foundation MRL Program under Grant No. DMR 76-00678.

*University Fellow; submitted in partial fulfillment of the requirements for the Ph.D.

below 4.1 K.[5] The crystal structure of $Hg_{3-\delta}AsF_6$ (δ = 0.18 at room temperature) consists of a host tetragonal array of AsF_6 octahedra and non-intersecting linear mercury chains running parallel to the $\vec{a}$ and $\vec{b}$ axes[6,7]. There are no chains along the $\vec{c}$-direction. The intrachain Hg-Hg distance is 2.66 Å. Since the unit cell parameters are a = b = 7.55 Å, the mercury chains are incommensurate with the three dimensional host lattice. The shortest distance between the chains along the $\vec{c}$-axis is 3.09 Å, and is comparable to the inter-atomic distance in mercury metal. We report here on the results of an initial investigation of the anisotropic thermopower of this novel compound.

EXPERIMENT

The thermoelectric power was measured on rectangular samples cut from the as-grown crystal. The samples were transferred in a controlled atmosphere dry box and sealed into a moisture free copper can sample holder for measurement. One end of sample was thermally anchored to a copper block where the sample temperature was measured. The other end of the sample was attached to a small heater which could create a temperature gradient (approximately 1 K to 3 K) across the sample. The copper electrical contacts were made by using spring pressure. All stray voltages were care-fully minimized, and the system was calibrated against a high purity (99.9999%) lead standard. The standard differential thermo-power measurement technique was applied.

RESULTS AND DISCUSSION

The anisotropic thermoelectric power obtained from single crystals of $Hg_{3-\delta}AsF_6$ is shown in Fig. 1 ($\vec{a}$ - $\vec{b}$ plane) and Fig. 2 ($\vec{c}$-axis). The thermopower is positive with overall magnitude of order of a few microvolts per degree; values typical of metallic behavior. At room temperature the results are 1.5 μV/K and 2.2 μV/K for the ($\vec{a}$ - $\vec{b}$) plane and $\vec{c}$-axis respectively. The anisotropy in the magnitude of the thermopower is relatively small compared to the electrical resistivity where the anisotropy is of order 100 to 1.

Above 220 K, the thermopower is approximately a linear function of temperature. The electron diffusion thermopower can be written as

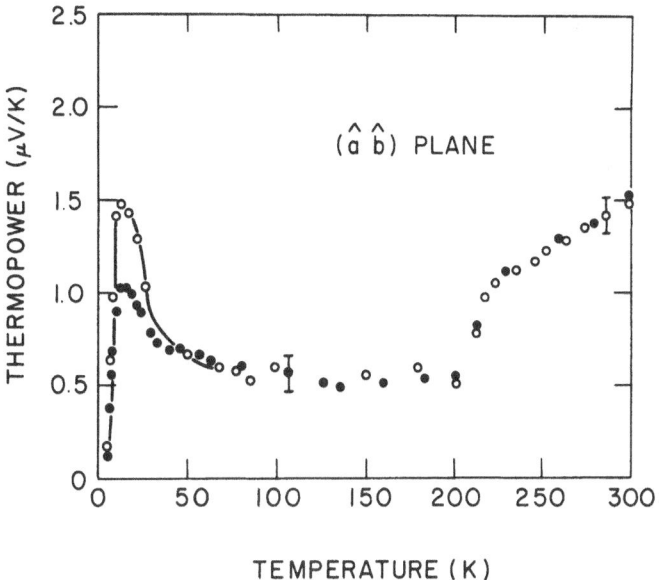

Fig. 1. Thermoelectric power of $Hg_{3-\delta}AsF_6$ single crystal ($\vec{a} - \vec{b}$ plane).

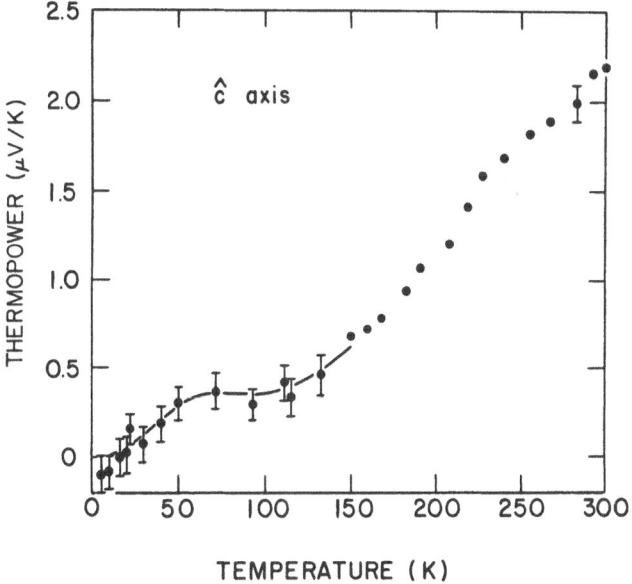

Fig. 2. Thermoelectric power of $Hg_{3-\delta}AsF_6$ single crystal ($\vec{c}$-axis).

$$S = \frac{\pi^2 k^2 T}{3e} \left[\frac{\partial \ln \sigma(\varepsilon)}{\partial \varepsilon} \right]_{E_F} . \tag{1}$$

For temperatures above the Debye temperature (Θ) and assuming energy independent scattering, eq. (1) can be expressed as

$$S = \frac{2\pi^2}{3} \cdot \frac{k^2 T}{e} N(o) \tag{2}$$

where $N(o)$ is the density of states (one sign of spin) at the Fermi energy. The relatively weak anisotropy observed in the thermopower suggests (eq. (1) and (2)) that the larger anisotropy in conductivity is predominantly the result of very strong scattering along the $\vec{c}$-axis. From the slope of the data we obtain $N(o) \simeq 0.14$ states/eV/carrier or 0.05 states/eV/Hg atom assuming one carrier for each AsF_6 anion. This value is to be compared with 0.08 states/eV/Hg atom obtained from the Knight shift.

At 220 K, the change in slope of S vs. T is suggestive of a phase transition in agreement with earlier observations[1,2]. The transition is evident in all directions; in the $\vec{a} - \vec{b}$ plane, the behavior is sharp while along the $\vec{c}$-axis the transition is more smooth. We have found no signature of the phase ordering phase transition discovered by elastic neutron scattering on very slow cooling through 120 K.[9]

The thermopower in the low temperature range is also aniso-tropic. In the $\vec{a} - \vec{b}$ plane, a maximum is found with peak value at 15 K. Measurement of four samples shows that the peak size varies from sample to sample in the range from 0.8 $\mu V/K$ to 1.5 $\mu V/K$. This variation and the overall peak structure are consistent with a phonon drag contribution which is typically sensitive to the impurity and/or defect concentration in the sample. In the $\vec{c}$-direction, there is a smaller peak which occurs near 60 K. Since phonon drag thermopower peaks typically appear at about $\frac{1}{10}$ to $\frac{1}{5}$ of Θ,[8] the structure observed at 15 K ($\vec{a} - \vec{b}$ plane) and 60 K ($\vec{c}$-axis) suggests an anisotropic phonon spectrum.

Inelastic neutron scattering studies[9] of the phonon spectrum have demonstrated the existence of one-dimensional (1 D) phonons originating from the incommensurate linear mercury chains, in addition to the usual three-dimensional (3 D) lattice phonons.

Moreover, specific heat measurements at low temperatures have shown that in addition to the usual T^3 contribution (3 D phonons), there exists a large specific heat contribution which is of lattice origin and approximately linear in T (1 D phonons). The magnitudes of the Debye temperatures are in reasonable agreement with those obtained directly from the phonon spectrum. Although a detailed theory is not available, the anisotropy observed in the phonon drag thermopower is in qualitative agreement with these observations. It is interesting that the $\vec{c}$-axis thermopower with peak near 60 K suggests stronger coupling to the 1 D phonons $(\Theta_{1D} \simeq 600$ K$)$ whereas the $\vec{a} - \vec{b}$ plane results with peak near 15 K suggest stronger coupling to the 3 D phonons $(\Theta_{3D} \simeq 80$ K$)$.

In summary, we have studied the anisotropic thermoelectric power of the linear chain mercury compound, $Hg_{3-\delta}AsF_6$, from 5 K to 300 K. The thermopower is small and typical of a metallic system. Anisotropy is observed over the entire temperature range. At high temperatures the results, when compared with the previously measured electrical conductivity, imply stronger scattering associated with transport along the $\vec{c}$-axis. Anisotropic phonon drag contributions are identified at low temperatures suggesting an anisotropic phonon spectrum. Comparison with the phonon spectrum obtained from inelastic neutron scattering and inferred from low temperature specific heat studies suggests that the $\vec{c}$-axis transport is strongly coupled to the 1 D phonons.

REFERENCES

1. B.D. Cutforth, W.R. Datars, A. van Schyndel and R.J. Gillespie, Solid State Commun. 21, 377 (1977).
2. C.K. Chiang, R. Spal, A. Denenstein, A.J. Heeger, N.D. Miro and A.G. MacDiarmid, Solid State Commun. 22, 293 (1977).
3. E.S. Koteles, W.R. Datars, B.D. Cutforth and R.J. Gillespie, Solid State Commun. 20, 1129 (1976).
4. D.L. Peebles, C.K. Chiang, M.J. Cohen, A.J. Heeger, N.D. Miro and A.G. MacDiarmid, Phys. Rev. B 15, 4607 (1977).
5. R. Spal, C.K. Chiang, A. Denenstein, A.J. Heeger, N.D. Miro, and A.G. MacDiarmid, to be published.
6. I.D. Brown, B.D. Cutforth, C.G. Davies, R.J. Gillespie, P.R. Ireland and J.E. Vekris, Can. J. Chem. 52, 791 (1974).
7. A.J. Schultz, J.M. Williams, N.D. Miro, A.G. MacDiarmid and A.J. Heeger, to be published.
8. See for example, R.D. Barnard, Thermoelectricity in Metals and Alloys, (Wiley, New York, 1972).

9.   J.M. Hastings, J.P. Pouget, G. Shirane, A.J. Heeger, N.D. Miro and A.G. MacDiarmid, to be published.

10.   T. Wei, A.F. Garito, C.K. Chiang, N.D. Miro and A.G. MacDiarmid, Phys. Rev. B (Accepted).

DISCUSSION

P.B. Allen: Did I hear you right, that the resistivity versus temperature in the A direction and B direction behaves as $T^{3/2}$ at low temperatures and exhibits no residual resistivity?

C.K. Chiang: Yes, in fact, that is fascinating, because crystal growing is so easy. I can do that without any temperature control (room temperature). Compared to people who try to get resistivity ratios of 60,000 this kind of work is trivial, but I still can get a resistivity ratio of 3,000 with no effort at all. That is a fascinating aspect of this compound, especially in view of an anisotropic compound.

SUMMARY REMARKS

B. R. Coles

Department of Physics
Imperial College
London, England

I am very honored to have been asked by the organizing committee to sum up but I am a little doubtful about my qualifications, since, although I measured an alloy thermoelectric power in 1951 which, I think, is even before Bill Pearson, I don't remember doing so in the last ten years. But then, of course, memories fade with age.

The other difficulty is that being asked to sum up a conference on thermoelectric power, at least this sort of conference, is rather like being asked to precis the New York Telephone Directory. There are, perhaps, overall classifications like the different boroughs, and one sees that there are some addresses which are more fashionable than others, but that's about the lot of it.

So I am going to be rather arbitrary, and I hope, for example, nobody will expect me, while it is still so fresh in your mind, to try to summarize anything you have heard this afternoon. The other thing, of course, that's remarkable, having said that, is that an outsider might have thought that the topic of our conference was a rather narrow one. But in fact what we are all impressed by is the range to which you expose yourself when you talk about thermopower in general.

Looking at the property, it's got rather special features. In condensed matter physics there are some properties that become a focus of study because they give a very specific type of information, and we can all think of our favorite examples (the de Haas-van Alphen effect is an obvious one). Then there are other properties which, although rather specialized, provide a vital test of new theoretical concepts developing in particular areas. There are other rather

specific properties of enormous range of applicability, like nuclear
magnetic resonance, where there is almost nothing in common between
some of the materials that people study in that way.

But thermopowers don't quite belong in any of these categories.
The trouble is that they are a bit like the book on elephants which
the small boy took back to the library the day after he got it out,
saying "this book tells me more about elephants than I want to know."
I did on several occasions today have the feeling that thermoelectric powers are a little too rich for the stomach of simple souls
like a condensed matter physicist. They should be left to the
biologist, to somebody who is used to getting more information than
he can cope with.

As far as the theory is concerned, I hope I won't be considered
unkind if I confess that I got a slight impression of the Scotsman
who was found on a Saturday night by a friend hunting around under
a street lamp for a sixpence he had dropped. The friend, having
joined the hunt, asked after a while if he was sure he had dropped
it there, and he said , "no, I dropped it down the street, but this
is where the light is." I got a feeling that the diffuse light of
theory is not shining in the areas where some of our problems seem
to be.

However, turning to the papers we have actually heard in the
last few days, I think we can see a number of components emerging
and a number of questions that crop up again and again.

There is evidence that thermopower measurements can provide
important tests of theoretical ideas that emerge in various areas.
For example, the papers on thermoelectric effects in superconductors
that we began with, showed that close consideration of current flow
in both bulk superconductors near T_c and in intermediate state materials at lower temperatures yield predictions which are best
tested by thermoelectric measurements. I was very struck, of
course, by the extent to which those measurements seemed to put
greater demands on both electrical and thermal hygiene than almost
any other measurements of thermoelectric effects.

A topic where I had rather hoped to find some nice tests of
theory is the possibility of thermoelectric effects associated with
crystalline field splitting of the states of impurity atoms in
metals, but that is a topic which, for some reason we did not hear
much about. Perhaps it is biding its time.

Well, today we've heard about the applications of thermoelectric
power to phase transitions and to defect studies, and, again in that
area, I felt that the richness of information that both the thermoelectric power and Hall effect gave one was almost more than my

tender digestion could cope with. The simple electrical resistance
told one rather a lot, and I was impressed by the bravery of those
who go on to do such measurements.

We also heard today about the characteristics of materials of
limited dimensionality. I won't try and summarize that. I would
like to comment on Dr. Park's study of critical phenomena, however,
because I was very pleased to see that Dr. Park supported my view
that, while neutrons are valuable they are extremely expensive, --
you have to build enormous spallation sources or high flux reactors
or something --, the probe that God gives one of magnetic order and
critical phenomena, that is, conduction electrons, are there all
the time, and are not only a lot cheaper but also at times even more
informative. Especially, of course, since rare earth metals don't
cooperate with neutrons but greedily absorb them.

In the Kondo problem, again, I think the thermoelectric power
was an early pointer to important effects, and it became later a
valuable test of theory. I was very glad to hear Dr. Fischer emphasize strongly the point that if you really had Kondo's original
Hamiltonian; that is to say, if there was only true exchange coupling between the conduction electrons and local moments, the giant
thermopowers would not be there; the essential feature, in fact, is
that the Kondo Hamiltonian is inadequate, and that we have to recognize that the effective exchange arises from a mixing term, the
Andersonian V_{kd}. I think the thermopower is perhaps the strongest
reminder of this limitation. The generals of ancient Rome had a
slave to whisper in their ears as they rode in triumph, "remember
thou art mortal;" the thermoelectric power is whispering to us
"remember that the JS.s Hamiltonian is not the whole story."

I hope that in the study of magnetic effects we shall begin to
see the role of thermopower in the study of interactions between
magnetic impurities, not only of the behavior of the isolated impurity, and I think Dr. Foiles' talk showed that it may prove a very
useful tool in distinguishing between spin-glass type freezing-in
of magnetic moments and long range magnetic order. I think that we
may see very many more studies of thermopower in critical concentration regions of magnetic solid solutions since scattering by finite
and by infinite magnetic clusters seems to have very distinct differences. We also, I think, will see a lot more coming along of
thermoelectric power studies in the exciting area of intermediate
valency alloys in compounds of rare earth metals. Scattering by
charge fluctuations is likely to be as important as scattering by
spin fluctuations.

Turning back almost to the beginning, we heard early on about
the present state of the art in normal, non-magnetic metals and
alloys, and I was impressed by Denis Greig's emphasizing explicitly

in his own work, and politely but implicitly in presenting the latest
flow of results from Dr. Vedernikov's laboratory, that we are still
really not able to draw any very specific conclusions about the
details of alloy electronic structures from measurements of their
thermoelectric powers. I was however reassured to find that some
of the vexing questions about the role of electron-electron and
electron-paramagnon scattering in resistivity are actually being
assisted by studies of thermoelectric powers rather than further
mystified.

Rather surprisingly, although perhaps that's because he is such
a convincing speaker, Dr. Enderby seemed to indicate that the picture
looks rather better for making sense of the variation with composition of thermopowers in liquid alloys than in solid alloys. If he
is right that one can get quite a long way with Boltzmann equation
formalisms, it is more difficult to make the case for a breakdown
of such formalisms in solid crystalline alloys.

On the question of phonon-drag, I was gratified to see that
although deviations from Matthiessen's rule may be bogging down, you
have a new field of endeavor, deviations from the Nordheim-Gorter
relationship, ready to take its place, and we will not be short of
deviationist activities.

Deviation is leading to my next topic because in his talk Father
Phil Taylor reminded us of our sins. He varied from being a priest
to an anthropologist, -- folk activities and ritual gestures on the
one hand and sins on the other, -- and whether we acknowledge and
bewail our manifold sins and wickednesses, I think most of us have
one confession to make, and that is a most encouraging one. It is
that really we worry about thermoelectric power not because we think
it is going to give us the ultimate answer but rather for the reason
that people climb Mt. Everest, because it is there, an intriguing
property with unexpected quirks. The thermoelectric power, more
than almost any other property of condensed matter, is a wonder
property; that is to say, you can always say, "I wonder what thermoelectric power I will get if I mix A and B, or melt X," or so on.
Even Matthias would hesitate, I think, to make bets on what the
results of such measurements might be.

To conclude, I think I'd like to pick out one point which impressed me particularly, and I am particularly glad to say that it
is a point which is getting special attention here at MSU; that is
the possibility of some very valuable results that may derive from
combining thermoelectric measurements with the application of high
magnetic fields. It is not completely a new idea. I know that
almost simultaneously, many years ago, Bill Pearson tried it on
some materials at Ottawa and I tried it on some other materials at

Carnegie Tech, but I didn't understand what I saw. The difference
now I think is that people are really beginning to make some valuable
suggestions about the interpretation of these results, and I see
that as a large growth area.

However, in referring to MSU at the end of these very arbitrary
remarks, I must also emphasize, of course, the contribution they have
made to thermoelectric power studies by holding this conference.
They have not only made it a lively, stimulating intellectual experience, they've also made it a very pleasant social experience.
The arrangements here have been extremely efficient, but not only
efficient, very warmhearted, and I'm sure we are all grateful for
that. I would like, in closing, to say on behalf of all of us,
thank you to the local committee and all the local boys who put in
a lot of work to give us a very fine time here.

PARTICIPANTS

P. B. Allen
S.U.N.Y.
Stony Brook, New York 11794

J. Bass
Michigan State University
East Lansing, Michigan 48824

R. Barnard
University of Salford
Salford, England

F. F. Bekker
Natuurkundig Laboratorium
der Universiteit van Amsterdam
Amsterdam, The Netherlands

G. Belanger
Universite de Montreal
Montreal, Canada

F. J. Blatt
Michigan State University
East Lansing, Michigan 48824

B. Blumenstock
Michigan State University
East Lansing, Michigan 48824

K. Böning
Universitat Munchen
Garching, Germany

R. I. Boughton
Northeastern University
Boston, Massachusetts 02181

R. R. Bourassa
University of Oklahoma
Norman, Oklahoma 73069

S. P. Bowen
Virginia Polytechnic Institute
and State University
Blacksburg, Virginia 24061

R. M. Boysel
Ohio State University
Columbus, Ohio 43210

C. Brook
8335 St. Louis Avenue
Skokie, Illinois 60076

P. M. Chaikin
U.C.L.A.
Los Angeles, California 90024

C. K. Chiang
University of Pennsylvania
Philadelphia, Pennsylvania 19104

J. Cleveland
Central Michigan University
Mount Pleasant, Michigan 48859

B. R. Coles
Imperial College
London, England

S. Crisp
University of Western Australia
Nedlands, Australia

W. R. Datars
McMaster University
Hamilton, Ontario
Canada

C. Dec
Energy Conversion Devices
Troy, Michigan 48067

J. E. Enderby
University of Bristol
England

C. M. Falco
Argonne National Lab.
Argonne, Illinois 60439

K. H. Fischer
Institut fur Festkorperforschung
Julich, Germany

R. Fletcher
Queens University
Kingston, Ontario
Canada

C. L. Foiles
Michigan State University
East Lansing, Michigan 48824

J. Garland
Ohio State University
Columbus, Ohio 43210

J. F. Goff
White Oak Laboratory
Silver Spring, Maryland 20910

D. Greig
University of Leeds
Leeds, England

A. M. Guénault
University of Lancaster
Lancaster, England

H.-J. Guntherodt
Universitat Basel
Switzerland

D. G. Hawksworth
Applied Superconductivity Lab.
University of Wisconsin
Madison, Wisconsin 53706

D. F. Heidel
Ohio State University
Columbus, Ohio 43210

W. G. Henry
Queens University
Kingston, Ontario
Canada

J. Ivory
Office of Naval Research
Branch Office
Chicago, Illinois 60605

F. Kus
McMaster University
Hamilton, Ontario
Canada

C. Leavens
National Research Council of
Canada
Ottawa, Ontario
Canada

K. H. Lee
University of Missouri
Columbia, Missouri 65201

S. K. Lyo
University of California
Los Angeles, California 90024

S. D. Mahanti
Michigan State University
East Lansing, Michigan 48824

R. F. Moreland
The University of Utah
Salt Lake City, Utah 84112

W. B. Muir
Eaton Laboratory
McGill University
Montreal, Canada

M. Müller
Der Universitat Basel
Switzerland

R. Newrock
University of Cincinnati
Cincinnati, Ohio 45221

J. Opsal
Michigan State University
East Lansing, Michigan 48824

R. D. Parks
University of Rochester
Rochester, New York 14627

J. C. Parlebas
University of California
Irvine, California 92717

W. B. Pearson
University of Waterloo
Canada

W. P. Pratt, Jr.
Michigan State University
East Lansing, Michigan 48824

S. Puri
Northeastern Illinois University
Chicago, Illinois 60625

D. J. Resnick
Ohio State University
Columbus, Ohio 43210

R. B. Roberts
C.S.I.R.O.
Sydney, Australia

J. B. Sampsell
Ohio State University
Columbus, Ohio 43210

P. A. Schroeder
Michigan State University
East Lansing, Michigan 48824

J. Sierro
University of Geneva
Switzerland

M. Snodgrass
Michigan State University
East Lansing, Michigan 48824

S. Steenwyk
Michigan State University
East Lansing, Michigan 48824

J. C. Swihart
University of Indiana
Bloomington, Indiana 47401

P. L. Taylor
Case Western Reserve
Cleveland, Ohio 44106

I. M. Templeton
National Research Council of
Canada
Ottawa, Ontario
Canada

B. Thaler
Michigan State University
East Lansing, Michigan 48824

C. Uher
Michigan State University
East Lansing, Michigan 48824

S. Ya Wang
University of Oklahoma
Norman, Oklahoma 73019

H. J. van Daal
Philips Research Laboratories
Eindhoven, The Netherlands

D. J. van Harlingen
Ohio State University
Columbus, Ohio 43210

J. B. van Zytveld
Calvin College
Grand Rapids, Michigan 49506

S. Woods
University of Alberta
Edmonton, Alberta
Canada

G. F. Zharkov
P. N. Lebedev Physical Institute
Moscow, U.S.S.R.

O. H. Zinke
University of Arkansas
Fayetteville, Arkansas 72701

Photograph of Participants

1. W. P. Pratt	23. K. H. Lee	46. R. Fletcher
2. F. F. Bekker	24. C. L. Foiles	47. K. Böning
3. J. Bass	25. C. Lee	48. J. B. Sampsell
4. P. L. Taylor	26. B. R. Coles	49. B. Blumenstock
5. J. C. Parlebas	27. A. M. Guénault	50. G. F. Zharkov
6. J. E. Enderby	28. W. B. Pearson	51. D. B. Heidel
7. W. B. Muir	29. P. M. Chaikin	52. C. Uher
8. B. Thaler	30. S. Ya Wang	53. M. Harrison
9. P. B. Allen	31. S. K. Lyo	54. S. Woods
10. C. Leavens	32. C. M. Falco	55. R. I. Boughton
11. G. Belanger	33. F. J. Blatt	56. O. H. Zinke
12. J. C. Swihart	35. R. M. Boysel	57. S. Puri
13. P. A. Schroeder	36. D. J. van Harlingen	58. J. Garland
14. K. H. Fischer	37. D. J. Resnick	59. R. Newrock
15. J. Ivory	38. C. K. Chiang	60. F. Kus
16. C. Brook	39. H.-J. Guntherodt	61. J. Cleveland
17. M. Müller	40. I. M. Templeton	62. S. Steenwyk
18. H. J. van Daal	41. C. Dec	63. W. G. Henry
19. D. Greig	42. J. F. Goff	64. W. R. Datars
20. R. Barnard	43. R. F. Moreland	65. S. Crisp
21. R. B. Roberts	44. S. P. Bowen	
22. J. Opsal	45. R. R. Bourassa